New Developments in Lipid–Protein Interactions and Receptor Function

NATO ASI Series

Advanced Science Institutes Series

A series presenting the results of activities sponsored by the NATO Science Committee, which aims at the dissemination of advanced scientific and technological knowledge, with a view to strengthening links between scientific communities.

The series is published by an international board of publishers in conjunction with the NATO Scientific Affairs Division

A	**Life Sciences**	Plenum Publishing Corporation
B	**Physics**	New York and London
C	**Mathematical and Physical Sciences**	Kluwer Academic Publishers
D	**Behavioral and Social Sciences**	Dordrecht, Boston, and London
E	**Applied Sciences**	
F	**Computer and Systems Sciences**	Springer-Verlag
G	**Ecological Sciences**	Berlin, Heidelberg, New York, London,
H	**Cell Biology**	Paris, Tokyo, Hong Kong, and Barcelona
I	**Global Environmental Change**	

Recent Volumes in this Series

Volume 243A —Biological Effects and Physics of Solar and Galactic Cosmic Radiation, Part A
edited by Charles E. Swenberg, Gerda Horneck, and E. G. Stassinopoulos

Volume 243B —Biological Effects and Physics of Solar and Galactic Cosmic Radiation, Part B
edited by Charles E. Swenberg, Gerda Horneck, and E. G. Stassinopoulos

Volume 244 —Forest Development in Cold Climates
edited by John Alden, J. Louise Mastrantonio, and Søren Ødum

Volume 245 —Biology of *Salmonella*
edited by Felipe Cabello, Carlos Hormaeche, Pasquale Mastroeni, and Letterio Bonina

Volume 246 —New Developments in Lipid–Protein Interactions and Receptor Function
edited by K. W. A. Wirtz, L. Packer, J. Å. Gustafsson, A. E. Evangelopoulos, and J. P. Changeux

Volume 247 —Bone Circulation and Vascularization in Normal and Pathological Conditions
edited by A. Schoutens, J. Arlet, J. W. M. Gardeniers, and S. P. F. Hughes

Series A: Life Sciences

New Developments in Lipid–Protein Interactions and Receptor Function

Edited by

K. W. A. Wirtz

University of Utrecht
Utrecht, The Netherlands

L. Packer

University of California
Berkeley, California

J. Å. Gustafsson

Karolinska Institute, NOVUM
Huddinge, Sweden

A. E. Evangelopoulos

The National Hellenic Research Foundation
Athens, Greece

and

J. P. Changeux

Institut Pasteur
Paris, France

Plenum Press
New York and London
Published in cooperation with NATO Scientific Affairs Division

Proceedings of a NATO Advanced Study Institute on
New Developments in Lipid-Protein Interactions and Receptor Function,
held August 16–27, 1992,
in Spetsai, Greece

NATO-PCO-DATA BASE

The electronic index to the NATO ASI Series provides full bibliographical references (with keywords and/or abstracts) to more than 30,000 contributions from international scientists published in all sections of the NATO ASI Series. Access to the NATO-PCO-DATA BASE is possible in two ways:

—via online FILE 128 (NATO-PCO-DATA BASE) hosted by ESRIN, Via Galileo Galilei, I-00044 Frascati, Italy

—via CD-ROM "NATO-PCO-DATA BASE" with user-friendly retrieval software in English, French, and German (©WTV GmbH and DATAWARE Technologies, Inc. 1989)

The CD-ROM can be ordered through any member of the Board of Publishers or through NATO-PCO, Overijse, Belgium.

Library of Congress Cataloging-in-Publication Data

New developments in lipid-protein interactions and receptor function
edited by K.W.A. Wirtz ... [et al.].
p. cm. -- (NATO ASI series. Series A, Life sciences ; v. 246.)
"Proceedings of a NATO Advanced Study Institute on New Developments in Lipid-Protein Interactions and Receptor Function, held August 16-27, 1992, in Spetsai, Greece"--Copr. p.
Includes bibliographical references and index.
ISBN 0-306-44521-2
1. Lipoproteins--Physiological effect--Congresses. 2. Membrane lipids--Congresses. 3. Membrane proteins--Congresses. 4. Cell receptors--Congresses. 5. Cellular signal transduction--Congresses. 6. Second messengers (Biochemistry)--Congresses. I. Wirtz, Karel W. A. II. NATO Advanced Study Institute on New Developments in Lipid-Protein Interactions and Receptor Function (1992 : Spetsai, Greece) III. Series.
QP552.L5N48 1993
574.87'6--dc20 93-11300

ISBN 0-306-44521-2

Printed in the United States of America

PREFACE

A NATO Advanced Study Institute on "New Developments in Lipid-Protein Interactions and Receptor Function" was held on the Island of Spetsai, Greece, from August 16–27, 1992. This Institute was organized to bring together researchers in the field of membrane organization and dynamics with those actively involved in studies on receptor function, signal transduction mechanisms and gene regulation.

Presentations and discussions focussed on the regulation of intracellular Ca^{2+}-levels, on the second messengers derived from inositol lipids and on the specific phospholipase C isozymes involved in these processes. A major focus was on G-proteins and the effect of lipid anchors on their function. These principles of regulation were further discussed in the context of receptors for acetylcholine, lysophosphatidic acid and low-density lipoproteins. In addition, various aspects of the genomic regulation of cell growth and differentiation by transcription factors were presented. These topics were put into perspective by discussing the most recent developments in lipid-protein interactions, protein insertion into membranes, membrane lipid organization and lipid dynamics as mediated by phospholipid transfer proteins.

This book presents the content of the major lectures and a selection of the most relevant posters. These proceedings offer a comprehensive account of the most important topics presented during the course of the Institute. The book is intended to make these proceedings accessible to a large audience.

The Editors

January 1993

CONTENTS

COMPONENTS OF SIGNAL TRANSDUCTION

RECEPTOR FUNCTION AND REGULATION

TRANSCRIPTION FACTORS AND GENE REGULATION

BIOSYNTHESIS, INSERTION AND FUNCTION OF MEMBRANE COMPONENTS

INTRACELLULAR Ca^{2+} STORES: WHY, WHERE AND HOW?

Jacopo Meldolesi, Alberto Ciardo, Antonello Villa, and Fabio Grohovaz

Department of Pharmacology, CNR Cytopharmacology and B. Ceccarelli Centers, S. Raffaele Scientific Institute, Via Olgettina 60, 20132 Milano, Italy

INTRODUCTION

Extensive experimental evidence, accumulated especially during the last decade, has demonstrated that, in the vast and highly diversified world of eukaryotic cells, the differences in terms of Ca^{2+} homeostasis are not drastic, but rather represent variations in a common theme. Individual actors responsible for specialized functions, for example the various types of voltage-gated Ca^{2+} channels of the plasmalemma, can be variously expressed and even miss altogether in some cells. Even in these cells, however, Ca^{2+} homeostasis exhibits general properties that can be summarized as follows: i. the cytosolic Ca^{2+} concentration, $[Ca^{2+}]_i$, is maintained around 10^{-7} M at rest, and can increase towards the low μM range after stimulation (see Carafoli, 1987; Pietrobon et al., 1990); ii. stimulated cells are all able to increase influx of Ca^{2+} across their plasma membrane although the density of the responsible channels, and the mechanisms of their regulation, can be markedly different (see Meldolesi et al., 1991); iii. all cells possess a Ca^{2+}-calmodulin-regulated Ca^{2+} ATPase in their plasma membrane that works as the high affinity-low capacity regulator of $[Ca^{2+}]_i$ (Carafoli, 1992); and iiii. all cells possess intracellular organelles (see Tsien and Tsien, 1990; Meldolesi et al., 1990) where Ca^{2+} is rapidly accumulated by another class of pumps (of the Sarcoplasmic-Endoplasmic Reticulum Ca^{2+}-ATPase (SERCA)-type see De Smedt et al., 1991) and from which Ca^{2+} is very rapidly released following activation of intracellular channels. The latter channels are usually referred to as the receptors for either the endogenous second messenger, inositol 1,4,5-trisphosphate (IP_3), or an exogenous high affinity ligand, the plant alkaloid ryanodine (Ry). In the present review we will concentrate on the rapidly exchanging Ca^{2+} stores. Our aim is the discussion of their essential properties, as revealed by the studies carried out during the last several years.

WHY?

This first question concerns the existence itself of the rapidly exchanging Ca^{2+} stores. As already mentioned, $[Ca^{2+}]_i$ (10^{-7} M) is much (in most cases over four orders of magnitude) lower than the concentration of Ca^{2+} in the extracellular medium (10^{-3} M). This

New Developments in Lipid-Protein Interactions and Receptor Function
Edited by K.W.A. Wirtz *et al.*, Plenum Press, 1993

coupled to the negative potential of the plasmalemma, makes the Ca^{2+} electrochemical gradient the steepest existing in biology. Wouldn't this driving force be sufficient to sustain $[Ca^{2+}]_i$ changes that occur after cell stimulation, via the simple activation of the Ca^{2+} channels of the plasmalemma? Why do cells need uptake into, and release from internal stores?

A neat answer to these questions comes from the present understanding of how do striated fibers work in both the skeletal and heart muscles. The differentiated function of these huge cells, i.e. contraction, requires to be fast and synchronously occurring throughout the cytoplasm. Were contractions sustained only by Ca^{2+} influx across the plasmalemma they would probably begin earlier, and be mechanically stronger, in the myofibrillae located superficially, and be progressively more and more weak and delayed the deeper we go inside the cytoplasm. This also because (important observation), due to the existance of so many high affinity Ca^{2+} binding molecules (especially proteins), diffusion of the cation is much (10-100 fold) slower throughout the cytoplasm (10-100 μm^2/sec) than in free solution (see Tsien and Tsien, 1990). In order to prevent delays and weakenings, the fibers possess an architecturally remarkable array of transverse (T) tubular infoldings of the plasmalemma that surround all myofibrillae at the level of the I-A band boundary, i.e. two T tubules/sarcomere. Being continuous with the plasmalemma the T tubules, on the one hand, contain extracellular (i.e., mM) concentrations of Ca^{2+}; on the other hand, participate in the plasmalemmal electrical events. Their Ca^{2+} channels (of the L type) are therefore activated with no delay when the fiber is depolarized. However, because of their astonishing length and narrow caliber, the exchange of the T tubule Ca^{2+} with the extracellular space cannot be fast. Thus, if large Ca^{2+} influx was indeed occurring at these sites, the tubule lumen would be quickly depleted of its cation. In order to cope with this problem, the abundant Ca^{2+} channel complement of the T tubule membrane has been switched from Ca^{2+} transport to a different purpose, i.e. sensing of the plasmalemma voltage and transfer of the activation to the intracellular Ca^{2+} channels, the Ry receptors, concentrated in the junctional membrane of the sarcoplasmic reticulum (SR). The junctional membrane is in fact directly attached to the T tubule membrane, to yield the well known triads and diads. This attachment is mediated by discrete bridges, the so called feet, now known to be composed of the coupled L channel - Ry receptor heterodymers.

The recently decoded meaning of the SR-T tubule interaction explains well both the structure and the physiology of the SR. This complex endomembrane system surrounds all myofibrillae as an anastomized water jacket, and pumps Ca^{2+} into its lumen because of its high complement of SERCA (over 100 fold higher than in most cells). Once in the SR lumen, however, Ca^{2+} does not remain evenly distributed, but is concentrated within the terminal cisternae where calsequestrin (CSQ), a Ca^{2+} storage (low affinity-high capacity) protein, is accumulated. This means that within the cisternal lumen the calcium content is very high (over 50 fold the extracellular concentration) and is continuously accumulated by the SERCA activity of the entire SR membrane system. The risk that under physiological conditions the activation of Ry receptors and the ensuing increased pumping of the surface ATPase could ultimately deplete the muscle of calcium is therefore highly unlikely because it is successfully competed by the intracellular SERCA pumps. The need of muscle fibers to express the SR has therefore finally received a convincing explanation.

What about other cells? Here both the general structure and the physiological needs are quantitatively different, yet the mechanisms apparently operate according to a logics similar to that of striated muscle. Direct plasmalemma-store coupling, analogous to the interaction of L-type channels with Ry receptors, might exist in both smooth muscles fibers and neurons, however not at the level of T tubules (that do not exist in those cells) but at the flat surface. Most of the stores are however separate and at some distance from the plasmalemma. Here their activation requires the diffusion of plasmalemma-generated second messenger(s), which apparently occurs at rates much faster than that of Ca^{2+} itself. The major advantage for the cell, however, does not concern rapidity but rather the discrete localization of the process at specific sites of the cytoplasm. This problem will be discussed in the following section.

WHERE ?

Cell topology, i.e. the study of the intracellular localization of specific components and events, has attracted considerable attention in recent years. Studies, however, have been focussed primarily on either macromolecules or specific ligands, whereas free ions are still widely believed to be evenly distributed within the cytosol and nuclear matrix. If one considers, however, that $[Ca^{2+}]_i$ (as discussed in the introduction) is nothing but the result of the dynamic equilibrium between pumps and channels, and that at different sites the concentration of these key molecules can vary, it follows that Ca^{2+} hot spots can be generated. A good example of ion heterogeneity, recently documented by the use of a low affinity Ca^{2+} indicator, is in the cytoplasmic rim immediately adjacent to the presynaptic membrane where, due to the high density in that membrane of N-type Ca^{2+} channels, $[Ca^{2+}]_i$ appears to rise after stimulation up to values 100 fold higher than in the rest of the cytoplasm (Llinas et al., 1992). This permits release of acetylcholine (and possibly of other classical neurotransmitters) to be a low affinity Ca^{2+}- dependent process. In the muscle, on the other hand, the SR functioning discussed in the previous section implies $[Ca^{2+}]_i$ changes to be larger at the I-A junctions, where triads are located, than in the rest of the sarcomere.

Another, more subtle but potentially even more interesting example of heterogeneity, also depending on intracellular Ca^{2+} stores, has been observed in the chicken cerebellum. Here the dendritic spines of Purkinje neurons, i.e. the structures to which impinging synapses are addressed and where synaptic plasticity takes place, have been recognized to be differently equipped for Ca^{2+} homeostasis with respect to the rest of the cytoplasm. The intraspinal cisternae are in fact rich of IP_3 receptors but lack both Ry receptors and the Ca^{2+} storage protein, CSQ, whereas the juxtaspine regions of dendrites are rich in all these components (Villa et al., 1991; Walton et al., 1991; Takei et al., 1992). Within the spines $[Ca^{2+}]_i$ changes of intracellular origin are therefore probably different (exclusively IP_3-dependent; more transient) from those in the dendritic stalk. Differences of this kind could be envisaged at other geometrically peculiar sites of neurons, for example the growth cones of axons and possibly also the axon terminals, where plasmalemma and intracellular Ca^{2+} transport processes are likely to operate in a coordinated fashion in order to control the Ca^{2+} homeostasis.

$[Ca^{2+}]_i$ heterogeneities are not the properties of only large and irregular cells, such as muscle fibers and neurons. Recently, in small chromaffin cells of the rat, the well known spontaneous $[Ca^{2+}]_i$ oscillations (Malgaroli et al., 1990) that ultimately invest the entire cytoplasm have been shown to originate from a discrete site, a sort of a pacemaker, where $[Ca^{2+}]_i$ does not remain stable but shows a pulsatile, up- and -down activity. Interestingly, such an activity was not observed in the cells of a line, named PC12, similar in origin (from a rat pheochromocytoma) and morphology to chromaffin cells, but incompetent to oscillate (D'Andrea et al., 1993). The way these $[Ca^{2+}]_i$ oscillations are generated is discussed below.

HOW?

This section is addressed to two main issues concerning mechanisms: how are intracellular Ca^{2+} stores established within the cell? and how do they function to control Ca^{2+} homeostasis?

Five years ago, when the cytological nature of intracellular Ca^{2+} stores begun to be discussed, consensus was existing only about the role of the SR in striated muscle fibers. In non muscle cells the field was split. Part of the scientists believed Ca^{2+} stores to coincide with the entire endoplasmic reticulum (ER) whereas for the others the stores were restricted to specific structures, indicated at that time as the calciosomes (see Volpe, 1988; Meldolesi et al., 1990). Although at that time calciosomes were preferentially envisaged as distinct organelles, analogous to the muscle SR, the possibility of their correspondence to specialized ER subcompartments was not excluded (see Volpe et al., 1988). The studies carried out in the meantime have clarified important aspects of the problem. First of all, detailed immunocytochemical and subcellular fractionation studies have revealed convincingly that specific structures specialized in Ca^{2+} storage and release exist not only in striated muscles, but also in various other cell types, including neurons (see above) eggs and smooth muscle fibers (Villa et al., 1991; Takei et al., 1992; Walton et al., 1991; McPherson et al., 1992; Villa et al., 1993). All these structures, including the muscle SR, are not completely distinct from the ER inasmuch as they contain specialized markers (such as the IP_3 and Ry receptors) together with general ER markers (e.g., the lumenal chaperons BiP and protein disulfide isomerase, Villa et al., 1991; Volpe et al., 1992). Moreover, physiological experiments carried out in a variety of cell types by the use of agents and drugs addressed to either one of the intracellular Ca^{2+} channels (IP_3 and Ry receptors) have yielded apparently conflicting results (discussed by Meldolesi et al., 1990; Sitia and Meldolesi, 1992). Activation of one of the channels was shown in fact to induce depletion of the Ca^{2+} store endowed with the other, however in only some cell types, whereas in different cell types the second store remained unaffected. These results suggest the co-localization of the two channels (or at the least, their segregation into lumenally continuous, adjacent membrane-bound structures) in the first group of cells; their segregation into discrete structures in the second group.

Taken together, these data lead us to conclude that Ca^{2+} stores correspond to ER subcompartments which however can be dual and function independently from each other. This conclusion fits well with the present understanding of the ER structure and function. Up to very recently, that endomembrane system was considered to be an uninterrupted network of membrane-bound elements (rough and smooth surfaced; the nuclear envelope) that, although different in shape, organization and distribution, were all lumenally continuous and destined therefore to function undissociately. The concept of the ER that emerges from recent studies in a variety of cell types (summarized by Sitia and Meldolesi, 1992) is in contrast that of a patchwork of specialized subcompartments, of which Ca^{2+} stores account for one (or two, see above). Specialization of the subcompartments appears to be the result of various processes, based however on the tendency of individual protein molecules (of both the membrane and the lumen) i., to interact with each other, and thus to form homoaggregates; ii., to bind specifically to other components either of the same or of adjacent structures (e.g., the cross-talk between the lumen and the limiting membrane; the coupling of the cytosolic surfaces of two membranes, as it occurs between SR and the T tubule at the junctional face of triads, see above; or of a single membrane to the cytoskeleton). These events should not be envisaged as static. Indeed they are probably the result of dynamic interactions in which the lumenal continuity with the rest of the ER can often be discontinued. Critical to the establishment of these interactions could be subtle heterogeneities in the microenvironment existing within the ER lumen, sustained by the flow of membranes recycling from various areas of the Golgi complex (for further details see Sitia and Meldolesi, 1992).

Also in relation to the functioning of the stores our ideas have changed considerably in recent years. Initially, the Ry receptors were believed to be expressed only by muscle fibers and to represent a specialization tool, necessary to sustain rapid contraction, whereas IP_3 receptors were believed to be expressed ubiquitously among cells. First the demonstration of abundant Ry receptors in various non muscle cells, such as neurons, eggs, secretory cells (see Tsien and Tsien, 1990); more recently the discovery of a new class of these receptors, which is widespread if not general, and is expressed not constitutively but under the control of growth factors, such as TGF-beta (Giannini et al., 1992); ultimately the demonstration that not only Ry, but also IP_3 receptors are not single but highly heterogeneous families of macromolecules (Sorrentino and Volpe, 1993; Meldolesi, 1992) have forced us to reconsider profoundly the entire issue. The present understanding is not at all complete, and various aspects remain obscure.

An important issue concerns the mechanisms of channel activation. Up until recently, the Ry receptor was believed to be activated either via the direct coupling with the surface L type channel described above, or by a rise of $[Ca^{2+}]_i$ within an appropriate range (peak 1-10 μM: Ca^{2+}- induced-Ca^{2+}-release), with inhibition at higher values. The IP_3 receptor, on the other hand, was believed to be sensitive only to the binding of its ligand. This black-and-white difference in the mechanisms of activation is no longer tenable. In fact, the Ry receptor has been shown to be bound with high affinity and activated by cyclic ADP-ribose, a metabolite of NAD with putative second messenger activity (Galione, 1992). Although up to now the generation of cyclic ADP-ribose has not been shown to be modified by cell activation, the possibility of a dual regulation (by both the putative second messenger and $[Ca^{2+}]$) of the Ry receptor needs to be kept into consideration. This type of regulation is apparently the case for the IP_3 receptor, for which however the positive modulation by $[Ca^{2+}]$ is recognized to occur in a narrower range and at lower levels (peak around 0.3 μM) compared to the Ry receptor (Bezprozvanny et al., 1991). Generalizations appear still premature, also in view of the already mentioned heterogeneity of the receptors. So far it is not yet known whether the newly discovered and widely expressed Ry receptor of Giannini et al. (1992) is regulated by Ca^{2+}-induced-Ca^{2+}-release; and whether the newly discovered isoforms of IP_3 receptors (II - IV types at the present time) differ from the classical type I isoform only for the affinity to the second messenger and the intracellular distribution, as discussed up to now (see Meldolesi 1992) or possibly also for their sensitivity to $[Ca^{2+}]$.

A final mention concerns $[Ca^{2+}]_i$ oscillations, a rythmyc activity that was recognized in various cell types as soon as the $[Ca^{2+}]_i$ measurement assays were developed at the single cell level. Up to recently two main possible candidates were available to explain the process: the IP_3 receptor, responding to oscillations of IP_3, generated via the rythmyc feed back inhibition of the generating enzyme, phospholipase C; and the Ry receptor, activated rythmically as a function of the $[Ca^{2+}]_i$ changes. Recent developments in rat chromaffin cells (a model particularly interesting because it exhibits oscillations also at rest, Malgaroli et al., 1990, whereas other cell types require moderate stimulation) appear to be of considerable interest. In these cells oscillations appear to be initiated by the pulsatile activity of a "pacemaker structure", presumably corresponding to a cell area where Ca^{2+} stores are concentrated. Pacemaker discharge increases $[Ca^{2+}]_i$ moderately, and this apparently sensitizes IP_3 receptors that become activable even at resting concentrations of the second messenger. Any moderate $[Ca^{2+}]_i$ rises, no matter the mechanism and origin, increase the probability of the oscillations, resulting in either their appearance (in the cells that were silent at rest) or in the increase of their frequency (when the cells were alredy active). In contrast, high $[Ca^{2+}]_i$ rises were inhibitory, as expected based on the present knowledge of IP_3 receptor function (see D'Andrea et al., 1993).

CONCLUSION

The progress occurred in the field of Ca^{2+} stores during the last decade has been certainly considerable, however important issues still remain open, concerning especially the heterogeneities of the receptors, their intracellular localization, their differential functional role. The present aim of our group is to further expand an integrated approach to the problem of Ca^{2+} stores where developments in various directions (immunocytochemistry; subcellular calcium and $[Ca^{2+}]$ measurements; electrophysiology etc.) might ultimately give rise to integrated pictures. Questions such as why, where and how Ca^{2+} stores work will then be given comprehensive and, hopefully, convincing answers.

REFERENCES

Bezprozvanny, I., Watras, J., and Ehrlich, B.E., 1991, Bell-shaped calcium-response curves of Ins(1,4,5)P_3- and calcium-gated channels from endoplasmic reticulum of cerebellum, *Nature* 351:751.

Carafoli, E., 1987, Intracellular calcium homeostasis, *Ann. Rev. Biochem.* 56:395.

Carafoli, E., 1992, The Ca^{2+} pump of the plasma membrane, *J. Biol. Chem.* 267:2115.

D'Andrea, P., Zacchetti, D., Meldolesi, J., and Grohovaz, F., 1993, Mechanism of $[Ca^{2+}]_i$ oscillations in rat chromaffin cells: the intracellular oscillator operates within a useful range of $[Ca^{2+}]_i$, submitted.

De Smedt, H., Eggermont, J.A., Wuytack, F., Parys, J.B., Van Den Bosch, L., Missiaen, L., Verbist, J., and Casteels, R., 1991, Isoform switching of the sarco(endo)plasmic reticulum Ca^{2+} pump during differentiation of BC3H1 myoblasts, *J. Biol. Chem.* 266:7092.

Galione, A., 1992, Calcium-induced calcium release and its modulation by cyclic ADP-ribose, *Trends Pharmacol. Sci.*, in press.

Giannini, G., Clementi, E., Ceci, R., Marziali, G., and Sorrentino, V., 1992, Expression of a ryanodine receptor-Ca^{2+} channel that is regulated by TGF-beta, *Science* 257:91.

Llinas, R., Sugimori, M., and Silver, R.B., 1992, Microdomains of high calcium concentration in a presynaptic terminal, *Science* 256:677.

Malgaroli, A., Fesce, R., and Meldolesi, J., 1990, Spontaneous $[Ca^{2+}]_i$ fluctuations in rat chromaffin cells do not require inositol 1,4,5-trisphosphate elevations but are generated by a caffeine- and ryanodine-sensitive intracellular Ca^{2+} store, *J. Biol. Chem.* 265:3005.

McPherson, S.M., McPherson, P.S., Mathews, L., Campbell, K.P., and Longo, F.J., 1992, Cortical localization of a calcium release channel in sea urchin eggs, *J. Cell Biol.* 116:1111.

Meldolesi, J., Madeddu, L., and Pozzan, T., 1990, Intracellular Ca^{2+} storage organelles in non muscle cells: heterogeneity and functional assignement, *Biochim. Biophys. Acta* 1055:130.

Meldolesi, J., Clementi, E., Fasolato, C., Zacchetti, D., and Pozzan, T., 1991, Ca^{2+} influx following receptor activation, *Trends Pharmacol. Sci.* 12:289.

Meldolesi, J., 1992, Multifarious IP_3 receptors, *Current Biology* 2:393.

Pietrobon, D., Di Virgilio, F., and Pozzan, T., 1990, Structural and functional aspects of calcium homeostasis in eukaryotic cells, *Europ. J. Biochem.* 193:599.

Sitia, R., and Meldolesi, J., 1992, The endoplasmic reticulum: a dynamic patchwork of specialized subregions, *Mol. Biol. Cell*, in press.

Sorrentino, V., and Volpe, P., 1993, Ryanodine receptors: how many and where, Trends Pharmacol. Sci, in press.

Takei, K., Stukenbrok, H., Metcalf, A., Mignery, G.A., Sudhof, T.C., Volpe, P., and De Camilli, P., 1992, Ca^{2+} stores in Purkinje neurons: endoplasmic reticulum subcompartments demonstrated by the heterogeneous distribution in the InsP_3 receptor, Ca^{2+}-ATPase, and calsequestrin, *J. Neurosci.* 12:489.

Tsien, R.W., and Tsien, R.Y., 1990, Calcium channels, stores, and oscillations, *Annu. Rev. Cell Biol.* 6:715.

Villa, A., Podini, P., Clegg, D.O., Pozzan, T., and Meldolesi, J., 1991, Intracellular Ca^{2+} stores in chicken Purkinje neurons, *J. Cell Biol.* 133:779.

Villa, A., Podini, P., Panzeri, M.C., Soling, H.D., Volpe, P., and Meldolesi, J., 1992, The endoplasmic-sarcoplasmic reticulum of smooth muscle fibers. Immunocytochemistry reveals specialized subcompartments differently equipped for the control of Ca^{2+} homeostasis, submitted.

Volpe, P., Krause, K.H., Hashimoto, S., Zorzato, F., Pozzan, T., and Meldolesi, J., 1988, "Calciosome", a cytoplasmic organelle: the inositol 1,4,5-trisphosphate-sensitive Ca^{2+} store of nonmuscle cells?, *Proc. Natl. Acad. Sci. USA* 85:1091.

Volpe, P., Villa, A., Podini, P., Martini, A., Nori, A., Panzeri, M.C., and Meldolesi, J., 1992, The endoplasmic reticulum-sarcoplasmic reticulum connection: distribution of endoplasmic reticulum markers in the sarcoplasmic reticulum of skeletal muscle fibers, *Proc. Natl. Acad. Sci. USA* 89:6142.

Walton, P.D., Airey, J.A., Sutko, J.L., Bech, C.F., Mignery, G.A., Sudhof, T.C., Deerinck, T.J., and Ellisman, M.H., 1991, Ryanodine and inositol trisphosphate receptors coexist in avian cerebellar Purkinje neurons, *J. Cell Biol.* 113:1145.

THE METABOLIC FATES AND CELLULAR FUNCTIONS OF MYOINOSITOL

Ian Batty[1], A. Nigel Carter[1], C. Peter Downes[1], Francisco Estevez[2], Daniel Sillence[1], Cyrus Vaziri[1]

[1] Department of Biochemistry
University of Dundee
Dundee DD1 4HN
[2] Facultat De Ciencas Medicas
University of Las Palmas
Gran Canaria

INTRODUCTION

Myo-inositol (Ins) is the predominant isomer of cyclohexane hexol that occurs in eukaryotic cells. The complexity of Ins metabolites that has emerged during the last several years suggests that distinct pathways of Ins metabolism may subserve a variety of cellular functions. These pathways are depicted in outline form in Figure 1.

Perhaps the best known function of Ins is as a precursor of phosphatidylinositol (PtdIns) which, besides its role as a substantial component of cell membranes, is required for the synthesis of PtdIns-glycan membrane protein anchors, diacylglycerol and inositol trisphosphate second messengers, and the recently described 3-phosphorylated inositol phospholipids. Although the calcium-mobilising function of inositol 1,4,5-trisphosphate ($Ins(1,4,5)P_3$) is well understood, little is known of the roles of other inositol polyphosphates such as the inositol pentakisphosphates and inositol hexakisphosphate (phytic acid). In this article we describe recent work from this laboratory which aims to provide a quantitative assessment of cellular Ins homeostatic mechanisms; to initiate studies of inositol polyphosphate metabolism in genetically tractable organisms; and to delineate the cellular functions of phosphatidylinositol 3,4,5-trisphosphate (PIP_3).

INOSITOL HOMEOSTASIS IN A CULTURED CELL LINE

Phosphatidylinositol 4,5-bisphosphate (PtdIns $(4,5)P_2$) is the immediate precursor of at least two, and perhaps more, intracellular messengers thought to mediate the action of many ligands which act through binding cell-surface receptors (Berridge and Irvine, 1989; Downes & Carter , 1991). The hydrolysis of PtdIns $(4,5)P_2$ by phospholipase C (PLC) releases $Ins(1,4,5)P_3$ and DAG. These signal molecules are inactivated by two separate pathways which ultimately yield the free inositol and CMP-phosphatidate from which PtdIns, and thus $PtdIns(4,5)P_2$, is re-synthesized. The final step in the dephosphorylation of $Ins(1,4,5)P_3$ to inositol is catalysed by inositol monophosphatase. This enzyme is potently inhibited by Li^+

New Developments in Lipid-Protein Interactions and Receptor Function
Edited by K.W.A. Wirtz *et al.*, Plenum Press, 1993

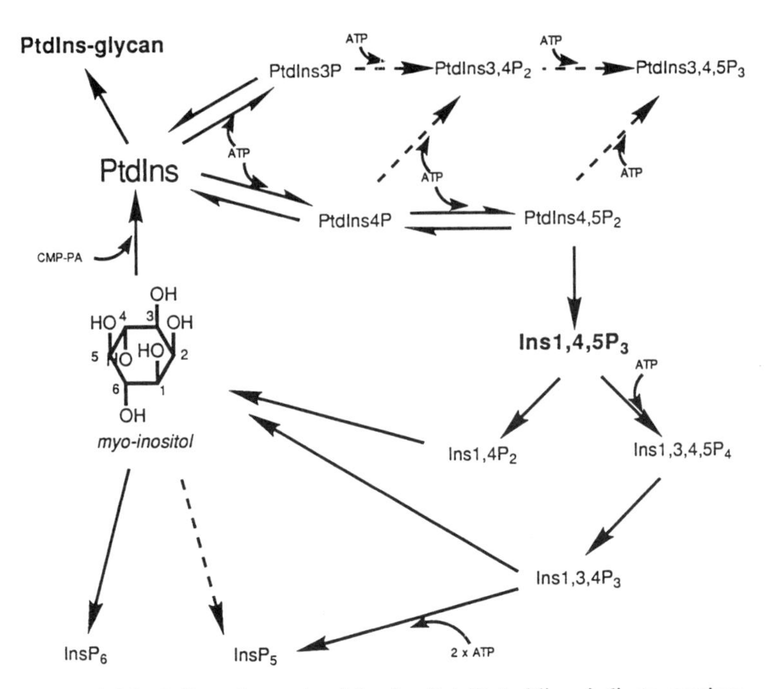

Figure 1. Metabolic pathways involving inositol. Dotted lines indicate reactions or pathways that are suspected, but not yet established. Parentheses are omitted from standard abbreviations to simplify the figure. Reproduced with permission from Eur.J.Biochem. 193, 1-18 (1990)

ions (Hallcher & Sherman, 1980). Inhibition of monophosphatase may lower cellular inositol concentrations thereby limiting phosphoinositide (PI) synthesis and impairing PI-dependent signalling (Berridge et al, 1989). As the inhibition of inositol monophosphatase by Li^+ is uncompetitive, the most effective action of Li^+ will be exerted in cells in which the PI-cycle is most active. Consequently, it has been suggested that inhibition of inositol monophosphatase may account for the therapeutic efficacy of Li^+ in the treatment of manic/depressive disorders, allowing selective down-regulation of signalling associated with abnormally active receptor systems (Berridge et al, 1982). In support of this, Li^+ ions have been shown to reduce cellular inositol concentrations, PI synthesis, and stimulated Ins(1,4,5)P_3 production and to impair associated cellular functions (Sherman, 1989; Nahorski et al, 1991). However, each of these actions is yet to be demonstrated in a causally related manner in a single identified cell type.

Such effects are not inevitable consequences of monophosphatase inhibition as cellular inositol concentrations are maintained not only by re-cycling of PI-derived inositol but also by synthesis de novo and by uptake from the extracellular environment (Sherman, 1989). The synthesis of inositol de novo from glucose-6-phosphate also involves the action of inositol monophosphatase and thus, may also be inhibited by Li^+, particularly in cells in which InsP concentrations are elevated through receptor activation. However, cells which are able to maintain inositol levels by uptake from their surroundings should be less vulnerale to Li^+.

We are currently studying the various factors which influence cellular inositol concentrations and the impact of their disruption on the signalling capability of PI-dependent mechanisms in cultured 132INI astrocytoma cells. The current status of these experiments is described below and summarised in Figure 2.

Sources of Cellular Inositol in 132INI Cells

The contributions made to cellular inositol homeostasis by synthesis de novo, uptake and inositol re-cycling in resting or stimulated cells are ill-defined. Thus, in preliminary studies we sought to establish the major sources of inositol in 132INI cells. The influence of extracellular inositol supply was studied by culture of cells for a period of days in medium ± inositol with appropriate, intervening medium changes, followed by labelling of confluent cell-monolayers to apparent isotopic equilibrium with tracer [^{3}H]-inositol and subsequent analysis of label distribution amongst cellular metabolites.

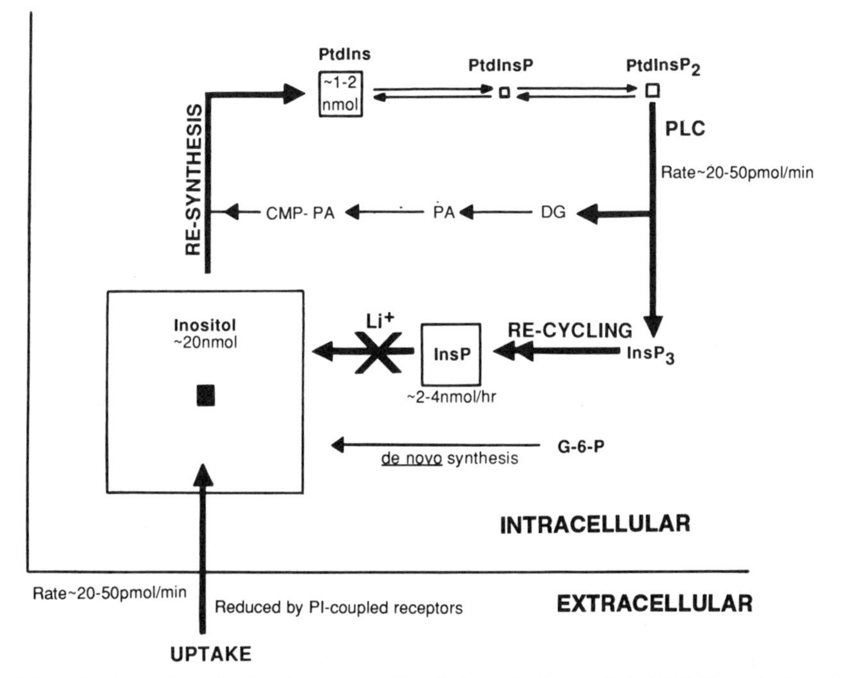

Figure 2. Basal and agonist-stimulated turnover of inositol metabolite pools in 1321 N1 cells. Box sizes are approximately representative of metabolite pool sizes. A single well of cells grown in inositol replete medium contains 20nmol of intracellular inositol. The filled box indicates the intracellular inositol pool size of inositol depleted cells.

Cells grown both ± inositol subsequently accumulated approximately 50% of the [^{3}H]-inositol intracellularly. As the intracellular volume is small compared to that of the culture medium, this implies that 132INI cells are able to accumulate inositol against a concentration gradient. However, in cells cultured in medium lacking inositol (depleted cells), the free [^{3}H]-inositol as a percentage of [^{3}H]-PtdIns was about 2 orders of magnitude lower than that in inositol replete cells. Preliminary estimates suggest PtdIns mass under both conditions is similar.

These simple observations show not only that these cells can accumulate inositol by uptake from the medium but also that this must be a major mechanism for maintaining resting inositol concentrations. By implication, synthesis of inositol de novo can have only a minimal role. In turn this raises further questions concerning: (i) the properties of the inositol uptake process in these cells; (ii) the capacity of this system relative to the rate of inositol re-cycling during stimulated PI-turnover; (iii) the relative effects on PI signalling of disrupting each of these processes.We have examined each of these issues in more detail as discussed below.

Inositol Uptake in 132INI Cells

The mechanisms by which different cells obtain inositol from the extracellular environment vary. Entry of inositol in to several types of cell appears to occur by a non-saturable, non-energy dependent diffusion-like process while other cells exhibit a relatively high affinity , sodium dependent transporter. Still further systems display both processes such that the dominant mechanism depends on the extracellular inositol concentration (Sherman, 1989; Nahorski et al, 1991)

We have studied inositol uptake into 132INI cells by following the intracellular accumulation of tracer [^{3}H]-inositol in the presence of varying extracellular inositol concentrations. At 1mM, [^{3}H]-inositol equilibrates between the intra- and extra-cellular compartments over 12-24 hours and over a wide range of extracellular concentrations, linear rates of intracellular [^{3}H]-inositol accumulation are measured over at least an initial 60-90 min. During this period, >98% of [^{3}H]-inositol uptaken remains within the free inositol pool in inositol replete cells. When measured under these conditions, the accumulation of intracellular label thus accurately reflects inositol uptake without complications of incorporation into other cellular metabolites.

Using this approach we have demonstrated that inositol uptake into 132INI cells occurs predominantly by a sodium dependent, saturable mechanism. The affinity of this system (Km ~40μM) is similar to that reported for several other cell types and at physiological inositol concentrations is likely to be 50-90% saturated (Sherman, 1989). The maximal rate of uptake is approximately 25-50 pmol/min/well cells which is of similar order to maximal against - stimulated rates of PI-hydrolysis measured under similar conditions (see below). Additional studies suggest that uptake by this mechanism may be subject to negative regulation by receptors which couple to PI-turnover as is also observed in parotid acinar cells (Torrens et al, 1991) although the significance of this is at present uncertain.

The Effects of Reduced Inositol Supply on PI-Signaling in 132INI Cells

As noted above, culture of 132INI cells in the absence of medium inositol markedly reduces cellular inositol but not PtdIns concentrations. In such inositol depleted cells, the re-synthesis of PtdIns during receptor activation of PLC may be dependent on the efficiency of inositol re-cycling. As this process is inhibited by Li^+ ions we have compared muscarinic receptor stimulated PI-turnover in inositol replete and depleted cells and the influence of Li^+ on this in the latter in order to establish the relative significance of inositol uptake vs re-cycling to continued signalling through this system.

In inositol replete cells a maximally effective concentration of carbachol (1mM) evoked a near linear accumulation of [^{3}H]-InsP over a 60-90 min incubation in the presence of 10 mM LiCl. implying a continuous uniform rate of PLC. Although the accumulation of [^{3}H]-InsP over this time was equivalent to 1-2 times the [^{3}H]PtdIns pool, (estimated at 1-2 nmol/well cells) the concentration of [^{3}H]PtdIns remained unchanged, reflecting effective re-synthesis from the much larger pool of free [^{3}H]-inositol. In contrast, in inositol depleted cells the initial rate of [^{3}H]-InsP accumulation declined after ~30min to a rate ~ 1-25% that seen initially and the concentration of [^{3}H]-PtdIns fell markedly, implying that in these cells PI synthesis is impaired, at least in the presence of Li^+ ions.

Further studies in inositol depleted cells showed that exposure to carbachol alone evoked increased concentrations of both [^{3}H]-InsP and [^{3}H]-inositol and a modest fall in the concentration of [^{3}H]PtdIns. Concentrations of each of these achieved new steady-states between 15-30 min which were then maintained for 60-75 min. In the presence of Li^+ however, the rise in [^{3}H]-inositol was prevented and the increase in [^{3}H]-InsP and fall in [^{3}H]-PtdIns were potentiated. Importantly however, although concentrations of [^{3}H]-PtdInsP responded similarly to those of [^{3}H]-PtdIns, concentrations of [^{3}H]-PtdInsP$_2$ (the immediate substrate for PLC) were influenced much less, if at all, by Li^+ and/or carbachol. These results suggest that despite a severe reduction in their free inositol concentration, these inositol depleted cells can sustain continued PI-turnover by efficient inositol re-cycling but when this is inhibited by Li^+ both PI re-synthesis and hydrolysis is disrupted.

We have quantified the extent to which PI-hydrolysis in inositol depleted cells is impaired by Li^+, again using [^{3}H]-InsP accumulation as an index of PLC activity. As noted previously, after carbachol stimulation for >30 min in the presence of Li^+ in these cells, [^{3}H]-InsP accumulation falls to a fraction of the initial rate such that close to a new steady-state is achieved. If, at this point, further muscarinic receptor-stimulated PLC activity is prevented by addition of atropine, the pre-accumulated concentration of [^{3}H]-InsP declines slowly. This shows that although inositol monophosphatase is inhibited by Li^+, this inhibition is incomplete. The rate of [^{3}H]-InsP removal after atropine addition is a measure of the residual monophosphatase activity but, more importantly, since the measure of this is made under close to steady-state conditions, this rate also gives a good approximation of the activity of PLC immediately prior to receptor blockade. When measured in this way after 60 min prior exposure to carbachol and Li^+, the rate of PLC is shown to have fallen to between ~15-35% of that observed during the initial 15 min stimulation (i.e. PLC activity is reduced by at least 65%). Additional studies are currently in progress to determine the impact of this on stimulated concentrations of the second messenger, Ins(1,4,5)P_3, although one would anticipate that these would be likely to fall in proportion to the severe attenuation of PLC activity.

Summary

The results of these studies demonstrate several important aspects of inositol homeostasis and its relationship to PI-signalling in 132INI cells. Firstly, resting cellular inositol concentrations are maintained predominantly by inositol uptake rather than synthesis. Secondly, since the maximal muscarinic receptor-stimulated rate of InsP accumulation in these cells is of the order of several tens of pmol/min/well cells, the kinetic parameters of this uptake process (Km ~ 40μM; Vmax~25-50 pmol/min/well cells) are both appropriate to efficient operation at physiological inositol concentrations and to import a large proportion of the inositol required to support stimulated PI-turnover. Thirdly, removal of medium inositol markedly reduces intracellular concentrations and, under these conditions, the effects of Li^+ on stimulated PLC activity and PtdIns concentration demonstrate the dependence of these cells on inostiol re-cycling. Fourthly, the latter is highly efficient and appears able to sustain persistent PI-turnover. These results also emphasize however, that Li^+ is likely to exert its most profound inhibitory influence on PI-signalling in cells which are unable to or inefficiently accumulate inositol from the extracellular environment and as a consequence of their limited inositol reserve are highly dependent on rapid inositol re-cycling. Finally, it is also notable that although the reduction in PLC activity by Li^+ in such cells is clearly associated with reduced cellular inositol, PtdIns and PtdInsP, this loss of activity is not readily attributed to a fall in PtdInsP_2 supply in 132INI cells. Previous studies have also been unable to show effects of Li^+on PtdInsP_2 concentrations although it has been argued that this may simply reflect the small proportion of this lipid associated with a hormone sensitive pool (Nahorski et al, 1991). We are currently employing pulse chase protocols which take advantage of the rapid rate of inositol uptake into the 132INI cells to investigate this point.

INOSITOL POLYPHOSPHATES

It will be of interest to know whether inositol depletion strategies, such as that described above, can be used to probe the putative signalling functions of other Ins metabolites. Some of these, the 3-phosphorylated inositol phosholipids, will be considered later, but we now want to consider the inositol polyphosphates (inositol tetrakis-, pentakis-

and hexakis- phosphates). Some of these compounds were amongst the first Ins metabolites to be detected in living organisms yet their functions, in the majority of cases, remain to be established (Downes and Macphee, 1990). Indeed, the importance of Ins polyphosphates is emphasised by the emerging information on their synthesis. Whilst these compounds have a ubiquitous distribution amongst eukaryotic cells, different organisms appear to have evolved distinct metabolic strategies for their biosynthesis. The seemingly most straightforward strategy, involving sequential phosphorylations starting from Ins itself, has been adopted by the slime mould *Dictyostelium discoideum* (Stephens and Irvine, 1990). By contrast, those mammalian cells that have been studied in detail appear to lack Ins kinase activities and probably synthesise inositol polyphosphates by a complex series of phosphorylations and dephosphorylations starting from Ins(1,4,5)P_3, the product of phospholipase C mediated cleavage of phosphatidylinositol bisphosphate (Shears, 1989). A similar strategy, but with a different set of intermediates, occurs in at least one plant, the unicellular alga, *Chlamydomonas eugametos* (Irvine *et.al.*, 1992).

In order to circumvent inherent difficulties in any coventional biochemical approaches to defining the fuctions of inositol polyphosphates, we have begun to define pathways of inositol polyphosphate biosynthesis in a genetically tractable organism, the budding yeast, *Saccharomyces cerevisiae*. This organism can synthesise inositol pentakisphosphate by the direct phosphorylation of Ins(1,4,5)P_3. The first step in this pathway is catalysed by a cytosolic kinase which specifically phosphorylates either the D-6 position to give Ins(1,4,5,6)P_4 or the 2-position to give Ins(1,2,4,5)P_4. This enzyme has now been purified greater than 1000-fold. The goal of this research is to isolate the gene encoding the yeast Ins(1,4,5)P_3 kinase and to use standard genetic approaches to understand its function and hence the functions of at least some inositol polyphosphates in this organism.

G-PROTEIN DEPENDENT REGULATION OF PHOSPHOLIPASE C IN TURKEY ERYTHROCYTES

Turkey erythrocyte ghosts have provided a uniquely responsive cell free system for the study of receptor and G-protein regulated phospholipase C (PLC) in native membranes. The PLC present in ghost membrane preparations is activated by P_{2y}-purinergic receptor agonists in a guanine nucleotide dependent manner (Harden et al, 1988; Boyer et al, 1989a). A 150kD PLC purified from turkey erythrocytes has been shown, by reconstitution with PLC depleted membranes, to be the G-protein regulated species of PLC in these cells (Morris et al, 1990a, 1990b). Receptor stimulated activation of turkey erythrocyte PLC (and of PLCβ isoenzymes in other cells) is mediated by a heterotrimeric guanine-nucleotide dependent regulatory protein (G-protein). A 43kD PLC activating G-protein α-subunit has recently been purified from turkey erythrocyte membranes (Waldo et al, 1991). Sequence analysis indicates this protein to be the avian homologue of a G-protein designated G_{11} (D. Maurice, A. Morris, G. Waldo, and T. K. Harden, personal communication).G_{11} is a member of the closely related G_q family of G-proteins (reviewed by Simon et al,1991), which have been shown to activate PLCβ (but not PLCγ) enzymes in vitro (Smrcka et al, 1991; Taylor et al, 1991).

Agonist occupied receptors catalyze the exchange of GDP bound to the α subunit of appropriate G-proteins for free GTP, a process which results in dissociation of the α from the βγ subunits. GTP liganded α subunits are able to regulate effector enzyme (eg PLC) activity. Hydrolysis of the GTP by an endogenous GTPase activity returns Gα to its inactive GDP liganded state. The GDP liganded α subunit recombines with βγ, thereby terminating the G-protein activation cycle. Non-hydrolyzable analogues of GTP (such as GTPγS) bind to, dissociate, and constitutively activate G-proteins in the absence of hormonal stimuli (reviewed by Gilman, 1987).

Bidirectional Signalling in Turkey Erythrocytes

The starting point for the work described here was an observation made by Boyer et al (1989b). These workers investigated the effects of exogenously added G-protein $\beta\gamma$ subunits on purinergic receptor regulated PLC activity in turkey erythrocyte ghosts. It was found that when reconstituted with turkey erythrocyte ghost "acceptor" membranes, $\beta\gamma$ subunits potentiated P_{2y}-purinergic agonist and GTP stimulated PLC activity. To explain this result, it was proposed that succesful coupling with the P_{2y} receptor required G_{11} to exist in its undissociated heterotrimeric ($\alpha\beta\gamma$) form, and that the ghost membranes contained an excess of free α_{11} over $\beta\gamma$. It was suggested therefore, that the reconstituted $\beta\gamma$ might be interacting with free α_{11} subunits in the membranes, and by mass action driving the formation of more G-protein heterotrimers available for coupling with the agonist occupied P_{2y}-receptor.

Hormonal activation of G-protein linked receptors, and the resultant dissociation of the G-protein heterotrimer into α and $\beta\gamma$ subunits, is a physiological event which is likely to elevate the free $\beta\gamma$ content of plasma membranes. Turkey erythrocyte membranes are known to contain both β-adrenergic receptors and adenosine receptors (Tolkovsky and Levitzki, 1978) which are linked to G_s (the G-protein responsible for mediating the hormonal stimulation of adenylyl cyclase). Selective dissociation of heterotrimeric G_s, by agonist and guanine-nucleotide, is likely to increase the free $\beta\gamma$ content of the plasma membranes, thereby providing a physiological correllate of the effects of adding exogenous $\beta\gamma$ to the system. Therefore, the potential modulation of P_{2y} receptor stimulated PLC by β-adrenoceptor and adenosine receptor activated G_s was investigated.

As would be predicted by the model outlined above, the β-adrenergic receptor agonist, isoproterenol, is able to increase the PLC response due to GTPγS and a maximally effective concentration of ADPβS (a P_{2y}-purinergic receptor agonist). However, isoproterenol also increases GTPγS stimulated PLC activity in the absence of ADPβS, indicating that stimulation of the β-adrenergic receptor does not simply modulate P_{2y}-purinergic receptor regulated PLC. Moreover, stimulation of adenosine receptors in the presence of GTPγS (with or without ADPβS) fails to modify PLC activity, indicating that the isoproterenol stimulation of PLC occurs independently of G_s (and adenylyl cyclase) activation.

Isoproterenol stimulated PLC activity exhibits an absolute requirement for GTP or GTPγS. Stimulation of PLC activity by GTPγS is preceded by a pronounced time lag, due to slow dissociation of GDP from the G-protein. Agonist occupied receptors increase the rate of exchange of G-protein bound GDP for free guanine nucleotide, thus reducing the lag phase which precedes activation of G-proteins by non-hydrolyzable guanine nucleotides. Both isoproterenol and ADPβS reduce the lag phase of PLC activation observed with GTPγS alone (Vaziri and Downes, 1992). Overall, these data are consistent with G-protein mediated activation of PLC by β-adrenergic receptors as well as P_{2y}-purinergic receptors in turkey erythrocytes.

The turkey erythrocyte β-adrenergic receptor is distinct from the mammalian $\beta1$, $\beta2$, and $\beta3$ receptor subtypes based on its primary sequence (Yarden et al, 1986), and pharmacological properties (Minneman et al, 1980). Nevertheless, a range of non-selective and $\beta1$ or $\beta2$ selective anatagonists inhibit isoproterenol stimulated adenylyl cyclase and PLC activities with the same rank order of potencies and comparable K_i values (C. V. and C. P. D., unpublished data). It is likely therefore, that a single β-adrenergic receptor subtype is able to stimulate both PLC and adenylyl cyclase activities in turkey erythrocytes. Surprisingly however, β-adrenergic stimuli are more potent at stimulating adenylyl cyclase activity. Isoproterenol stimulates adenylyl cyclase with an EC_{50} of 4.4nM. By contrast, isoproterenol stimulates PLC activity with an EC_{50} of 126nM. If, as seems likely, a single β-receptor subtype elicits both responses, it would appear that a greater degree of receptor occupancy is required for maximal stimulation of PLC than is necessary for maximal stimulation of adenylyl cyclase activity.

Some fundamental questions arising from the observations described above relate to the identities of the β-adrenergic receptor linked G-protein, and the PLC. The rate of activation of PLC due to combined maximally effective doses of isoproterenol and ADPβS (in the presence of GTPγS) is only approximately 70% of a calculated fully additive response to maximally effective concentrations of isoproterenol or ADPβS alone.This indicates that the activated β-adrenergic and P_{2y}-purinergic receptors converge upon common downstream components that are rate limiting under conditions of supraphysiological stimulation. These common components are likely to be G-protein and/or PLC, as discussed below.

As discussed above, G-proteins of the $G_{q/11}$ family have been shown to stimulate purified PLCβ isozymes from bovine brain (Smrcka et al, 1991), bovine liver (Taylor et al, 1991) and turkey erythrocytes (Waldo et al, 1991). Antisera raised against a decapeptide common to the C-termini of G_q and G_{11}, recognize a single protein band of 43kD on Western blots of turkey erythrocyte membranes (C. V. and C. P. D., unpublished data). By analogy with other G-proteins (eg G_s, G_i, and transducin), the C-terminal region of $G_{q/11}$ is believed to be important for interaction with activated receptors. Preincubation of turkey erythrocyte ghosts with antisera to $G_{q/11}$ inhibits subsequent activation of PLC by β-adrenergic as well as by P_{2y}-purinergic receptor agonists. However, stimulation of PLC by AlF_4^-, which directly binds to and activates GDP-liganded G-proteins, is unaffected by antibody pretreatment, indicating that the antisera do not impair function of the PLC interactive region of turkey erythrocyte G_{11}(C. V. and C. P. D., manuscript in preparation). These data demonstrate that a $G_{q/11}$ like G-protein mediates activation of PLC by P_{2y}-purinergic and by β-adrenergic receptors.

The PLC present in turkey erythrocyte ghosts is uncoupled from activation by G_{11} following treatment of the ghosts with Mg^{2+}-free buffers. However, purified 150kD PLC from turkey erythrocyte cytosol has been shown to reconstitute ADPβS and GTPγS dependent PLC activity, when combined with the uncoupled ghosts (Morris et al, 1990b). Likewise, purified PLC reconstitutes isoproterenol stimulated activity when combined with uncoupled ghost preparations in the presence of GTPγS. The isoproterenol dose dependency of reconstituted PLC is identical to that of isoproterenol stimulated PLC in fully coupled ghosts (Vaziri and Downes, 1992). These experiments demonstrate that a single species of purified PLC can be regulated by both P_{2y}-purinergic and β-adrenergic receptors.

In conclusion, the results described here suggest that the turkey erythrocyte β-adrenergic receptor can stimulate adenylyl cyclase or PLC activities via direct interactions with G_s and G_{11} respectively. Furthermore, it would appear that the β-adrenergic and P_{2y}-purinergic receptors in these cells interact with a common G-protein, G_{11}, to stimulate the same species of PLC. Previous studies have indicated that purified receptors can interact with different G-proteins when reconstituted into lipid vesicles (Rubenstein et al, 1991). Similarly, expression of Leutenizing Hormone (LH)-receptors in cells lacking endogenous receptors for LH confers hCG (an LH receptor agonist) dependent production of cAMP as well as IP_3 , apparently due to direct interaction of the activated LH-receptor with G_s and G_q respectively (Gudderman et al, 1992). It is likely that turkey erythrocytes are an example of a cell in which β-adrenergic stimulation of PLC and adenylyl cyclase activities reflects physiological interaction of a single multipotential receptor with different G-proteins. Fig.3 illustrates the multiplicity of intracellular second messengers generated in turkey erythrocytes, due to activation of different signal transduction pathways in response to diverse hormonal stimuli.

REGULATION OF PHOSPHATIDYLINOSITOL 3-KINASE IN PC12 CELLS

Recent work has established that, in addition to the well known inositol phospholipids considered earlier, cells possess 3 further lipids each having an additional monoester phosphate group in the 3-position of the inositol ring. The occurrence of these lipids reflects the activity of phosphatidylinositol 3-kinase (PtdIns 3-kinase). PtdIns 3-kinase has been

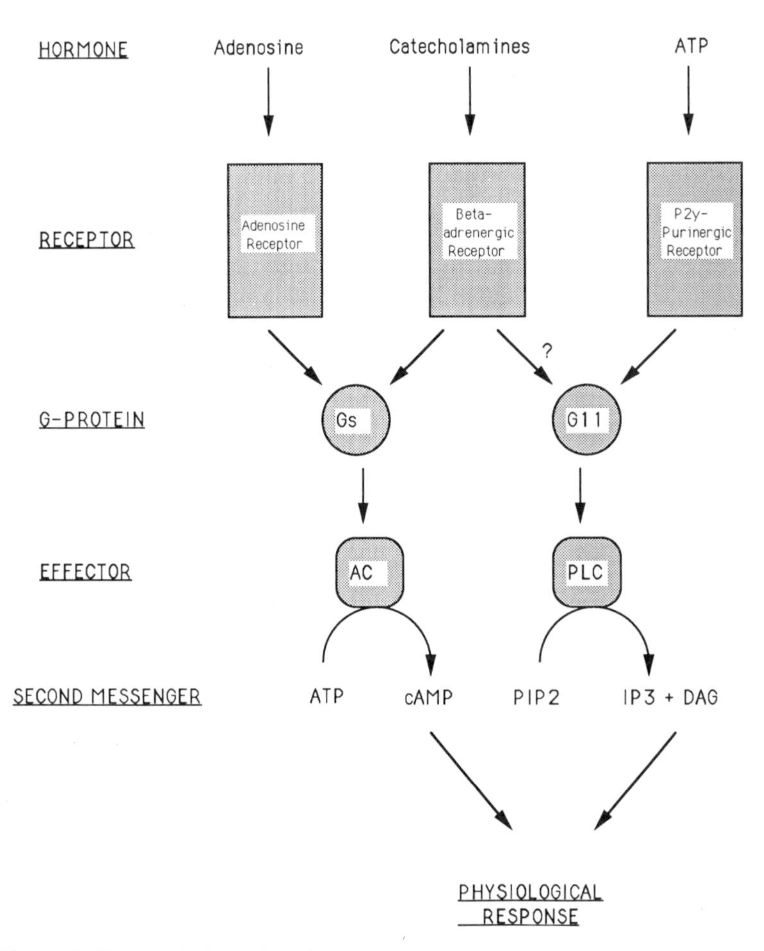

Figure 3. Hormonal stimulation of PLC and adenylyl cyclase in turkey erythrocytes.

shown to be activated by a number of growth factor receptors possessing a ligand activated tyrosine kinase activity, in a growth factor dependent fashion. These include the receptors for platelet derived growth factor, colony stimulating factor 1, stem cell factor, and insulin. PtdIns 3-kinase is also involved in the actions of certain oncoproteins that are constitutively active tyrosine kinases. It has been shown to form a complex with and appears to be activated by $pp60^{v\text{-}src}$, polyoma virus middle T antigen/$pp60^{c\text{-}src}$ complex, and $pp160^{gag\text{-}abl}$ (Downes and Carter, 1991; Cantley et al, 1991).The intracellular products of PtdIns 3-kinase that rise acutely in all of these situations, phosphatidylinositol 3,4-bisphosphate ($PtdIns(3,4)P_2$) and phosphatidylinositol 3,4,5-trisphosphate ($PtdIns(3,4,5)P_3$), have, therefore, been proposed as novel second messengers involved in mitogenic signal transduction, and the aetiology of cellular transformation leading to oncogenesis.

The product of the TRKA protooncogene, $gp140^{trkA}$, is a receptor tyrosine kinase that has been shown to be activated by nerve growth factor (NGF) in rat pheochromocytoma (PC12) cells. $gp140^{trkA}$ appears to be the high affinity receptor for NGF, although the functional receptor may require oligomerisation involving the previously characterised low affinity p75 NGF receptor (see Downes and Carter, 1991). The actions of NGF may be mediated through similar mechanisms to those operating for other growth factors, such as epidermal growth factor (EGF), whose receptors possess a ligand activated tyrosine kinase domain. Immediate functional targets of the NGF receptor tyrosine kinase, however, have yet to be identified.

Rat pheochromocytoma (PC12) cells have been used extensively as a model of the biogenesis and maintenance of sympathetic neurons. This stems from the fact that PC12 cells respond to nerve growth factor (NGF) by cessation of growth, extension of neurites and assumption of many of the biochemical characteristics of sympathetic neurons. The signal transduction processes initiated by NGF are poorly understood, although a role for cAMP has been proposed on the basis that both dibutyryl cAMP and forskolin stimulate neurite outgrowth in PC12 cells. However, the morphology induced by these agents is distinct from that induced by NGF. More recently a number of groups have detected a small increase in tyrosine phosphorylation of phospholipase Cγ1 in response to NGF, with a concomitant increase in inositol phosphate levels (reviewed by Cantley et al, 1991).

Studies on the effects of inducible oncogenes in PC12 cells has revealed that oncogenic $p21^{ras}$ and $pp60^{v\text{-}src}$ do not transform PC12 cells, as they do in fibroblasts, but instead promote neurite outgrowth in a manner reminiscent of the effects of NGF. One small distinction is that the oncogenic responses are not density inhibited as is the case for NGF. Microinjection of neutralising antibodies to $pp60^{v\text{-}src}$ does not inhibit the response to $p21^{ras}$, however microinjection of neutralising antibodies to $p21^{ras}$ does inhibit the response to $pp60^{v\text{-}src}$, implying that both src and ras are required for neurite outgrowth, and that src is upstream of ras. An extension of this work reported recently used miroinjection of the same antibodies to show that the neurite outgrowth response to NGF could also be blocked with either anti-$p21^{ras}$ or anti-$pp60^{src}$, thus establishing these proteins as important for the NGF response in PC12 cells (Bar-Sagi and Feramisco, 1985; Kremer et al, 1991; Thomas et al, 1991).

We have reported that PtdIns 3-kinase is activated acutely in PC12 cells stimulated with NGF (Carter and Downes, 1992 and Table 1). Remarkably, the levels of the product(s) of this reaction can be elevated by as much as 25-30 fold above basal.

Table 1. Effects of growth factor stimulation of dual [^{3}H]inositol / [^{32}P]Pi labelled PC12 cells on PtdIns(3,4)P_2 and PtdIns(3,4,5)P_3.

	[^{3}H] dpm		[^{32}P] dpm	
Growth factor	PtdIns(3,4)P_2	PtdIns(3,4,5)P_3	PtdIns(3,4)P_2	PtdIns(3,4,5)P_3
Control	378 +/- 110	41 +/- 26	321 +/- 101	124 +/- 86
NGF 5mins	5275 +/- 70	7727 +/- 250	6949 +/- 941	2229 +/- 26
EGF 1min	8949 +/- 15	8707 +/- 78	11157 +/- 538	2262 +/- 124

It seems likely that this response represents a prominent signal transduction pathway utilised by NGF, and as there appear to be essential roles for $p21^{ras}$ and $p60^{src}$ in the neurite forming action of NGF, we have been investigating the possibility that PtdIns 3-kinase participates with these proteins in the co-ordination of neurite outgrowth.

Recent studies have established that the dominant-negative mutant of $p21^{ras}$, where Ser17 is mutated to an Asn, can block the effects of NGF on PC12 cells (Thomas et al, 1992; Wood et al, 1992), in particular the activation of the mitogen activated protein kinase (MAP kinase). However, in insulin stimulated A14 cells (NIH 3T3 cells that overexpress the human insulin receptor), where the $p21^{ras}$Asn17 blocks MAP kinase activation, the PtdIns 3-kinase response appears not to be impaired (de Vries-Smits et al, 1992), so if PtdIns 3-kinase is prominent in this response then it lies upstream of the activation of $p21^{ras}$, or alternatively on a parallel signal transduction pathway. As there is also evidence that $p60^{src}$ lies upstream of $p21^{ras}$ in PC12 cells, then one possiblity is that PtdIns 3-kinase and $p60^{src}$ are co-operating in some way.

NGF induces accumulation of PtdIns 3-kinase in antiphosphotyrosine immunoprecipitates in PC12 cells, and western blots show a tyrosine phosphorylated protein of ~60 KDa from NGF stimulated cells (Carter and Downes, 1992). We reasoned that PtdIns 3-kinase and $pp60^{src}$ might interact directly and, therefore, looked for complexes containing these 2 proteins by immunoprecipitating members of the src family tyrosine kinases with the antibody cst-1, which recognises $pp60^{c\text{-}src}$, $p60^{c\text{-}yes}$ and $p56^{fyn}$. NGF induced an increase of PtdIns 3-kinase activity in these immunoprecipitates which exhibited a similar time course to the activity found in antiphosphotyrosine immunoprecipitates.

Using more specific antibodies we attenpted to determine which member(s) of the src-family of tyrosine kinases could co-immunoprecipitate PtdIns 3-kinase activity. Table 2 illustrates the results of an experiment utilising antibodies specific for $pp60^{c\text{-}src}$, $pp60^{c\text{-}yes}$ and $p56^{fyn}$. The only immunoprecipitates which showed an NGF dependent increase in PtdIns 3-kinase activity were those for $pp60^{c\text{-}src}$, with no increases being found in either $pp60^{c\text{-}yes}$ or $p56^{fyn}$. This suggests a specific interaction between a PtdIns 3-kinase dependent signal transduction pathway and $pp60^{c\text{-}src}$.This may suggest a direct, presumably functional association between $p60^{src}$ and PtdIns 3-kinase. Alternatively each of these components could be associating with an additional protein (for example the NGF receptor itself), hence causing antibodies directed against $pp60^{c\text{-}src}$ to immunoprecipitate PtdIns 3-kinase activity.

Table 2. Recovery of PtdIns 3-kinase activity in anti-src family tyrosine kinase immuno–precipitates following 5 minute NGF stimulation of PC12 cells

Antibody specificity	+/- NGF	PI 3-kinase activity [^{32}P]-PI 3-P per assay (dpm)
src+yes+fyn 0 mins	-	3721 +/- 132
src+yes+fyn 5 mins	-	3133 +/- 304
src 5 mins	+	11097 +/- 686
yes 5mins	+	3159 +/- 749
fyn 5mins	+	3879 +/- 200

In conclusion, several lines of evidence indicate that PtdIns(3,4,5)P_3 is a second messenger with important roles in mediating cellular responses to growth factors and oncogenic transformation. Our results also implicate this lipid signal molecule in the molecular basis of neuronal differentiation and survival induced by NGF. Figure 4 summarises some recent findings on NGF signal transduction pathways in PC12 cells.

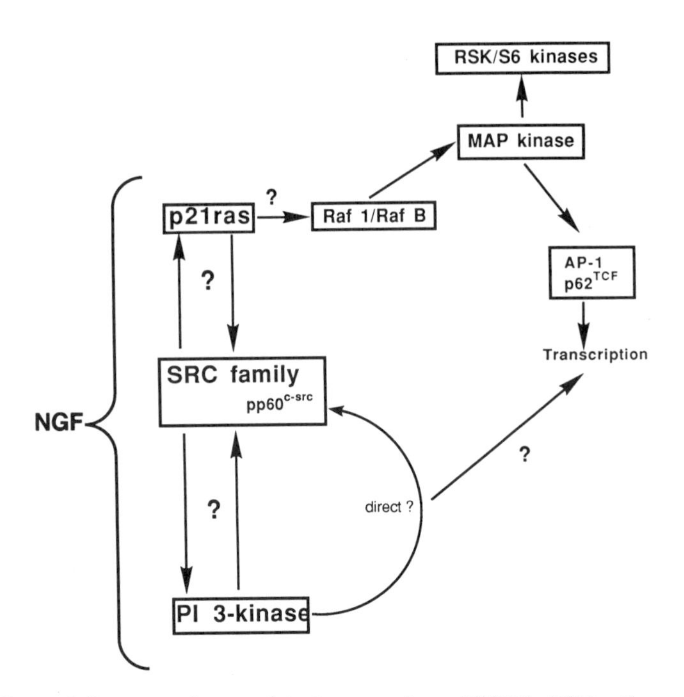

Figure 4. Summary of some of the known actions of NGF in PC12 cells

ACKNOWLEDGEMENTS

We are very grateful to Dr S. Courtneidge, EMBL, Heidelberg, Germany, for the gift of antibodies to src-family tyrosine kinases. D.S. and C.V. were supported by research studentships and I.B. and A.N.C.by Project Grants awarded by the Medical Research Council of Great Britain.

REFERENCES

Bar-Sagi, D. and Feramisco, J.R.(1985) Cell 42, 841-848.

Berridge, M. J. & Irvine, R. F. (1989) Nature 341, 197-205.

Berridge, M. J., Downes, C. P. & Hanley, M. R. (1982) Biochem. J. 206, 587-595.

Berridge, M. J., Downes, C. P. & Hanley, M. R. (1989) Cell 59 411-419.

Boyer, J. L., Downes, C. P. and Harden, T. K. (1989a) J. Biol. Chem. 264, 884-890

Boyer, J. L., Waldo, G. L., Evans, T., Northup, J. K., Downes, C. P. and Harden, T. K. (1989b) J. Biol. Chem. 264, 13917-13922.

Cantley, L.C., Auger, K.R., Carpenter, C, Duckworth, B., Graziani, A., Kapeller, R. and Soltoff, S. (1991) Cell 64, 281-302.

Carter, A.N. and Downes, C.P. (1992) J.Biol.Chem., 267, *in press.*

Downes, C. P. & Carter, A. N. (1991) Cellular Signalling 3, 501-513.

Downes, C.P. and Macphee, C.H. (1990) Eur.J.Biochem. 193, 1-18.

Gilman, A. G. (1987) Ann. Rev. Biochem. 56, 615-649

Gudermann, T., Birnbaumer, M. and Birnbaumer, L. (1992) J. Biol. Chem. 267, 4479-4488.

Hallcher, L. M. & Sherman, W. R. (1980) J. Biol. Chem. 255, 10896-10901.

Harden, T. K., Hawkins, P. T., Stephens, L., Boyer, J. L. and Downes, C. P. (1988) Biochem. J. 252, 583-593.

Irvine, R.F., Letcher, A.J., Stephens, L.R. and Musgrave, A. (1992) Biochem.J. 281, 261-266.

Kremer, N.E., D`Arcangelo, G., Thomas, S.M., DeMarco, M., Brugge, J.S., Halegoua, S. (1991) J.Cell Biol. 115, 809-819.

Minneman, K. P., Weiland, G. A. and Molinoff, P. B. (1979) Mol. Pharmacol. 17, 1-7.

Morris, A. J., Waldo, G. L., Downes, C. P. and Harden, T. K. (1990a) J. Biol. Chem. 265, 13501-13507.

Morris, A. J., Waldo, G. L., Downes, C.P. and Harden, T. K. (1990b) J. Biol. Chem. 265, 13508-13513.

Nahorski, S. R., Ragan, C. I. & Challiss, R. A. J. (1991) Trends Pharmacol. Sci. 12, 297-303.

Rubenstein, R. C., Linder, M. E. and Ross, E. M. (1991) Biochemistry 30, 10769-10777.

Shears, S.B. (1989) Biochem J. 260, 313-324.

Sherman, W. R. (1989) in Inositol Lipids in Cell Signalling (eds. R. H. Michell, A.M. Drummond & C. P. Downes) pp 39-79, Academic Press.

Simon, M. I., Strathmann, M. P. and Gautam, N, (1991) Science 252, 802-808.

Smrcka, A. V., Hepler, J. R., Brown, K. O. and Sternweis, P. C. (1991) Science 251, 804-807.

Stephens, L.R. and Irvine, R.F. (1990) Nature 346, 580-583.

Taylor, S.J., Chae, H. Z., Rhee, S. G. and Exton, J. H. (1991) Nature 350, 516-518.

Thomas, S.M., DeMarco, M., D`Arcangelo, G., Halegoua, S. and Brugge, J.S. (1992) Cell 68, 1031-1040.

Thomas, S.M., Hayes, M., D`Arcangelo, G., Armstrong, R.C., Meyer, B., Zilberstein, A., Brugge, J.S., and Halegoua, S. (1991) Mol.Cell Biol. 11, 4739-4750.

Tolkovsky, A. M. and Levitzki, A. (1978) Biochemistry 17, 3795-3810.

Torrens, Y., Diett, M., Beaujovan, J. C. & Glowinski, J. (1991) J. Pharmacol.Exp. Ther. 258, 639-646.

Vaziri, C. and Downes, C. P. (1992) Biochem. J. 284, 917-922.

de Vries-Smits, A.M.M., Burgering, B.M.T., Leevers, S.J., Marshall, C.J.and Bos, J.L.(1992) Nature 357, 602-604.

Waldo, G. L., Boyer, J. L., Morris, A. J. and Harden, T. K. H. (1991) J. Biol.Chem. 266, 14217-14225.

Wood, K.W., Sarnecki, C., Roberts, T.M. and Blenis, J. (1992) Cell 68, 1040-1050.

Yarden, Y., Rodriguez, H., Wong, S. K. F., Brandt, D. R., May, D. C., Burnier, J., Harkins, R. N., Chen, E. Y., Ramachandran, J., Ullrich, A. and Ross, E. M. (1986) Proc. Natl. Acad. Sci. USA 83, 6795-6799.

ACTIVATION MECHANISMS OF PHOSPHOLIPASE C ISOZYMES

Sue Goo Rhee

Laboratory of Biochemistry
National Heart, Lung and Blood Institute
National Institutes of Health
Bethesda, Maryland 20892 U.S.A.

INTRODUCTION

On binding to their cell surface receptors, many extracellular signaling molecules including hormones, peptide growth factors, neurotransmitters and immunoglobulins, elicit intracellular responses by activating inositol phospholipid-specific phospholipase C (PLC).[1] Activated PLC catalyzes the hydrolysis of phosphatidylinositol 4,5-bisphosphate (PIP_2) to generate diacylglycerol and inositol 1,4,5-trisphosphate (IP_3). Diacylglycerol is the physiological activator of protein kinase C (PKC) and IP_3 induces the release of Ca^{2+} from internal stores.[1] This bifurcating pathway constitutes the cornerstone of a transmembrane signal transduction mechanism that is now known to regulate a large array of cellular processes, including metabolism, secretion, contraction, neural activity, and proliferation.

Direct protein isolation and molecular cloning studies have revealed the existence of multiple PLC isozymes in mammalian tissues. The various PLC isoforms appear to be activated by different receptors and different mechanisms and to interact differently with inhibitory mechanisms such as those mediated by cAMP-dependent protein kinase (PKA) and PKC. Here we summarize our current knowledge on PLC with emphasis on the primary structures and activation mechanisms of the various isoforms.

PLC ISOFORMS

Several distinct PLC enzymes have been purified from a variety of mammalian tissues ([2] and refs. therein), and a total of 16 amino acid sequences -- 14 mammalian enzymes and 2 Drosophila enzymes -- have been deduced from the nucleotide sequences of their corresponding cDNAs ([3] and refs. therein). Comparison of deduced amino acid sequences has indicated that the PLCs can be divided into three types -- PLC-ß, PLC-γ, and PLC-δ -- and that each type contains more than one subtype; subtypes are designated by adding Arabic numerals after the Greek letters as in PLC-ß1 and PLC-ß2, for example. Three mammalian (ß1, ß2, and ß3) and two Drosophilia (norpA and P21) subtypes are known for PLC-ß, whereas two PLC-γ1 subtypes (γ1 and γ2) and four

ACTIVATION OF PLC-γ ISOZYMES

Polypeptide growth factors such as platelet-derived growth factor (PDGF), epidermal growth factor (EGF), fibroblast growth factor (FGF), colony-stimulating factor (CSF-1), nerve growth factor (NGF), and insulin mediate their pleiotropic actions by binding to and activating cell surface receptors that have a similar molecular topology, including a cytoplasmic region that contains a tyrosine kinase domain. Despite the structural similarities between their receptors, the participation of polypeptide growth factors in PI signaling does not appear to be universal: binding of PDGF, EGF, and NGF to their respective receptors induces PI turnover; in contrast, insulin and CSF-1 appear to have no effect on PI turnover.

Growth factor-induced stimulation of PLC appears to be independent of G proteins and to require the intrinsic tyrosine kinase activity of the receptors. Treatment of a number of cell types with EGF, PDGF, or NGF led to an increase in the phosphorylation of PLC-γ1 -- but not of PLC-ß1 or PLC-δ1 -- with the increased phosphorylation occurring on both serine and tyrosine residues ([3] and refs. threin). EGF-induced tyrosine phosphorylation of PLC-γ1, which is mediated directly by the EGF receptor tyrosine kinase, occurs rapidly and correlates well with stimulation of PIP_2 hydrolysis. Treatment of cells with EGF, PDGF, or NGF also promotes the association of PLC-γ1 with the cognate receptor.[3] Thus, antibodies to either PLC-γ1 or growth factor receptors immunoprecipitate both proteins. The receptor-PLC-γ1 association is mediated by a high-affinity interaction between the SH2 domains of PLC-γ1 and a specific tyrosine-autophosphorylated site of the receptor.[4] Among the five autophosphorylated sites at the carboxyterminus of the EGF receptor, Tyr992 has been identified as the high-affinity binding site of the EGF receptor for PLC-γ1 SH2 domains.[23] In the FGF receptor, Tyr766 and its flanking sequences have been identified as the major binding site for PLC-γ1.[20]

Association of growth factor receptors with PLC-γ1 precedes tyrosine phosphorylation of PLC-γ1 by the receptor tyrosine kinase. The major sites of PLC-γ1 phosphorylated by the receptors for EGF, PDGF, and NGF appear to be identical and are Tyr771, Tyr783, and Tyr1254.[21] The presence of SH2 domains in PLC-γ1 has been shown to facilitate PLC-γ1 phosphorylation by lowering the apparent K_m of substrate, not by increasing the V_{max}.[22] The role of tyrosine phosphorylation was investigated by substituting Phe for Tyr at the three sites of PLC-γ1 and expressing the mutant enzymes in NIH 3T3 cells.[21] Tyr783, and to a lesser extent Tyr1254, were shown to be essential for PDGF-stimulated inositol phosphate formation in intact cells. Like the wild-type enzyme, PLC-γ1 substituted with Phe at Tyr783 associated with the PDGF receptor and was phosphorylated at serine residue in response to PDGF. These results suggest that phosphorylation of Tyr783 is essential for PLC-γ1 activation, and that neither the association of PLC-γ1 with the receptor nor its phosphorylation on serine residues is sufficient to account for PDGF-induced activation of PLC-γ1. However, the experiments with cells expressing mutant EGF receptors that lack Tyr992 indicate that both tyrosine phosphorylation of PLC-γ1 and the SH2-mediated interaction with the activated EGF receptor are necessary for PLC-γ1 activation _in vivo_.[23]

In vitro studies indicate that phosphorylated and unphosphorylated PLC-γ exhibit similar activities under standard assay conditions.[21] However, the unphosphorylated enzyme was found to be selectively inhibited in the presence of a micellar concentration of Triton X-100[24] or in the presence of the small soluble actin-binding protein profilin, which also shows a high affinity for PIP_2.[25] Because only the phosphorylated enzyme catalyzes the hydrolysis of profilin-bound PIP_2, thereby releasing profilin and altering actin polymerization, profilin was suggested as the link between transmembrane signaling and cellular responses such as changes in shape and increased motility.[25]

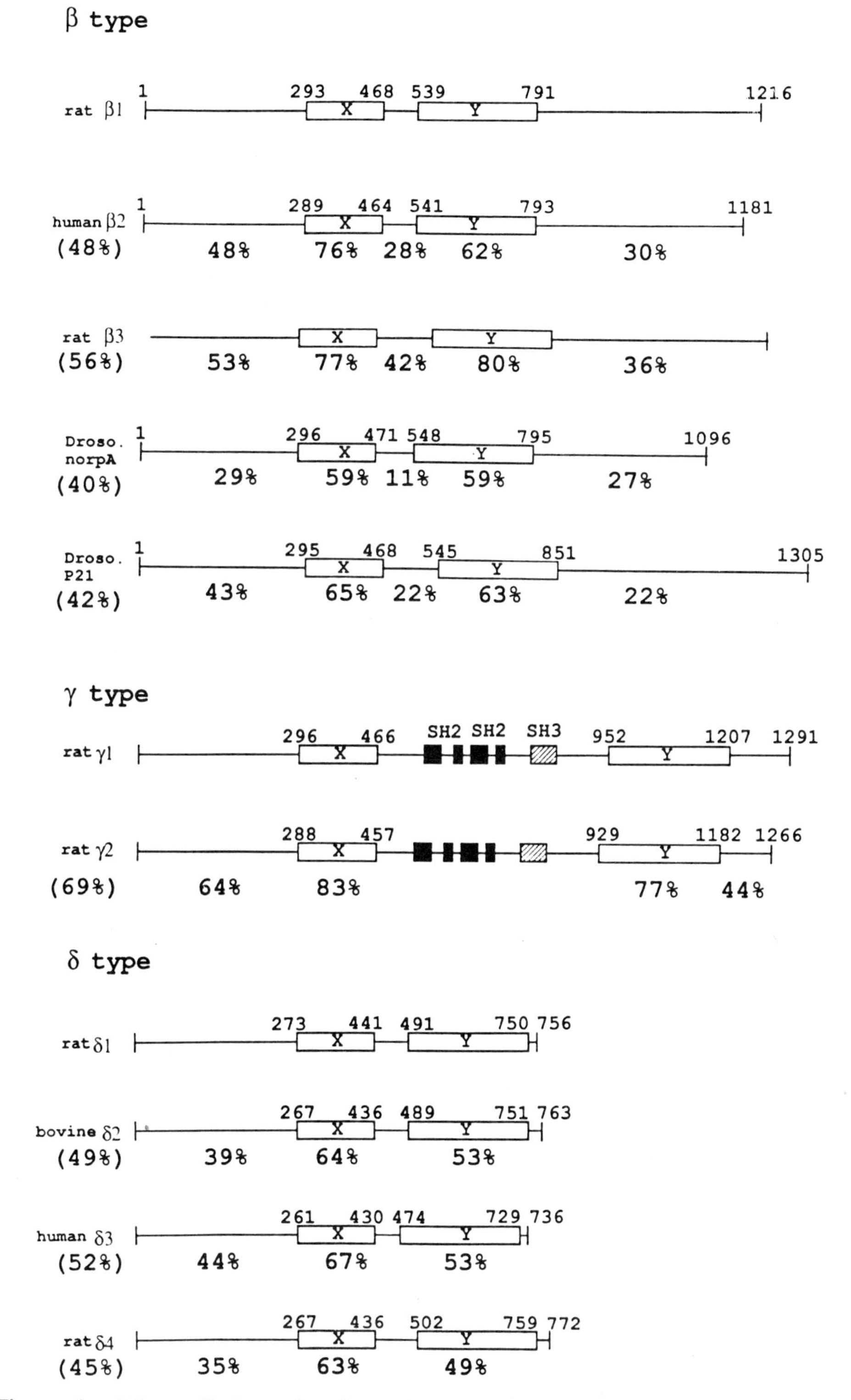

Figure 1. Linear display of amino acid sequences deduced from cDNA of PLC isozymes, highlighting the difference between the three types of PLCs (ß, γ, and δ types) and the similarity between the subtypes.

Little is known about the molecular events that occur after the tyrosine phosphorylation of PLC-γ1. It is possible that the SH2 domains of tyrosine-phosphorylated PLC-γ1 interact intramolecularly with PLC-γ1 tyrosine phosphates. Such an intramolecular interaction may elicit a conformational change that allows the SH3 domain to bind to the membrane cytoskeleton and position the putative catalytic X and Y domains at the cytoplasmic face of the cell membrane. Although there is no direct support for this model, it is consistent with the proposed role of the SH3 domain[26] and with the observation that EGF or PDGF treatment of cells induces the translocation of PLC-γ1 from a predominantly cytosolic localization to membrane fractions.[27]

The tyrosine phosphorylation and activation of PLC-γ1 can also be achieved through the action of nonreceptor protein tyrosine kinase (PTKs) in response to ligation of certain cell surface receptors in leukocytes ([28] and refs. therein). The T cell antigen receptor (TCR) complex functions to recognize antigen and to transduce signals across the plasma membrane. The TCR complex is composed of at least seven polypeptides, including the three polypeptides that are collectively termed CD3. Although none of the components of the TCR complex is a protein kinase, ligation of the TCR complex activates PTK activity. Candidates for the TCR-regulated PTK include the products of fyn and lck, both of which are members of the src family. Recently, TCR stimulation was shown to result in a rapid and transient phosphorylation of PLC-γ1 on both serine and tyrosine residues.[29] Two-dimensional phosphopeptide map analysis revealed that the major sites of tyrosine phosphorylation in PLC-γ1 in activated T cells are the same as those phosphorylated in cells treated with PDGF or EGF.[29] Tyrosine phosphorylation of PLC-γ1 by nonreceptor PTKs was also shown to be involved in the elevated PIP_2 hydrolysis seen in response to ligation or cross-linking of the membrane IgM in B lymphocytes, the high-affinity IgE receptor ($Fc_{\epsilon}RI$) in basophilic leukemia cells, and IgG receptors (FcγRI and FcγRII) in monocytic cells.[28,30]

PLC-γ2 also appears to be activated by nonreceptor PTKs. Our recent data (unpublished) suggest that stimulation of the TCR, membrane IgM, $Fc_{\epsilon}RI$ or FcγRs elicits the tyrosine phosphorylation of PLC-γ2. However, the ratio of PLC-γ2 to PLC-γ1 concentrations and the relative extents of phosphorylation differed significantly between the respective cells. Two residues, Tyr753 and Tyr759, were identified as phosphorylation sites in PLC-γ2. In PLC-γ1, Tyr771, Tyr783, and Tyr1254 were phosphorylated and phosphorylation of Tyr783 was shown to be essential for enzyme activation. The sequence surrounding Tyr759 in PLC-γ2 is similar to that surrounding PLC-γ1 Tyr783, and the relative location of Tyr753 in PLC-γ2 is approximately equivalent to that of PLC-γ1 Try771. Tyrosine phosphorylation was not observed in the carboxyl-terminal region of PLC-γ2.

As an example of nonreceptor PTK-dependent activation of PLC-γ, the activation of PLC-γ1 in response to TCR stimulation is shown on the right side of Figure 2B. The nature of the interaction between PTKs and leukocyte receptors is not known. The depicted interaction of PLC-γ1 with the nonreceptor PTK through the PLC-γ1 SH2 domains and the PTK phosphotyrosine residues was patterned on the interaction of PLC-γ1 with growth factor receptors.

ACTIVATION OF PLC-δ ISOZYMES

Neither the receptors nor the transducer that are coupled to any of the PLC-δ members is known.

REFERENCES

1. R.S. Rana and L.E. Hokin, Role of phosphoinositides in transmembrane signaling, *Physiol. Rev.* 70:115-164 (1990).

2. S.G. Rhee, P.G. Suh, S.H. Ryu, and S.Y. Lee, Studies of inositol phospholipid-specific phospholipase C, *Science* 244:546-550 (1989).
3. S.G. Rhee, and K.D. Choi, Multiple forms of phospholipase C isozymes and their activation mechanisms, *in:* "Advances in Second Messenger and Phosphoprotein Research," J.W. Putney, Jr., ed., Raven Press, New York, pp. 35-60 (1992).
4. A.C. Koch, D.A. Anderson, M.F. Moran, C. Ellis, and T. Pawson, T., SH2 and SH3 domains: Elements that control interactions of cytoplasmic signaling proteins, *Science* 252:668-674 (1991).
5. S.P. Srivastava, N.Q. Chen, Y.X. Liu, and J.L. Holtzman, Purification and characterization of a new isozyme of thiol:protein-disulfide oxidoreductase from rat hepatic microsomes, *J. Biol. Chem.* 266:20337-20344 (1991).
6. M.I. Simon, M.P. Strathmann, and N. Gautam, Diversity of G proteins in signal transduction, *Science* 252:802-808 (1991).
7. A.V. Smrcka, J.R. Hepler, K.O. Brown, and P.O. Sternweis, Regulation of polyphosphoinositide-specific phospholipase C activity by purified G_q, *Science* 250:804-807 (1991).
8. S.J. Taylor, H.Z. Chae, S.G. Rhee, and J.H. Exton, Activation of the ß1 isozyme of phospholipase C by α subunits of the G_q class of G protein, *Nature* 350:516-518 (1991).
9. J.L. Blank, A.H. Ross, and J.H. Exton, Purification and characterization of two G-proteins that activate the ß1 isozyme of phosphoinositide-specific phospholipase C, *J. Biol. Chem.* 266:18206-18216 (1991).
10. G. Waldo, J. Boyer, A. Morris, and T.K. Harden, Purification of an AlF_4^- and G-protein ßγ-subunit-regulated phospholipase C-activating protein, *J. Biol. Chem.* 266:14217-14225 (1991).
11. A. Shenker, P. Goldsmith, C.G. Unson, and A.M. Spiegel, The G protein coupled to the thromboxane A_2 receptor in human platelets is a member of the novel G_q family, *J. Biol. Chem.* 266:9309-9313 (1991).
12. R.L. Wange, A.V. Smrcka, P.C. Sternweis, and J.H. Exton, Photoaffinity labeling of two rat liver plasma membrane proteins with $[^{32}P]\gamma$-azidoanilido GTP in response to vasopressin, *J. Biol. Chem.* 266:11409-11412 (1991).
13. S. Gutowski, A. Smrcka, L. Nowak, D. Wu, M. Simon, and P.C. Sternweis, Antibodies to the α_q subfamily of guanine nucleotide-binding regulatory protein α subunits attenuate activation of phosphatidylinositol 4,5-bisphosphate hydrolysis by hormones, *J. Biol. Chem.* 266:20519-20524 (1991).
14. S.K.-F. Wong, E.M. Parker, and E.M. Ross, Chimeric muscarinic cholinergic:ß-adrenergic receptors that activate G_s in response to muscarinic agonists, *J. Biol. Chem.* 265:6219-6224 (1990).
15. D. Wu, C.H. Lee, S.G. Rhee, and M.I. Simon, Activation of phospholipase C by the α subunits of the G_q and G_{11} proteins in transfected Cos-7 cells, *J. Biol. Chem.* 267:1811-1817 (1992).
16. D. Park, D.-Y. Jhon, R. Kriz, J. Knopf, and S.G. Rhee, Cloning, sequencing, expression, and G_q-independent activation of phospholipase C-ß2, *J. Biol. Chem.*, in press (1992).
17. C.H. Lee, D. Park, D. Wu, S.G. Rhee, and M.I. Simon, Members of the G_q α subunit gene family activate phospholipase C-ß isozymes, *J. Biol. Chem.*, in press (1992).

18. G. Berstein, J.L. Blank, D.-Y. Jhon, J.H. Exton, S.G. Rhee, and E.M. Ross, Phospholipase C-ß1 is a GTPase activating protein (GAP) for $G_{q/11}$, its physiologic regulator, *Cell*, in press (1992).
19. D. Rotin, B. Margolis, M. Mohammadi, R. Daly, G. Daum, N. Li, W. Burgess, E.H. Fischer, A. Ullrich, and J. Schlessinger, SH2 domains prevent tyrosine dephosphorylation of the EGF receptor: Identification of Tyr992 as the high-affinity binding site for SH2 domains of phospholipase C-γ, *EMBO J.* 11:559-567 (1992).
20. M. Mohammadi, A.M. Honegger, D. Rotin, R. Fischer, F. Bellot, W. Li, C.A. Dionne, M. Jaye, M. Rubinstein, and J. Schlessinger, A tyrosine-phosphorylated carboxy-terminal peptide of the fibroblast growth factor receptor (Flg) is a binding site for the SH2 domain of phospholipase C-γ1, *Mol. Cell. Biol.* 11:5068-5078 (1991).
21. H.K. Kim, J.W. Kim, A. Zilberstein, B. Margolis, C.K. Kim, J. Schlessinger, and S.G. Rhee, PDGF stimulation of inositol phospholipid hydrolysis requires PLC-γ1 phosphorylation on tyrosine residues 783 and 1254, *Cell* 65:435-441 (1991).
22. D. Rotin, A.M. Honegger, B.L. Margolis, A. Ullrich, and J. Schlessinger, Presence of SH2 domains of phospholipase Cγ1 enhances substrate phosphorylation by increasing the affinity towards the EGF-receptor, *J. Biol. Chem.*, in press (1992).
23. Q.C. Vega, C. Cochet, O. Filhol, C.P. Chang, S.G. Rhee, and G.N. Gill, A site of tyrosine phosphorylation in the C terminus of the epidermal growth factor receptor is required to activate phospholipase C, *Mol. Cell. Biol.* 12:128-135 (1992).
24. M.I. Wahl, G.A. Jones, S. Nishibe, S.G. Rhee, and G. Carpenter, Growth factor stimulation of phospholipase C-γ1 activity: Comparative properties of control and activated enzymes, *J. Biol. Chem.*, in press (1992).
25. P.J. Goldschmidt-Clermont, J.W. Kim, L.M. Machesky, S.G. Rhee, and T.D. Pollard, Regulation of phospholipase C-γ1 by profilin and tyrosine phosphorylation, *Science* 251:1231-1233 (1991).
26. D.G. Drubin, J. Mulholland, Z. Zhu, and D. Botstein, Homology of a yeast actin-binding protein to signal transduction proteins and myosin-I, *Nature* 343:288-290 (1990).
27. G. Todderud, M.I. Wahl, S.G. Rhee, and G. Carpenter, Stimulation of phospholipase C-γ1 membrane association by epidermal growth factor, *Science* 249:296-299 (1990).
28. S.G. Rhee, D.J. Park, and D. Park, Regulation of phospholipase C isozymes, *in* "Cellular and Molecular Mechanisms of Inflammation: Signal Transduction," C.G. Cochrane and M.A. Gimbrone, eds., Academic Press, New York, in press (1992).
29. D.J. Park, H.W. Rho, and S.G. Rhee, CD3 stimulation causes phosphorylation of phospholipase C-γ1 on serine and threonine residues in a human T-cell line, *Proc. Natl. Acad. Sci. U.S.A.* 88:5453-5456 (1991).
30. F. Liao, H.S. Shin, and S.G. Rhee, Tyrosine phosphorylation of phospholipase C-γ1 induced by cross-linking of the high-affinity or low-affinity Fc receptor for IgG in U937 cells, *Proc. Natl. Acad. Sci. U.S.A.*, in press (1992).

TUMOR NECROSIS FACTOR CYTOTOXICITY IS ASSOCIATED WITH ACTIVATION OF CELLULAR PHOSPHOLIPASES

Dirk De Valck, Rudi Beyaert, Frans Van Roy, and Walter Fiers

Laboratory of Molecular Biology
Gent University
K.L. Ledeganckstraat 35
B-9000 Gent, Belgium

INTRODUCTION

Tumor Necrosis Factor (TNF), a cytokine primarily produced by activated macrophages, exerts a broad range of activities on different cell types, mostly related to inflammation and immunomodulation (reviewed in Camussi et al., 1991; Fiers, 1991; Aggarwal and Vilček, 1992). Among these, the most interesting feature of TNF is its selective toxicity for many tumor cells, leaving normal cells unaffected. Remarkably, the selective killing of transformed cells is often much more pronounced in combination with synergizing agents [e.g. interferon-γ (Williamson et al., 1983), LiCl (Beyaert et al., 1989)], making this molecule a good candidate for future cancer therapy development.

So far, the molecular mechanism of TNF action on cells in culture is largely unknown. The classical cell line for testing TNF cytotoxicity as well as TNF-mediated gene induction (e.g. interleukin-6) is the murine fibrosarcoma cell line L929. The cytotoxic action of TNF is nucleus independent, and the effect is even considerably enhanced by transcription or translation inhibitors (Ruff and Gifford, 1981).

TNF binds as a trimer to target cells via either of two different, high affinity receptors (reviewed in Loetscher et al., 1991). The TNF receptors can be regulated by protein kinases since activation of protein kinase C with phorbol ester, downregulates the receptors, while activation of protein kinase A upregulates the receptors (Scheurich et al., 1989). Signal transduction to the intracellular space occurs by receptor clustering (Shalaby et al., 1990; Tartaglia et al., 1991), while the complexes are internalized and degraded. Furthermore, TNF receptors might be coupled to a G-protein, as both TNF-mediated interleukin-6 induction and TNF cytotoxicity are inhibited by pertussis toxin treatment in intact cells, and as the non-hydrolyzable GTP-analogue GTPγS is able to enhance the TNF signal in permeabilized cells (Imamura et al., 1988; Earl et al., 1990). Activation of cellular phospholipases (PLA_2, PLC, PLD) with generation of important lipid second messengers, is also believed to occur upon TNF-treatment (Suffys et al., 1987 and 1991; Beyaert et al., submitted; De Valck et al., 1993).

New Developments in Lipid-Protein Interactions and Receptor Function
Edited by K.W.A. Wirtz *et al.*, Plenum Press, 1993

TNF induces the transcription of a number of genes by a rapid, but transient induction of the transcription factors c-fos and c-jun, and by a more lasting activation of NFκB (Lin et al., 1987; Osborn et al., 1989). Furthermore, there is evidence for the involvement of the mitochondria, where the electron transport chain becomes disrupted at the ubisemiquinone site, with generation of highly reactive superoxide radicals (Lancaster et al., 1989; Schulze-Osthoff et al., 1992). This is most probably a key step in the cytotoxic action of TNF as superoxide radicals can cause irreversible cell damage by lipid peroxidation, protein degradation and DNA fragmentation. The different TNF-induced signaling pathways discussed in this introduction are illustrated in Figure 1.

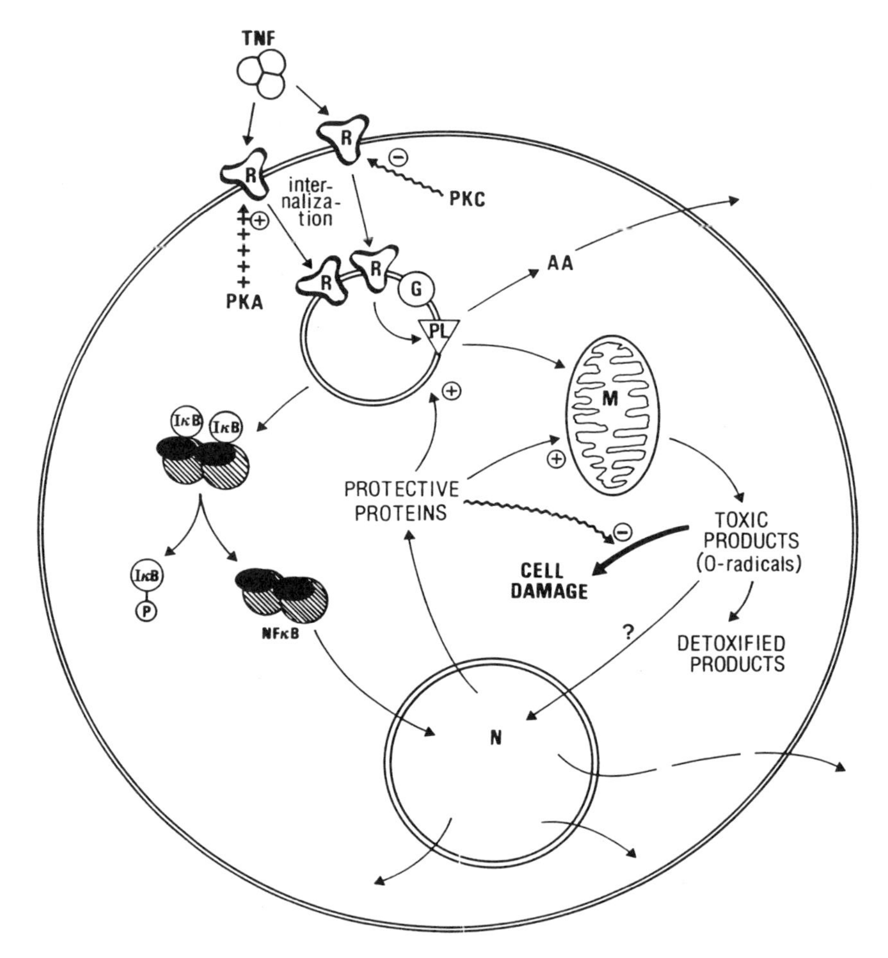

Figure 1. Model for the mechanism of action of TNF. The trimeric TNF interacts with its receptor (R) and by clustering causes internalization. The receptor itself is negatively regulated by protein kinase C and positively by protein kinase A. Internalized, clustered TNF receptor complexes transmit one or more signals leading to activation of phospholipases (PL), which might be mediated by a G-protein (G), and disruption of the electron transport system with formation of radicals in the mitochondria (M). These reactive oxygen species may eventually kill the cell and lead to the activation of the transcription factor NFκB, which induces the transcription of a number of genes in the nucleus (N).

TNF ACTIVATES CELLULAR PHOSPHOLIPASES

Activation of an Arachidonic Acid-selective Phospholipase A_2

TNF-sensitive cells release arachidonic acid into the medium upon TNF treatment. This release is already significant from 2-4 h on and precedes cell lysis (Suffys et al., 1991). As arachidonic acid is mainly incorporated in the sn-2 position of phospholipids, this TNF-induced release suggests the activation of a phospholipase A_2 (PLA_2). Studies with various PLA_2-inhibitors (such as quinacrine, dibucaine, dexamethasone) also point in the same direction, as they inhibit TNF-mediated arachidonic acid release and protect cells against TNF cytotoxicity (Hepburn et al., 1987; Suffys et al., 1987). Although steroids such as dexamethasone are known to inhibit PLA_2 by the induction of lipocortins, we could find no evidence for an involvement of lipocortin in TNF-induced signaling (Beyaert et al., 1990).

When the cells contain the appropriate enzymes, arachidonic acid may be converted to prostaglandins by cyclooxygenases, and to other eicosanoids as thromboxanes or leukotrienes by lipoxygenases. However, inhibitor studies revealed that there is no reason to believe that these metabolites play a role in cell killing. In addition, evidence could also be obtained against an involvement of arachidonic acid as such in TNF cytotoxicity (Suffys et al., 1987 and 1991).

Activation of a Phosphatidylinositol-specific Phospholipase C

In 1989, our laboratory reported that LiCl increases considerably the cytotoxic activity of TNF towards some transformed cell lines such as L929 (Beyaert et al., 1989). Since LiCl is a potent inhibitor of inositol phosphatases, this initial observation prompted us to investigate whether the inositol lipid cycle was affected by TNF (Figure 2).

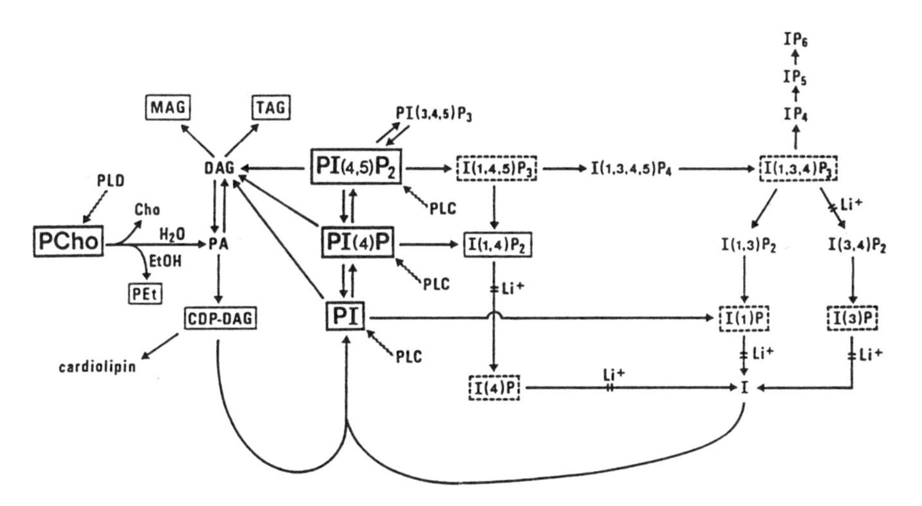

Figure 2. Simplified scheme of the inositol lipid cycle. Phosphatidylinositol lipids (PI, PI(4)P, PI(4,5)P_2) can be cleaved by PLC to produce DAG and IP_n. The latter are further metabolized to inositol (I) through several Li^+-sensitive and Li^+-insensitive steps. Resynthesis of PI occurs through phosphorylation of DAG to PA and reaction of CDP-DAG with inositol. Another way to generate PA is by PLD-mediated hydrolysis of phosphatidylcholine (PC). Alternative pathways to metabolize DAG are the synthesis of triacylglycerols (TAG) and monoacylglycerols (MAG).

When intracellular inositol is depleted by LiCl, TNF induces a transient increase of CDP-diacylglycerol (CDP-DAG) in L929 cells (Beyaert et al., submitted). However, this finding does not allow to distinguisch between upstream PLD or PLC activation, since activation of both phospholipases can result in PA and subsequent CDP-DAG formation.

Treatment of L929 cells with the combination of TNF and LiCl leads to the prolonged accumulation of inositol phosphates (IP_n, i.e. IP_1, IP_2 and IP_3; Beyaert et al., submitted). Characterization of IP_2 isomers by HPLC analysis revealed that the TNF + LiCl-induced increase in inositol phosphate levels was due to activation of a PLC and not a PLD. The TNF + LiCl-induced increase in IP_3 further suggests a role for intracellular Ca^{2+}-mobilization in TNF cytotoxicity.

Activation of a Phosphatidylcholine specific Phospholipase D

Besides the well-documented activation of intracellular PLA_2 and PLC upon agonist stimulation, evidence is rapidly growing to indicate that also PLD plays a role in second messenger generation (reviewed in Dennis et al., 1991). The primary product which is formed upon PLD activation is PA. Lyso-PA, has been shown to be a Ca^{2+}-mobilizing stimulus, and to posses growth factor-like activity (Moolenaar et al., 1986; Jalink et al., 1990). Furthermore, PA can be converted to a well documented protein kinase C activator, viz. DAG, by the action of a PA phosphohydrolase (Billah et al., 1989). Although PA and DAG can result from PLD activity, these products are not specific indicators for PLD activation. Indeed, PLC activity and subsequent DAG kinase activity can also lead to the formation of DAG and PA, respectively (Dennis et al., 1991).

In contrast to the formation of PA and DAG, the transphosphatidylation of phospholipids using a primary alcohol as acceptor molecule has been shown to be a unique feature of PLD (Yang et al., 1967; Figure 3). Therefore TNF-sensitive L929 cells were prelabeled with [^{14}C]palmitic acid or [^{14}C]lysophosphatidylcholine, and treated with TNF in the presence of 0.5% ethanol. Lipids were extracted from the cells, and phosphatidylethanol (PEt) which was formed due to PLD-catalyzed transphosphatidylation, was then separated from other lipid metabolites using thin layer chromatography.

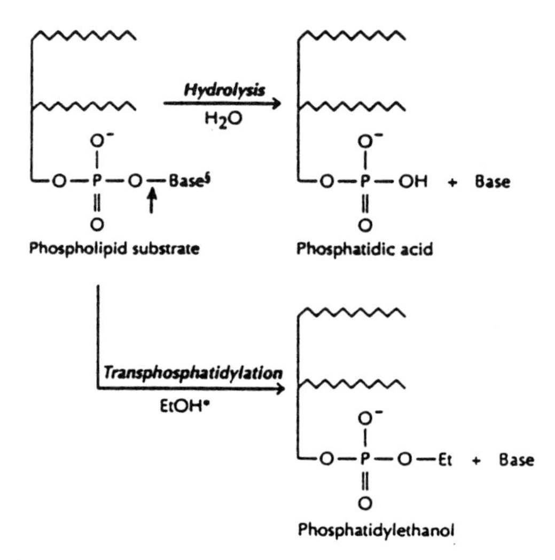

Figure 3. Schematic diagram of PLD-catalyzed transphosphatidylation (Shukla and Halenda, 1991).

In L929 cells, TNF-induced PEt production was first detectable 3-4 h after TNF addition and then gradually increased as a function of time (Figure 4). Measurement of the TNF-induced release of radioactivity from parallel cell cultures labeled with [^{3}H]uridine showed that TNF-induced PLD activity preceded the onset of cell lysis only by 1 h. TNF-induced PLD activity was dose dependent, with an EC_{50} value of approximately 250 IU/ml (data not shown). Although PLD-mediated hydrolysis of lipids induces the formation of PA and DAG, we were unable to detect these metabolites upon TNF-treatment. In contrast, a marked elevation of MAG and TAG levels was observed (data not shown), indicating that PA, via DAG, is rapidly metabolized into these neutral lipids.

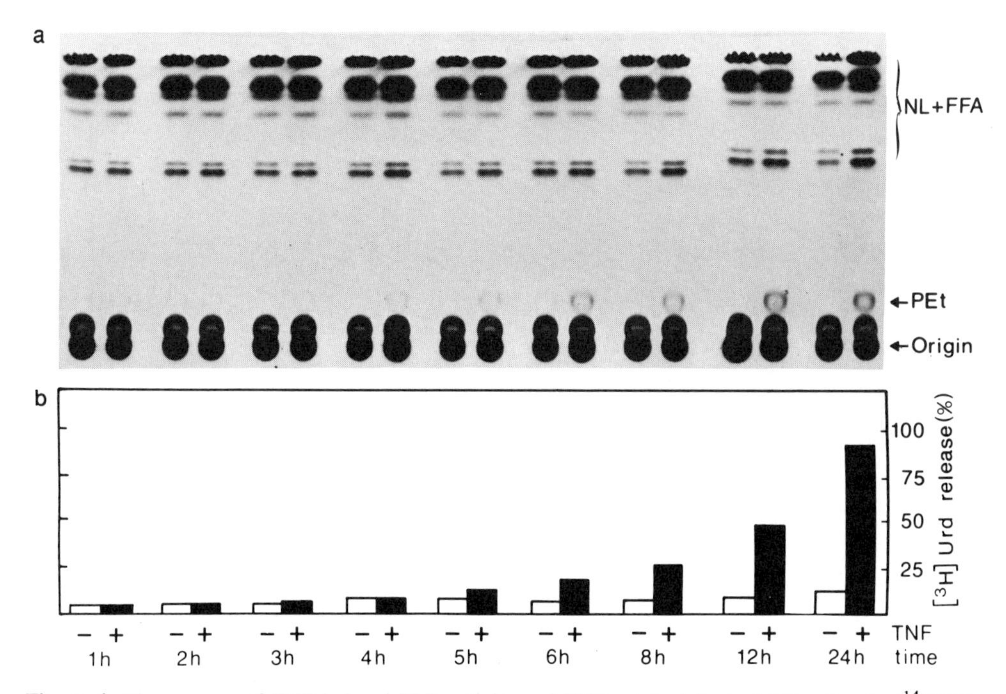

Figure 4. Time course of TNF-induced PLD activity and TNF cytotoxicity in L929 cells. A, [^{14}C]PEt formation by PLD-catalyzed transphosphatidylation on cells, prelabeled with [^{14}C]palmitic acid, in the presence of 0.5% ethanol. B, percentage of radioactivity released in the supernatant of cells prelabeled with [^{3}H]uridine ([^{3}H]Urd). Cells were either untreated (-; white boxes) or incubated with 5000 IU TNF/ml (+; black boxes) for the times indicated.

In addition to L929 cells, TNF induced PLD activity in the TNF-sensitive WEHI164cl13 and U937 cells, but failed to do so in TNF resistant L929r2 cells. As this type of resistant subclone becomes sensitive to TNF cytotoxicity in the presence of RNA or protein synthesis inhibitors (Vanhaesebroeck et al., 1991), we examined TNF-induced PLD activity after addition of the transcription inhibitor actinomycin D. Under these conditions, TNF-induced PLD activity could be restored (data not shown). These results clearly demonstrate that TNF-mediated PLD activation is restricted to cells which are sensitive to the cytotoxic effect of TNF.

In order to investigate the location of PLD in the TNF-signaling pathway, we tested the effect of drugs, known to modulate TNF cytotoxicity at different biochemical steps, on TNF-induced PLD activity. The following drugs were previously shown to inhibit TNF cytotoxicity: the G-protein inhibitor pertussis toxin (Suffys et al., 1987; Imamura et al., 1988; Earl et al., 1990); the PLA_2 inhibitor dexamethasone (Hepburn et al., 1987; Suffys et al., 1987); the serine-type protease inhibitor N_α-p-tosyl-L-arginine methyl ester (TosArgOMe; Ruggiero et al., 1987; Suffys et al., 1988); the monoamine oxidase inhibitor pargyline (Beyaert, unpublished results); the mitochondrial electron

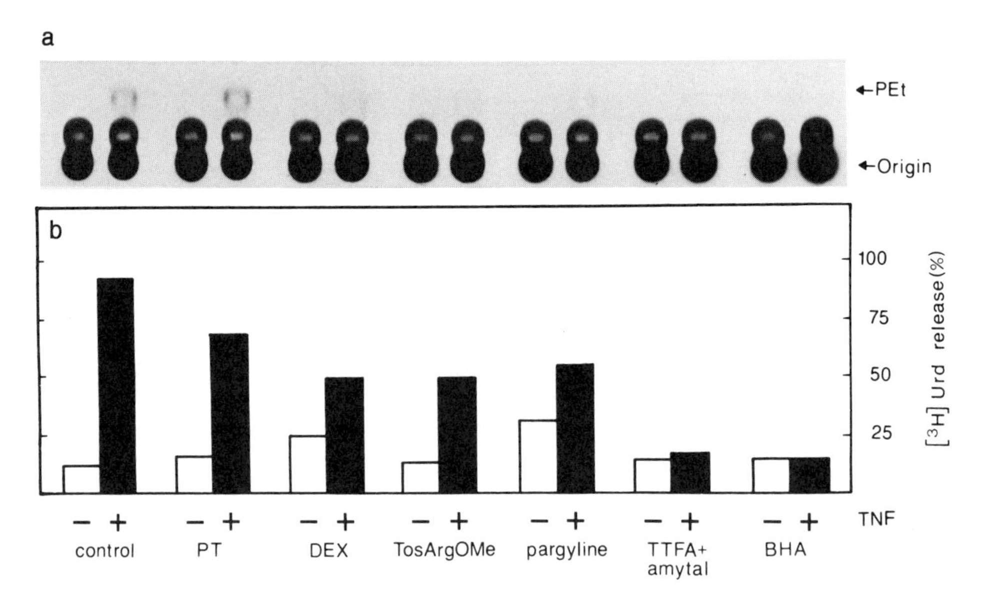

Figure 5. Effect of TNF cytotoxicity-inhibiting drugs on TNF-induced PLD activity in L929 cells. Cells prelabeled with [^{14}C]palmitic acid (A) or [^{3}H]uridine (B) were untreated (-) or were incubated with 5000 IU TNF/ml (+) for 18 h in the presence of pertussis toxin (PT; 1 μg/ml), dexamethasone (DEX; 50 μM), N_α-p-tosyl-L-arginine methyl ester (TosArgOMe; 5 mM), pargyline (3 mM), thenoyltrifluoroacetone (TTFA; 150 μM) + amytal (40 μg/ml), or butylated hydroxyanisole (BHA; 200 μM). All drugs were added 1 h before TNF.

transport inhibitors thenoyltrifluoroacetone and amytal, and the antioxidant butylated hydroxyanisole (Schulze-Osthoff et al., 1992). When TNF-induced PLD activity and TNF cytotoxicity were assayed in parallel, both activities were equally inhibited by the different drugs tested (Figure 5). Also when TNF cytotoxicity was enhanced by cotreatment with the inositol phosphatase inhibitor LiCl (Beyaert et al., 1989) or the protein kinase inhibitor staurosporine (Beyaert et al., submitted), potentiation of TNF-induced PLD activity could be found (data not shown). These results suggest that TNF-induced PLD activity occurs at a rather late step in the TNF-signaling pathway, as it is preceded by several other signaling reactions.

A major question which arises is whether PLD is specifically activated upon TNF stimulation or merely the result of cell killing per se. Therefore, we tested the effect of a number of cytotoxic agents, different from TNF, on PLD activity. The agents used included menadione (a synthetic quinone), paraquat (a bipyridyl herbicide), as well as dibromothymoquinone and daunorubicine (both quinone-containing antitumor antibiotics). All compounds killed L929 cells within the same incubation time and to approximately the same extent as TNF (Figure 6). Nevertheless, only cells killed by TNF showed PLD activation. These results demonstrate that PLD activation is specific for cell killing induced by TNF, and that it is not merely the result of cell death.

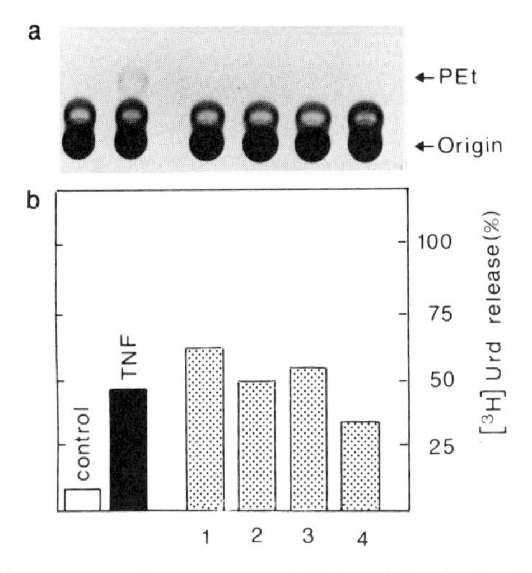

Figure 6. Effect of different cytotoxic reagents on PLD activity in L929 cells. Cells prelabeled with [^{14}C]palmitic acid (A) or [^{3}H]uridine (B) were untreated (control; white boxes), or were incubated for 18 h with TNF (5000 IU/ml; black boxes), or with different cytotoxic reagents (stippled boxes) including menadione (50 μg/ml; lane 1), paraquat (2 mg/ml; lane 2), dibromothymoquinone (100 μM; lane 3), or daunorubicine (100 μM; lane 4).

CONCLUSION

Although the exact molecular mechanism of TNF cytotoxicity is at present still unclear, we obtained evidence for TNF-induced activation of different phospholipases. Since the time kinetics for the PLA_2, PLC and PLD species involved are very much alike, intensive cross-talk might exist. Further attempts to characterize the exact biological roles of these enzymes in TNF-induced cytotoxicity are in progress.

ACKNOWLEDGEMENTS

The authors thank F. Van Houtte and W. Drijvers for technical assistance. D.D.V. holds a fellowship from the *Belgisch Werk tegen Kanker*. R.B. is a Research Assistant, and F.V.R. a Research Director with the *NFWO*. Research was supported by the *IUAP*, the *FGWO*, the *ASLK* and the *Nationale Loterij*.

REFERENCES

Aggarwal, B.B. and Vilček, J. (eds) (1992) Tumor necrosis factors: Structure, function, and mechanism of action, Marcel Dekker, New York.

Beyaert, R., Vanhaesebroeck, B., Suffys, P., Van Roy, F., and Fiers, W. (1989) Lithium chloride potentiates tumor necrosis factor-mediated cytotoxicity in vitro and in vivo. *Proc. Natl. Acad. Sci. USA* 86, 9494-9498.

Beyaert, R., Suffys, P., Van Roy, F., and Fiers, W. (1990) Inhibition by glucocorticoids of tumor necrosis factor-mediated cytotoxicity. Evidence against lipocortin involvement. *FEBS Lett.* 262, 93-96.

Billah, M.M., Eckel, S., Mullmann, T.J., Egan, R.W., and Siegel, M.I. (1989) Phosphatidylcholine hydrolysis by phospholipase D determines phosphatidate and diglyceride levels in chemotactic peptide-stimulated human neutrophils. Involvement of phosphatidate phosphohydrolase in signal transduction. *J. Biol. Chem.* 64, 17069-17077.

Camussi, G., Albano, E., Tetta, C., and Bussolino, F. (1991) The molecular action of tumor necrosis factor-α. *J. Biochem.* 202, 3-14.

Dennis, E.A., Rhee, S.G., Billah, M.M., and Hannun, Y.A. (1991) Role of phospholipases in generating lipid second messengers in signal transduction. *FASEB J.* 5, 2068-2077.

De Valck, D., Beyaert, R., Van Roy, F., and Fiers, W. (1993) Tumor necrosis factor cytotoxicity is associated with phospholipase D activation. *Eur. J. Biochem.* 212, 491-497.

Earl, C.Q., Stadel, J.M., and Anzano, M.A. (1990) Tumor necrosis factor-mediated biological activities involve a G-protein-dependent mechanism. *J. Biol. Resp. Modif.* 9, 361-367.

Fiers, W. (1991) Tumor necrosis factor. Characterization at the molecular, cellular and in vivo level. *FEBS Lett.* 285, 199-212.

Hepburn, A., Boeynaems, J.M., Fiers, W., and Dumont, J.E. (1987) Modulation of tumor necrosis factor-α cytotoxicity in L929 cells by bacterial toxins, hydrocortisone and inhibitors of arachidonic acid metabolism. *Biochem. Biophys. Res. Commun.* 149, 815-822.

Imamura, K., Sherman, M.L., Spriggs, D., and Kufe, D. (1988) Effect of tumor necrosis factor on GTP binding and GTPase activity in HL-60 and L929 cells. *J. Biol. Chem.* 263, 10247-10253.

Jalink, K., van Corven, E.J., and Moolenaar, W.H. (1990) Lysophosphatidic acid, but not phosphatidic acid, is a potent Ca^{2+}-mobilizing stimulus for fibroblasts. Evidence for an extracellular site of action. *J. Biol. Chem.* 265, 12232-12239.

Lancaster Jr., J.R., Laster, S.M., and Gooding, L.R. (1989) Inhibition of target cell mitochondrial electron transfer by tumor necrosis factor. *FEBS Lett.* 248, 169-174.

Loetscher, H., Steinmetz, M., and Lesslauer, W. (1991) Tumor necrosis factor: Receptors and inhibitors. *Cancer Cells* 3, 221-226.

Moolenaar, W.H., Kruijer, W., Tilly, B.C., Verlaan, I., Bierman, A.J., and de Laat, S.W. (1986) Growth factor-like action of phosphatidic acid. *Nature* 323, 171-173.

Osborn, L., Kunkel, S., and Nabel, G.J. (1989) Tumor necrosis factor α and interleukin 1 stimulate the human immonodeficiency virus enhancer by activation of the nuclear factor κB. *Proc. Natl. Acad. Sci.* USA 86, 2336-2340.

Ruff, M.R. and Gifford, G.E. (1981) Tumor necrosis factor, in *Lymphokines* (Pick E., ed) vol. 2, pp. 235-272, Academic Press, New York.

Ruggiero, V., Johnson, S.E., and Baglioni, C. (1987) Protection from tumor necrosis factor cytotoxicity by protease inhibitors. *cell. Immunol.* 107, 317-325.

Scheurich, P., Kobrich, G., and Pfizenmaier, K. (1989) Antagonistic control of tumor necrosis factor receptors by protein kinases A and C. Enhancement of TNF receptor synthesis by protein kinase A and transmodulation of receptors by protein kinase C. *J. Exp. Med.* 170, 947-958.

Schulze-Osthoff, K., Bakker, A.C., Vanhaesebroeck, B., Beyaert, R., Jacob, W.A., and Fiers, W. (1992) Cytotoxic activity of tumor necrosis factor is mediated by early damage of mitochondrial functions. Evidence for the involvement of mitochondrial radical generation. *J. Biol. Chem.* 267, 5317-5323.

Shalaby, M.R., Sundan, A., Loetscher, H., Brockhaus, M., Lesslauer, W., and Espevik, T. (1990) Binding and regulation of cellular functions by monoclonal antibodies against human tumor necrosis factor receptors. *J. Exp. Med.* 172, 1517-1520.

Shukla, S.D. and Halenda, S.P. (1991) Phospholipase D in cell signalling and its relationship to phospholipase C. *Life Sci.* 48, 851-866.

Suffys, P., Beyaert, R., Van Roy, F., and Fiers, W. (1987) Reduced tumour necrosis factor-induced cytotoxicity by inhibitors of the arachidonic acid metabolism. *Biochem. Biophys. Res. Commun.* 149, 735-743.

Suffys, P., Beyaert, R., Van Roy, F., and Fiers, W. (1988) Involvement of a serine protease in tumour-necrosis-factor-mediated cytotoxicity. *Eur. J. Biochem.* 178, 257-265.

Suffys, P., Beyaert, R., De Valck, D., Vanhaesebroeck, B., Van Roy, F., and Fiers, W. (1991) Tumour-necrosis-factor-mediated cytotoxicity is correlated with phospholipase-A_2 activity, but not with arachidonic acid release *per se*. *Eur. J. Biochem.* 195, 465-475.

Tartaglia, L.A., Weber, R.F., Figari, I.S., Reynolds, C.,Palladino Jr., M.A., and Goeddel, D.V. (1991) The two different receptors for tumor necrosis factor mediate distinct cellular responses. *Proc. Natl. Acad. Sci. USA* 88, 9292-9296.

Vanhaesebroeck, B., Van Bladel, S., Lenaerts, A., Suffys, P., Beyaert, R., Lucas, R., Van Roy, F., and Fiers, W. (1991) Two discrete types of tumor necrosis factor-resistant cells derived from the same cell line. *Cancer Res.* 51, 2469-2477.

Williamson, B.D., Carswell, E.A., Rubin, B.Y, Prendergast, J.S., and Old, L.J. (1983) Human tumor necrosis factor produced by human B-cell lines: Synergistic cytotoxic interaction with human interferon. *Proc. Natl. Acad. Sci. USA* 80, 5397-5401.

Yang, S.F., Freer, S., and Benson A.A. (1967) Transphosphatidylation by phospholipase D. *J. Biol. Chem.* 242, 477-484.

RELATIONSHIP BETWEEN PHOSPHOLIPASE C, TYROSINE-KINASES AND CYTOSKELETON IN THROMBIN-STIMULATED PLATELETS

Christine Guinebault, Monique Plantavid, Bernard Payrastre, Pascal Grondin, Claire Racaud-Sultan, Gérard Mauco, Monique Breton, and Hugues Chap

INSERM Unité 326
Hôpital Purpan
31059 Toulouse Cedex, France

INTRODUCTION

Phospholipase C (PLC)-mediated inositol lipid hydrolysis is a critical step in stimulation of platelets by thrombin. However, mechanisms of PLC activation are not yet well understood and might be multiple since several isoenzymes have been described in platelets (Banno et al., 1992) as in other cells (Suh et al., 1988). PLC-γ1 has been shown to be regulated by growth factor receptors containing tyrosine-kinase such as epidermal growth factor or platelet-derived growth factor receptors (Rhee and Choi, 1992) and also by non-receptor tyrosine-kinases (Park et al., 1991). This tyrosine phosphorylation-mediated activation could allow either an allosteric activation (Wahl et al., 1992) or an appropriate localization near the plasma membranes where PLC substrates are located. Several reports have recently shown that another isoenzyme, PLC-β1 is regulated by α subunits of a novel trimeric G-protein family known as G_q (Taylor and Exton, 1991; Smrcka et al., 1991). In contrast, a soluble PLC present in HL60 granulocytes has been shown to be stimulated by $\beta\gamma$ subunits of G proteins (Camps et al., 1992). In platelets, there is good evidence that thrombin mainly stimulated PLC activity *via* G proteins, without excluding other pathways (Haslam and Davidson, 1984; Hrbolich et al., 1987). It is also interesting to note that a member of the novel Gq family has been identified in platelets (Shenker et al., 1991). Recently, the structure of the receptor for thrombin has been characterized (Vu et al., 1991). It belongs to the class of seven-transmembrane domain receptors coupled to trimeric G proteins, suggesting the implication of the G protein pathway with activation of PLC in thrombin-stimulated platelets.

On the other hand, there is now compelling evidence that thrombin is able to activate tyrosine-kinases in platelets (Ferrel and Martin, 1988; Golden and Brugge, 1989). It is also known that platelets express high levels of $pp60^{src}$ and other kinases of the *src* family, like *lyn*, *fyn* and *yes* (Golden et al., 1986; Huang et al., 1991) and possess both PLC-γ1 and PLC-γ2 isoenzymes (Banno et al., 1992).

New Developments in Lipid-Protein Interactions and Receptor Function
Edited by K.W.A. Wirtz *et al.*, Plenum Press, 1993

Altogether, these observations raise the question of a possible involvement of these two isoenzymes in thrombin-activated platelets and their activation by tyrosine-kinases. Furthermore, PLC activation may be involved in the thrombin-induced reorganization of cytoskeleton since some cytoskeletal proteins, such as profilin or gelsolin, specifically interact with phosphatidylinositol 4,5-bisphosphate (PtdIns(4,5)P_2) (Janmey and Stossel, 1987; Goldschmidt-Clermont et al., 1990). We describe here the inhibitory effect of the specific tyrosine-kinase inhibitor, tyrphostin AG-213 on phosphatidic acid (PtdOH) production in thrombin-treated platelets. Furthermore, an increase of PLC activity was measured in a pool of proteins immunoprecipated from thrombin-stimulated platelets with an antiphosphotyrosine antibody, compared to the same immunoprecipitate obtained from resting platelets. We have also shown that thrombin was able to increase PLC activity in the cytoskeleton as well as in the antiphosphotyrosine immunoprecipitate obtained from this cytoskeleton. Finally, these results seem to be confirmed by a preliminary experiment of Western blotting showing both the presence of PLC-γ1 in this immunoprecipitate and an increase of its content upon thrombin stimulation.

EXPERIMENTAL PROCEDURES

Platelet Preparation and Activation

Prelabelled [^{32}P]orthophosphate platelets were isolated and their lipids extracted and analysed by HPLC as described by Sultan et al. (1990). When specified, platelets were preincubated for 5 min in the presence or in the absence of 100 μM tyrphostin AG-213 (A. Levitzki) and subsequently stimulated with human thrombin (Sigma, Saint Louis, MO, USA) at 37°C for 5 min.

Cytoskeleton Extraction

Cytoskeletons were isolated from either resting or thrombin-stimulated platelets as described by Grondin et al. (1991).

Immunoprecipitation

Cytoskeletons and whole platelets were solubilized with sodium dodecyl sulfate (0.1 %). Solubilized cytoskeletons and platelet lysates were then incubated with an antiphosphotyrosine antibody coupled to agarose (Oncogene Science) as previously described by Payrastre et al. (1990). Immunopurified proteins were then assayed for PLC activity exactly as described by Grondin et al. (1991) or probed by immunoblotting with an anti-PLC-γ1 antibody (1/3000, Zymed Laboratories, Inc.). Goat anti-mouse immunoglobulins conjugated to alkaline phosphatase were used as secondary antibodies.

RESULTS AND DISCUSSION

Inhibition of PtdOH Production by Tyrphostin AG-213

When [^{32}P]-labelled platelets were stimulated by thrombin (O.7 IU/ml), their pre-treatment by tyrphostin AG-213 (Gazit et al., 1989) significantly decreased (36% $\pm$ 5.6

inhibition, n = 3) the labelling of PtdOH (Fig. 1). Since, in platelets, PtdOH is a good reflect of PLC activation, our results may suggest a partial inhibition by tyrphostin of the PtdIns(4,5)P_2 hydrolysis stimulated by thrombin. We have confirmed this hypothesis by measuring inositol phosphate production in thrombin-stimulated [^{3}H]-*myo*-inositol prelabelled platelets and incubated in the absence or in the presence of tyrphostin (not shown). This inhibition, only partial, might indicate that several PLC isoenzymes could be implicated in the inositol lipid hydrolysis upon thrombin treatment; one would probably belong to the PLC-β family, activated *via* the trimeric Gq proteins and at least a second one would be regulated by tyrosine-kinases; in this case, the isoenzyme would likely be a PLC-γ.

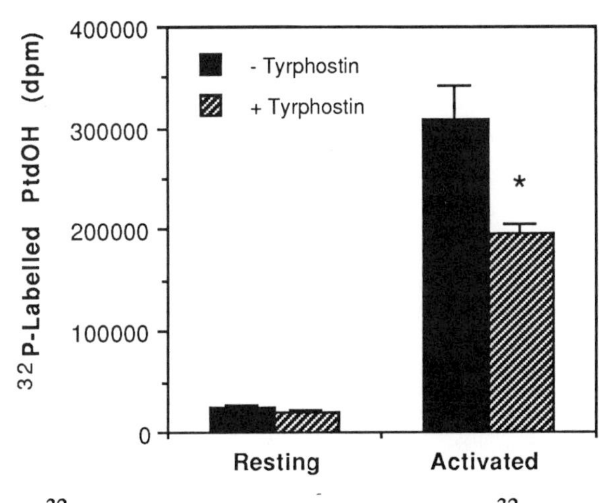

Figure 1. Inhibition of ^{32}P-PtdOH formation by tyrphostin treatment. [^{32}P]-Labelled platelets (2 x 10^9 cells/ml) were preincubated in the absence or in the presence of 100 μM tyrphostin AG-213 for 5 min and then activated or not with 0.7 IU/ml of human thrombin. Lipids were extracted, deacylated and subsequently analyzed by HPLC technique. Results are expressed in dpm of PtdOH produced $\pm$ S.E.M. (n = 3).

Association of PLC with the Cytoskeleton

Activation of platelets by thrombin (1 IU/ml; 5 min) leads to an increase of total cytoskeleton proteins compared with resting platelets (264 versus 948 μg/1.5 x 10^9 platelets) and also an increase of PLC activity (3.2 $\pm$ 0.7 versus 53.0 $\pm$ 7.5 pmol of [^{3}H]-IP_3 released/min/1.5 x 10^9 platelets) (Fig. 2a and Fig. 2b). As a consequence, the specific activity of PLC associated with cytoskeleton increased upon thrombin stimulation (x 4.5).

This result raises two questions : is there a translocation of a PLC to the cytoskeleton after thrombin treatment or is there an activation of a PLC already associated with cytoskeleton ?

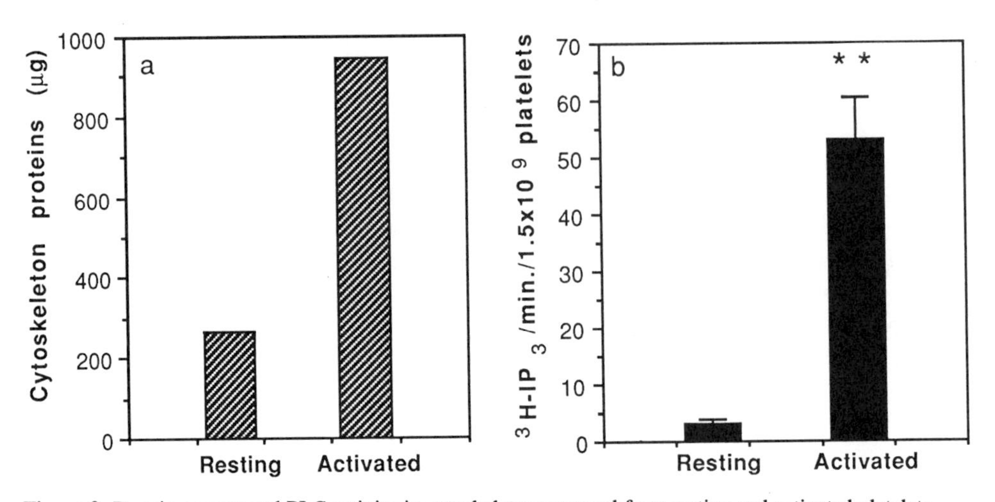

Figure 2. Protein content and PLC activity in cytoskeleton prepared from resting and activated platelets.
a: Protein content. Results from one experiment representative of three identical experiments.
b: Phospholipase C activity. Results are expressed in pmol of $[^3H]$-IP_3 released per min and per 1.5×10^9 platelets ± S.E.M. (n = 6).

Antiphosphotyrosine Immunoprecipitation of PLC Activity from Whole Platelets and Cytoskeleton

When phosphotyrosyl proteins obtained from platelet lysate or solubilized cytoskeleton were immunoprecipitated with an antiphosphotyrosine antibody, PLC activity was present in both immunoprecipitates and its activity was approximately 2.5 fold higher after thrombin stimulation (Table 1).

The recovery of PLC activity in antiphosphotyrosine immunoprecipitates does not obligatory mean a direct tyrosine phosphorylation of this enzyme. Actually, PLC may only be associated to a tyrosine phosphorylated protein. However, our results strongly

Table 1. PLC activity measured in the antiphosphotyrosine immunoprecipitate obtained from whole platelets and cytoskeleton. Results are expressed in pmol of $[^3H]$-IP_3 produced in 10 min per 1×10^9 platelets.

	Platelet homogenate		Platelet cytoskeleton	
Activity	Resting	Activated	Resting	Activated
$[^3H]$ IP_3 (pmol)	0.76	2.0	0.61	1.47
	(x 2.63)		(x 2.41)	

suggest that the PLC activity detected in these antiphosphotyrosine immunoprecipitates may be related to a PLC-γ isoenzyme.

In a preliminary experiment, using an anti-PLC-γ1 antibody, we have detected the presence of PLC-γ1 in the pool of immunoprecipitated proteins obtained from platelet cytoskeleton. Furthermore, thrombin treatment increased PLC-γ1 detection (not shown).

This preliminary result suggests an association of PLC-γ1 with platelet cytoskeleton. This is in agreement with a previous study showing a colocalization of PLC-γ1 with actin microfilaments in rat embryo fibroblasts (Mc Bride et al., 1991). The increase of PLC activity observed in antiphosphotyrosine immunoprecipitate obtained from platelet lysate and platelet cytoskeleton after thrombin treatment suggests that a PLC activation *via* tyrosine-kinases plays a role in thrombin-mediated inositol lipid hydrolysis. Further investigations concerning the presence of PLC-γ1 in platelet cytoskeleton are now in progress.

Acknowledgements

The authors would like to thank Dr. A. Levitzki for providing tyrphostin AG-213 and Y. Jonquière for correcting the English manuscript. This work was supported by a grant from Association pour la Recherche sur le Cancer (France).

REFERENCES

Banno, Y., Nakashima, T., Kumada, T., Ebisawa, K., Nonomura, Y., and Nozawa, Y., 1992, Effects of gelsolin on human platelet cytosolic phosphoinositide-phospholipase C isoenzymes, *J. Biol. Chem.* 267:6488-6494.

Camps, M., Hou, C., Sidiropoulos, D., Stock, J.B., Jakobs, K.H., and Gierschik, P., 1992, Stimulation of phospholipase C by guanine-nucleotide-binding protein ßγ subunits, *Eur. J. Biochem.* 206:821-831.

Ferrell, J.E., and Martin, G.S., 1988, Platelet tyrosine-specific protein phosphorylation is regulated by thrombin, *Mol. Cell. Biol.* 8:3603-3610.

Gazit, A., Yaish, P., Gilson, C., and Levitzki, A., 1989, Tyrphostins I : synthesis and biological activity of protein tyrosine kinase inhibitors, *J. Med. Chem.* 32:2344-2352.

Golden, A., and Brugge, J.S., 1989, Thrombin treatment induces rapid changes in tyrosine phosphorylation in platelets, *Proc. Natl. Acad. Sci. USA* 86:901-905.

Goldschmidt-Clermont, P.J., Machesky, L.M., Baldassare, J.J., and Pollard, T.D., 1990, Regulation of phospholipase C-γ by profilin and tyrosine phosphorylation, *Science* 247:1575-1578.

Grondin, P., Plantavid, M., Sultan, C., Breton, M., Mauco, G., and Chap, H., 1991, Interaction of $pp60^{c\text{-}src}$, phospholipase C, inositol-lipid, and diacylglycerol kinases with the cytoskeletons of thrombin-stimulated platelets, *J. Biol. Chem.* 266:15705-15709.

Haslam, R.J., and Davidson, M.M.L., 1984, Receptor-induced diacylglycerol formation in permeabilized platelets ; possible role for a GTP-binding protein, *J. Recept. Res.* 4:605-629.

Hrbolich, J.K., Culty, M. and Haslam, R.J., 1987, Activation of phospholipase C associated with isolated rabbit platelet membranes by guanosine 5'-[γ-thio] triphosphate and by thrombin in the presence of GTP, *Biochem. J.* 243:457-465.

Huang, M.-H., Bolen, J.B., Barnwell, J.W., Shattil, S.J., and Brugge, J.S., 1991, Membrane glycoprotein IV (CD36) is physically associated with the Fyn, Lyn and Yes protein, *Proc. Natl. Acad. Sci. USA* 88:7844-7848.

Janmey, P.A., and Stossel, T.P., 1987, Modulation of gelsolin function by phosphatidylinositol 4,5-bisphosphate, *Nature* 325:362-364.

Mc Bride, K., Rhee, S.G., and Jaken, S., 1991, Immunocytochemical localization of phospholipase C-γ in rat embryo fibroblast, *Proc. Natl. Acad. Sci. USA* 88:7111-7115.

Park, D.J., Rho, H.W., and Rhee, S.G., 1991, CD_3 stimulation causes phosphorylation of phospholipase C-γ1 on serine and tyrosine residues in human T-cell line, *Proc. Natl. Acad. Sci. USA* 88:5453-5456.

Payrastre, B., Plantavid, M., Breton, M., Chambaz, E.M., and Chap, H., 1990, Relationship between phosphoinositide kinase activities and protein tyrosine phosphorylation in plasma membrane, *Biochem. J.* 272:665-670.

Rhee, S.G., and Choi, K.D., 1992, Regulation of inositol phopsholipid-specific phospholipase C isoenzyme, *J. Biol. Chem.* 267:12393-12396.

Shenker, A., Goldsmith, P., Unson, C.G., and Spiegel, A.M., 1991, The G protein coupled to the thromboxane A_2 receptor in human platelets is a member of the novel Gq family, *J. Biol. Chem.* 266:9309-9313.

Smrcka, A.V., Hepler, J.R., Brown, K.O., and Sternweis, P.C., 1991, Regulation of polyphosphoinositide-sepcific phospholipase C activity by purified Gq, *Science* 251:804-807.

Suh, P.-G., Ryu, S.H., Moon, K.H., Suh, H.W., and Rhee, S.G., 1988, Cloning and sequence of multiple forms of phospholipase C, *Cell* 54, 161-169.

Sultan, C., Breton, M., Mauco, G., Grondin, P., Plantavid, M., and Chap, H., 1990, The novel inositol lipid phosphatidylinositol 3,4-bisphosphate is produced by human blood platelets upon thrombin stimulation, *Biochem. J.* 269:831-834.

Taylor, S.J., and Exton, J.H., 1991, Two α subunits of the Gq class of G proteins stimulate phosphoinositide phospholipase C-ß1 activity , *FEBS Lett.* 286:214-216.

Vu, T.-K. H., Hung, D.T., Wheaton, V.I., and Coughlin, S.R., 1991, Molecular cloning of a functional thrombin receptor reveals a novel proteolytic mechanism of receptor activation, *Cell* 64:1057-1068.

Wahl, M.I., Jones, G.A., Nishibe, S., Rhee, S.G., and Carpenter, G., 1992, Growth factor stimulation of phospholipase C-γ1 activity, *J. Biol. Chem.* 267, 10447-10456.

LIPID MODIFICATIONS OF GTP-BINDING REGULATORY PROTEINS

Patrick J. Casey

Section of Cell Growth, Regulation and Oncogenesis and
Department of Biochemistry
Duke University Medical Center
Durham, NC 27710

INTRODUCTION

Modification by lipid groups is being increasingly appreciated as a mechanism to promote the interaction of a wide variety of proteins with cell membranes and, in many cases, to impart functional properties to these proteins. There are three basic classes of lipid-modified proteins, although others may exist. The first class are the acylated proteins, which contain probably the most common lipid modification. The lipid group is most often a saturated fatty acyl group, specifically myristoyl (14-carbon) or palmitoyl (16-carbon), and each of these groups are found on defined sites in proteins (Towler et al, 1988; James, and Olson, 1990). The glycosyl phosphatidylinositol (GPI) anchored proteins comprise the second class; these proteins contain the most complex lipid modification from both structural, as well as biosynthetic, standpoints. The modifying group contains both sugar and lipid components, and this modification is confined to a discrete subset of proteins whose ultimate destination is the external face of the cell's plasma membrane (Doering et al, 1990). Proteins in the third class of modified proteins contain the the most recent lipid modification identified. These so-called prenylated proteins are modified via addition of either the farnesyl (15-carbon) or the geranylgeranyl (20-carbon) isoprenoids to conserved cysteine residue(s) near or at the COOH-terminus of proteins (Casey, 1992; Maltese, 1990). Additionally, most if not all prenylated proteins are subject to further modifications at the COOH-terminus which apparently impart additional membrane-interaction and/or regulatory properties to these proteins. This review will be concerned with protein acylation and prenylation, since these modifications occur on GTP-binding regulatory proteins (G proteins).

The heterotrimeric G proteins, comprised of α, β and γ subunits, are predominately associated with the inner surface of the plasma membrane, where they serve to transduce signals from membrane receptors (Gilman, 1987). It is the GTP-binding α subunit, of which many subtypes exist, that defines the oligomer and, in most cases, carries the signal to the effector molecule that is regulated. The βγ subunits, which also exist in multiple forms, function as a complex. Individual forms of the βγ complex may be shared among different α subunits. The basic features of G protein signal transduction are described in the model in Figure 1. Activation of the G protein by an appropriately liganded receptor stimulates the exchange of GTP for GDP on the α subunit. This activation is accompanied by dissociation of α-GTP from βγ. The interaction of α-GTP (or, in some cases, βγ) with a membrane-associated effector protein produces the intracellular response (e.g. generation of a second messenger such as cAMP).

The association of the G protein subunits with the plasma membrane is crucial in ensuring efficient signal transduction (Gilman, 1987; Spiegel et al, 1991). However, the molecular basis of G protein association with cellular membranes was unclear until recently. Although most G proteins are similar to integral membrane proteins in that they require detergents to be solubilized, none of the subunits' polypeptide sequences contain sufficient hydrophobic sequences to account for this membrane interaction. The important clues for the interaction of the G proteins with membranes came from studies designed to detect possible lipid modifications of the proteins (Buss et al, 1987; Yamane et al, 1990; Mumby et al, 1990).

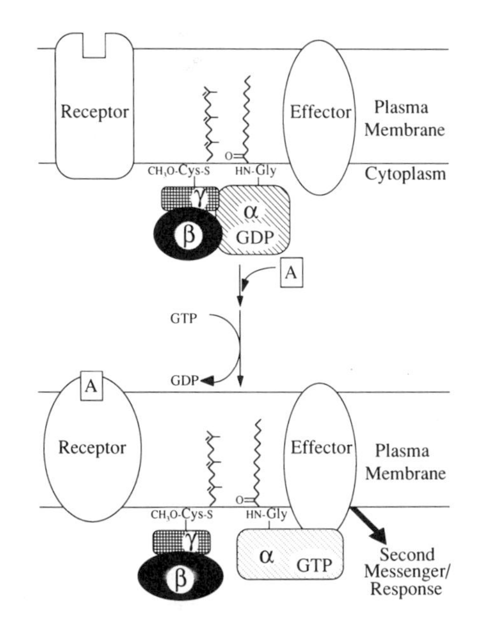

Figure 1. Model of G protein transmembrane signal transduction. Agonist (A) binding to its appropriate receptor stimulates activation of the G protein by GTP binding. The lipid groups associated with the G protein subunits, shown as α subunit myristoylation and γ subunit prenylation, contribute to the association of these proteins with the membrane. See text for further details.

Two of the above-mentioned classes of lipid modifications are now known to be involved in G protein-membrane interactions; acylation and prenylation. Acylation (specifically myristoylation) of a number of G protein α subunits mediates the interaction of these proteins with the plasma membrane (Mumby et al, 1990; Jones et al, 1990), and, additionally, plays a role in their interactions with the $\beta\gamma$ complex (Linder et al, 1991). The more recent discovery that G protein γ subunits are modified by isoprenoids has helped define the mechanism by which this subunit is membrane-associated (Simonds et al, 1991; Yamane et al, 1990; Muntz et al, 1992; Maltese, and Robishaw, 1990). The first section of this chapter covers current knowledge on myristoylation of G protein α subunits, while the second part deals with the the prenylation of the G protein γ subunits.

MYRISTOYLATION OF G PROTEIN α SUBUNITS

Analysis of the α_i and α_o subtypes of G protein α subunits by both direct chemical as well as metabolic labelling approaches revealed the presence of myristic acid on the proteins (Buss et al, 1987). The nature of the linkage was found to be base-resistant but acid-labile. This result suggested that, similar to other known myristoylated proteins (Gordon et al, 1991), this fatty acid was linked via an amide bond by a process termed N-myristoylation. Subsequent studies confirmed that these G protein α subunits, including α_{i1}, α_{i2}, α_{i3}, and α_o subtypes, are N-myristoyated at a conserved NH_2-terminal glycine residue. These observations helped explain previous studies that showed that the NH_2-terminus of G protein α subunits was important in their interaction with membranes and with the $\beta\gamma$ complex (Neer et al, 1988; Navon, and Fung, 1987).

Enzymology of Myristoylation

The addition of the myristoyl group to an NH_2-terminal glycine residue of a protein is a co-translational event (Wilcox et al, 1987). Since the bond formed to the protein is a very stable amide linkage, the reaction is believed to be essentially irreversible. Thus, myristoylation is apparently a life-of-the-protein modification. The reaction is catalyzed by an enzyme termed myristoyl-CoA:protein N-myristoyltransferase (NMT)(Rudnick et al, 1990). This enzyme has a pronounced substrate specificity for myristoyl-CoA as well as an absolute requirement for an NH_2-terminal glycine on the substrate protein (or peptide, as the enzyme can recognize them also) (Towler et al, 1988). Although no strict consensus sequence for myristoylation exists, analysis of >100 peptides has revealed some preferences of the enzyme for amino acid residues at certain positions (Gordon et al, 1991). For example, a serine at position 5 promotes high-affinity interactions of NMT with the peptide, and basic residues at positions 7 and 8 are preferred over others.

Fatty acyl specificity studies conducted with NMT have shown that some heteroatom analogs of myristoyl-CoA, most notably those with oxygen-for-methylene substitutions at discrete points along the acyl chain, can be accepted as substrates by the enzyme (Gordon et al, 1991). Since the use of these analogs in the enzymatic reaction results in the formation of myristoylated products with reduced hydrophobicity, the analogs have proven useful in more detailed analysis of the requirement for the hydrophobic nature of the acyl chain in the properties of the modified proteins. Results from these studies have indicated that the membrane asociation of some myristoylated proteins are much more sensitive than others to alterations of the hydrophobicity in the attached acyl chain, suggesting that the ability of the myristoyl group to target proteins to membranes is not solely due to the ability of the acyl chain to insert into lipid bilayers.

Myristoylation and Membrane Association of G Protein α Subunits

The discovery of the myristoylation of G protein α subunits prompted a number of studies designed to define the role of the modification in the membrane association and functional activities of these proteins (Linder et al, 1991; Mumby et al, 1990; Jones et al, 1990). A popular technique to begin to address this problem has been transient expression of G protein α subunits in mammalian cells. This technique, coupled with metabolic labelling studies utilizing [^{3}H]myristate, has been used to demonstrate the covalent attachment of myristic acid to the G protein α subunits designated α_{i1}, α_{i2}, α_{i3}, α_o, α_t and α_z, while the α_s subtype was unable to be myristoylted (Mumby et al, 1990; Jones et al, 1990). Site directed mutagenesis was used to assess the role of myristoylation in the membrane targeting of the modified proteins. The NH_2-terminal glycine residues (the site of myristoylation) of α_{i1} and α_o were changed from glycine to an alanine and the cDNAs encoding the mutant α subunits transfected into cells. Expression of these cDNAs produced α subunits which did not incorporate [^{3}H]myristate and also did not associate with celllular membranes, whereas transfections with wild-type α subunits produced myristoylated proteins which did associate with membranes (Mumby et al, 1990; Jones et al, 1990). While both the wild-type and mutant α subunits produced in these cells could interact with βγ subunits, the nonmyristoylated mutant α subunit required much larger amounts of βγ to observe the interaction than did the normal, myristoylated, protein (Jones et al, 1990).

A unique approach to studying functional properties imparted to G protein α subunits by myristoylation involved the co-expression of the NMT and specific α subunits in *E. coli*, which allows production of myristoylated proteins in this organism which normally lacks this capacity (Linder et al, 1991). Thus, α_o proteins in both myristoylated and nonmyristoylated form could be produced in and purified from *E. coli*. Analysis of these two forms of recombinant α_o by a variety of techniques demonstrated that the myristoylated, recombinant protein exhibited normal interactions with the G protein βγ complex, while the nonmyristoylated counterpart had a very poor affinity for βγ (Linder et al, 1991). Taken together, the studies noted above indicate that myristoylation of G protein α subunits affects not only the membrane interactions of these polypeptides, but also controls subunit (i.e. α-βγ) interactions of G proteins.

While the myristoyl group almost certainly plays a role in the attachment of the α subunit to the plasma membrane, it is still not clear whether this interaction is driven by myristoyl insertion into the lipid bilayer, as illustrated in Fig. 1, or perhaps mediated by additional membrane protein(s). In this regard, it is noteworthy that α-subunits purified from bovine brain bind to artificial phospholipid vesicles poorly unless the vesicles contain the βγ complex (Sternweis, 1986). One conclusion from this data is that the membrane association of the α subunit involves more than the hydrophobic interaction provided by the NH_2-terminal myristate, and that protein-protein interactions are also important. Additional data in support of this model comes from the use of the heteroatom-substituted myristate analogs mentioned above in the modification of α subunits. Incorporation of 11-oxa myristate into α_i and α_o resulted in some redistribution of these proteins to the soluble fraction of cells, while similar experiments conducted with α_z and α_t showed that they associated equally well with membranes whether myristate or the 11-oxa analog was attached (Mumby et al, 1990). Thus, while the myristoyl group is most certainly involved in directing the acylated α subunits to the membrane, it seems likely that myristate is not the only signal required (Gordon et al, 1991). Analysis of the additional determinants on the α subunit that participate in membrane targeting of these proteins is likely to be a fruitful area of exploration.

At least two α subunits do not appear to be dependent on myristoylation for membrane attachment. These two proteins are α_s, the protein which mediates stimulation of adenylyl cyclase, and α_t, which functions in visual signaling. While α_s has not yet been shown to incorporate any fatty acid, α_t can be myristoylated when its cDNA is expressed in mammalian cells (Mumby et al, 1990) even though no myristate was detected by chemical analysis of the protein purified from retina (Buss et al, 1987). It is possible that the reason for this discrepancy is that the environment of α_t in the retinal cell does not facilitate myristoylation, however α_t may also normally (i.e. in the retina) be modified by a fatty acid other than myristate.

PRENYLATION OF G PROTEIN γ SUBUNITS

As noted in the "Introduction", the G protein β and γ subunits are tightly associated and function as a complex in G protein-mediated signal transduction (and see Figure 1). Like the α subunits, amino acid sequences of both β and γ do not contain hydrophobic domains which could account for their tight interaction with membranes. The finding that the G protein γ subunits are prenylated has helped explain the hydrophobic properties of this subunit complex and suggests a mechanism for the association of these proteins with membranes.

Processing of G protein γ subunits

The G protein γ subunits are members of a family of proteins distinguished by a COOH-terminal Cys-AAX motif (where A and X signify any of several amino acids). It is now appreciated that the presence of this motif directs the processing of these polypeptides by a series of three post-translational events (Casey, 1992; Maltese, 1990; Clarke, 1992). This processing is illustrated in Figure 2. The initial step in the processing is prenylation, which involves the enzymatic addition of either a 15-carbon, farnesyl, or 20-carbon, geranylgeranyl, isoprenoid to the cysteine residue in the Cys-AAX motif. This reaction occurs in the cytosol, and two distinct enzymes have been identified which attach either the geranylgeranyl or farnesyl group (Reiss et al, 1990; Moomaw and Casey, 1992). The identity of the terminal residue ("X") of the Cys-AAX motif determines which isoprenoid will be attached. If X is a leucine residue, a protein geranylgeranyltransferase (PGGT) adds a geranylgeranyl group, whereas a methionine, serine or glutamine residue at this position signals addition of a farnesyl group by a protein farnesyltransferase (PFT). The isoprenoid substrates for these enzymes are the corresponding diphosphates (i.e. farnesyl diphosphate and geranylgeranyl diphosphate) which are produced in the cholesterol biosynthetic pathway from mevalonate (Goldstein, and Brown, 1990).

Processing subsequent to the addition of the isoprenoid involves the proteolytic removal of the three terminal amino acids (the "AAX") (Ashby et al, 1992; Hrycyna, and Clarke, 1992) followed by methylation of the now carboxyl-terminal, prenylated cysteine residue (Gutierrez et al, 1989; Stephenson, and Clarke, 1990). The enzymes that catalyze these two processing steps have been localized to the microsomal membrane fraction, although neither enzyme has been purified. Available information on the protease indicates it is an endoprotease, cleaving the prenylated protein on the COOH-terminal side of the prenylated cysteine residue. The specific methyltransferase then apparently recognizes the free carboxyl group of the prenylated cysteine residue and modifies it in a S-adenosylmethionine-dependent reaction. The net result of these processing steps is a modified COOH-terminus which, in most cases, serves to direct and/or anchor these proteins to their specific cellular membrane(s) (Glomset et al, 1991).

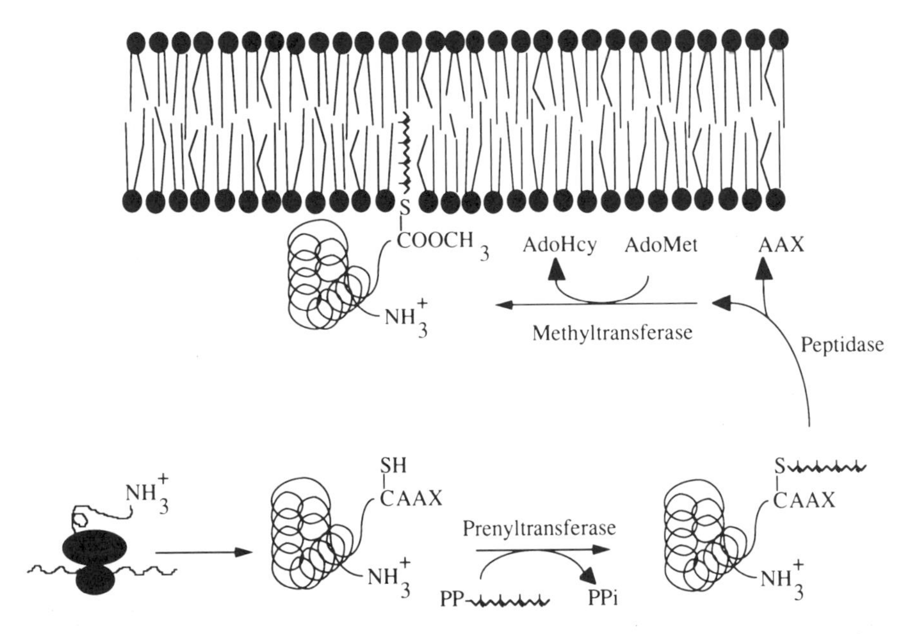

Figure 2. Carboxyl-terminal processing of prenylated γ subunits. The proteins are synthesized as soluble precursors containing a cysteine residue fourth from the COOH-terminus (the so-called "CAAX-motif"). Prenylation is followed by a proteolytic removal of the final three amino acids and methylation of the now-free carboxyl group of the modified cysteine residue. See text for further details. Adapted from Casey, 1992.

Specific G protein γ subunits are modified by either farnesyl or geranylgeranyl isoprenoids (Lai et al, 1990; Yamane et al, 1990; Maltese, and Robishaw, 1990; Mumby et al, 1990). The known non-retinal γ subunits contain COOH-terminal leucine residues and are geranylgeranyl modified, while the COOH-terminus of the γ subunit of the retinal

G protein, transducin, is a serine residue and thus this protein is farnesylated. Why two different prenyl groups are attached to these highly homologous proteins is not yet clesr, but may reflect affinity differenes observed for retinal versus nonretinal βγ complexes for membranes (Hancock et al, 1991).

Prenylation and Membrane Association of G Protein γ Subunits

Studies utilizing transfection of cDNAs encoding γ subunits into mammalian cells similar to those described above for studying the effects of α subunit myristoylation have been carried out to examine the influence of prenylation on γ subunit localization. Here also wild-type and mutant γ subunits were expressed, in this case the mutation was a Cys-to-Ser change in the Cys-AAX motif which completely blocks the processing of these proteins (Casey et al, 1989; Hancock et al, 1989). When the cDNA encoding the major non-retinal form of the γ subunit was transfected into mammalian cells and the expressed protein was examined by metabolic labelling and distribution analysis, it was shown to be prenylated and associated with the membranes (Muntz et al, 1992). However, transfection of a cDNA encoding the same γ subunit with a Cys-to-Ser substitution in the Cys-AAX domain produced a γ subunit which couldn't get prenylated and no longer associated with the plasma membrane (Muntz et al, 1992; Simonds et al, 1991). Additionally, inhibiting the synthesis of mevalonate (the precursor of isoprenoids) in cells coexpressing β and γ subunits resulted in redristribution of a fraction of the γ (and β, see below) subunits from the membrane to the cytoplasmic fraction (Muntz et al, 1992).

The influence γ subunit prenylation on the interaction of the polypeptide with the β subunit was also examined by cDNA expression studies. Thus, coexpression of a β subunit with an unprenylated (Cys-to-Ser mutant) γ subunit resulted in the shift of a substantial amount of β subunit to the cytoplasmic fraction, which is where the unprenylated γ subunit accumulates (Muntz et al, 1992). This results suggests that the β subunit can interact with an unprenylated γ subunit. Additional evidence supporting the conclusion that β subunits can form a complex with γ subunits regardless of their prenylation status comes from protease treatment studies on the protein produced in cDNA expression systems. A characteristic 23 kDa fragment of the β subunit is generated upon limited trypsin treatment only when this polypeptide is associated with the γ subunit (Schmidt, and Neer, 1991). In cells overexpressing a β subunit with either wild-type γ or that with the Cys-to-Ser mutant, the 23 kDa fragment was observed upon trypsin treatment, even though in the latter case the βγ analyzed was from the soluble fraction (Muntz et al, 1992). These studies also suggest that the properties of γ subunit influence the cellular localization of the β subunit.

The retinal G protein, transducin, is apparently unique in that it contains a γ subunit which is modified by the 15-carbon, farnesyl isoprenoid (Lai et al, 1990). Surprisingly, both farnesylated and unfarnesylated γ can be found associated with β subunits in the retina; the unfarnesylated γ apparently is truncated just before the prenylated cysteine residue (Fukada et al, 1990; Ohguro et al, 1991; Fukada et al, 1989). Since βγ subunit complexes containing both forms of the γ subunit are membrane-associated, it seems that determinants other than the farnesyl group mediate binding of transducin βγ to the retinal membrane (Fukada et al, 1990; Fukada et al, 1989). Since the transducin βγ complex is the only known one which can be released from the membrane without detergent, the mechanism of its membrane association may be quite different than for the $\beta\gamma$ complex from other G proteins.

While its membrane association does not appear to require a prenylated γ subunit, transducin $\beta\gamma$ does need the modification for efficient signal transduction (Fukada et al, 1990). Thus, the ability of light-activated rhodopsin to stimulate GTP binding to the α subunit of transducin was 30-fold more efficient in the presence of $\beta\gamma$ containing a farnesylated γ subunit than with that containing the unfarnesylated subunit (Fukada et al, 1990). This study further highlights the importance of prenylation in G protein signaling.

CONCLUDING REMARKS

The discoveries that G protein subunits are covalently modified by lipid groups has vastly improved our understanding of how these protein interact with cell membranes. Nonetheless, many of the molecular details of these interactions remain unclear. For example, the contribution of as yet unidentified (or unappreciated) proteins on specific membranes that play roles in directing or stabilizing G protein subunits to that particular membrane needs to be more fully explored. Also, the extent to which these lipid modifications impart functional properties onto their hosts has not been clearly established. Recent advances in the isolation and cloning of many of the specific enzymes involved in these modifications have provided the tools to begin a more detailed examination of these and other questions on lipid modifications of proteins. We can certainly look forward to more surprises as the impact of these processes on the biology of signal transduction becomes clarified.

ACKNOWLEDGEMENTS

Thanks are due Julia Thissen for helpful comments on this manuscript and Joyce Higgins for the preparation of the artwork. Work from the author's laboratory was supported by research grants from the National Science Foundation and American Cancer Society and an Established Investigator Award from the American Heart Association

REFERENCES

Ashby, M.N., King, D.S., and Rine, J. (1992) Endoproteolytic processing of a farnesylated peptide in vitro, *Proc.Natl.Acad.Sci.USA* 89:4613.

Buss, J.E., Mumby, S.M., Casey, P.J., Gilman, A.G., and Sefton, B.M. (1987 Myristoylated alpha subunits of guanine nucleotide-binding regulatory proteins, *Proc.Natl.Acad.Sci.USA* 84:7493.

Casey, P.J. (1992) Biochemistry of protein prenylation, *J.Lipid Res.* in press.

Casey, P.J., Solski, P.A., Der, C.J., and Buss, J.E. (1989) p21ras is modified by a farnesyl isoprenoid, *Proc.Natl.Acad.Sci.USA* 86:8323.

Clarke, S. (1992) Protein isoprenylation and methylation at carboxyl-terminal cysteine residues, *Ann.Rev.Biochem.* 61:355.

Doering, T.L., Masterson, W.J., Hart, G.W., and Englund, P.T. (1990) Minireview: Biosynthesis of glycosyl phosphatidylinositol membrane anchors, *J.Biol.Chem.* 265:611.

Fukada, Y., Ohguro, H., Saito, T., Yoshizawa, T., and Akino, T. (1989) $\beta\gamma$-subunit of bovine transducin of two components with distinctive gamma subunits, *J. Biol. Chem.* 264:5937.

Fukada, Y., Takao, T., Ohguro, H., Yoshizawa, T., Akino, T., and Shimonishi, Y. (1990) Farnesylated gamma subunit of photoreceptor G protein indispensable for GTP-binding, *Nature* 346:658.

Gilman, A.G. (1987) G proteins: Transducers of receptor-generated signals, *Ann.Rev.Biochem.* 56:615.

Glomset, J., Gelb, M., and Farnsworth, C. (1991) The prenylation of proteins, *Curr. Opinion Lipidology* 2:118.

Goldstein, J.L., and Brown, M.S. (1990) Regulation of the mevalonate pathway, *Nature* 343:425.

Gordon, J.I., Duronio, R.J., Rudnick, D.A., Adams, S.P., and Goke, G.W. (1991) Protein N-myristoylation, *J.Biol.Chem.* 266:8647.

Gutierrez, L., Magee, A.I., Marshall, C.J., and Hancock, J.F. (1989) Post-translational processing of p21ras is two-step and involves carboxyl-methylation and carboxy-terminal proteolysis, *EMBO J.* 8:1093.

Hancock, J.F., Cadwallader, K., and Marshall, C.J. (1991) Methylation and proteolysis are essential for efficient membrane binding of prenylated p21K-ras(B), *EMBO J.* 10:641.

Hancock, J.F., Magee, A.I., Childs, J.E., and Marshall, C.J. (1989) All ras proteins are polyisoprenylated but only some are palymitoylated, *Cell* 57:1167.

Hrycyna, C.H., and Clarke, S. (1992) Maturation of isoprenylated proteins in *Saccharomyces cerevisiae*: Multiple activities catalyze the cleavage of the three carboxyl-terminal amino acids from farnesylated substrates in vitro, *J.Biol.Chem.* 267:10457.

James, G., and Olson, E.N (1990) Fatty acylated proteins as components of intracellular signaling pathways, *Biochemistry* 29:2623.

Jones, T.L.Z., Simonds, W.F., Merendino, J.J., Brann, M.R., and Spiegel, A.M. (1990) Myristoylation of an inhibitory GTP-binding protein alpha subunit is essential for its membrane attachment, *Proc. Natl. Acad. Sci.* 87:568.

Lai, R.K., Perez-Sala, D., Canada, F.J., and Rando, R.R. (1990) The γ subunit of transducin is farnesylated, *Proc.Natl.Acad.Sci.USA* 87:7673.

Linder, M.E., Pang, I-H., Duronio, R.J., Gordon, J.I., Sternweis, P.C., and Gilman, A.G. (1991) Lipid modifications of G protein subunits:Myristoylation of $G_{0\alpha}$ increases its affinity for $\beta\gamma$, *J.Biol.Chem.* 266:4654.

Maltese, W.A. (1990) Posttranslational modification of proteins by isoprenoids in mammalian cells, *FASEB J.* 4:3319.

Maltese, W.A., and Robishaw, J.D. (1990) Isoprenylation of C-terminal cysteine in a G protein gamma subunit, *J.Biol.Chem.* 265:18071.

Moomaw, J.F., and Casey, P.J. (1992) Mammalian protein geranylgeranyltransferase: Subunit composition and metal requirements, *J.Biol.Chem.* 267:17438.

Mumby, S.M., Casey, P.J., Gilman, A.G., Gutowski, S., and Sternweis, P.C. (1990) G protein gamma subunits contain a 20-carbon isoprenoid, *Proc.Natl.Acad.Sci.USA* 87:5873.

Mumby, S.M., Heukeroth, R.O., Gordon, J.I., and Gilman, A.G. (1990) G protein alpha subunit expression, myristoylation, and membrane association in COS cells, *Proc.Natl.Acad.Sci.USA* 87:728.

Muntz, KH., Sternweis, PC., Gilman, AG., and Mumby, SM. (1992) Influence of gamma subunit prenylation on association of guanine nucleotide-binding regulatory proteins with membranes, *Mol.Biol.Cell* 3:49.

Navon, S.E., and Fung, B.K.-K. (1987) Characterization of transducin from bovine retinal rod outer segment. Participation of the amino terminal region of T_α in subunit interactions, *J.Biol.Chem.* 262:15746.

Neer, E.J., Pulsifer, L., and Wolf, L.G. (1988) The amino terminus of G protein alpha subunits is required for interaction with βγ, *J.Biol.Chem.* 263:8996.

Ohguro, H., Fukada, Y., Takao, T., Shimonishi, Y., Yoshizawa, T., and Akino, T. (1991) Carboxyl methylation and farnesylation of transducin gamma subunit synergistically enhance its coupling with meterhodopsin II, *EMBO* 10:3669.

Reiss, Y., Goldstein, J.L., Seabra, M.C., Casey, P.J., and Brown, M.S. (1990) Inhibition of purified p21ras farnesyl:protein transferase by Cys-AAX tetrapeptides, *Cell* 62:81.

Rudnick, D.A., McWherter, C.A., Adams, S.P., Ropson, I.J., Duronio, R.J., and Gord4on, J.I. (1990) Structural and functional studies of *Saccharomyces cerevisiae* myristoyl-CoA:protein N-myristoyltransferase produced in *Escherichia coli*, *J. Biol. Chem.* 265:13370.

Schmidt, C.J., and Neer, E.J. (1991) In vitro synthesis of G protein βγ dimers, *J.Biol.Chem.* 266:4538.

Simonds, W.F., Butrynski, J.E., Gautam, N., Unson, C.G., and Spiegel, A.M. (1991) G protein βγ dimers: Membrane targeting requires subunit coexpression and intact gamma CAAX domain, *J.Biol.Chem.* 266:5363.

Spiegel, A.M., Backlund, P.S., Butyrinski, J.E., Jones, T.L.Z., and Simonds, W.F. (1991) The G protein connection: molecular basis of membrane association, *Trends Biochem.Sci.* 16:338.

Stephenson, R.C., and Clarke, S. (1990) Identification of a C-terminal protein carboxyl methyltransferase in rat liver membranes utilizing a synthetic farnesyl cysteine-containing peptide substrate, *J.Biol.Chem.* 265:16248.

Sternweis, P.C. (1986) The purified alpha subunits of Go and Gi from bovine brain require βγ for association with phospholipid vesicles, *J.Biol.Chem.* 261:631.

Towler, D.A., Gordon, J.I., Adams, S.P., and Glaser, L. (1988) The biology and enzymology of eukaryotic protein acylation, *Ann.Rev.Biochem.* 57:69.

Wilcox, C., Hu, J.-S., and Olson, E.N. (1987) Acylation of proteins with myristic acid occurs cotranslationally, *Science* 238:1275.

Yamane, H.K., Farnsworth, C.C., Xie, H., Howald, W., Fung, B.K.-K., Clarke, S., Gelb, M.H., and Glomset, J.A. (1990) Brain G protein gamma subunits contain an all-trans-geranylgeranyl-cysteine methyl ester at their carboxyl termini, *Proc.Natl.Acad.Sci.USA* 87:5868.

LYSOPHOSPHATIDIC ACID AS A LIPID MEDIATOR: SIGNAL TRANSDUCTION AND RECEPTOR IDENTIFICATION

Wouter Moolenaar, Rob van der Bend, Emile van Corven, Kees Jalink, Thomas Eichholtz and Wim van Blitterswijk

Division of Cellular Biochemistry, The Netherlands Cancer Institute Plesmanlaan 121, 1066 CX Amsterdam, The Netherlands

INTRODUCTION

Phospholipids are currently attracting much interest in studies on cellular signalling and growth control. It is now well established that the breakdown products of several plasma membrane phospholipids act as signal molecules, i.e. as intracellular second messengers or as agonists that modulate cell function. Well known examples of phospholipid-derived signaling molecules include diacylglycerol, inositol trisphosphate and prostaglandins, which are all rapidly generated upon receptor stimulation. The simplest natural phospholipid, lysophosphatidic acid (LPA or monoacylglycerol-3-phosphate), is a particularly intriguing case in that it not only serves a critical precursor role in de novo lipid biosynthesis but also shows striking hormone- and growth factor-like activities when added exogenously to appropriate target cells, as if LPA binds to and thereby activates its own G-protein-coupled receptor(s). Although LPA's lipid precursor role has been known for many decades, the possibility that this simple phospholipid may have an additional role as a signaling molecule is being appreciated only since a few years. In this chapter, we briefly summarize our recent findings on the various actions of LPA, with particular emphasis on the signal transduction pathways triggered by LPA and the identification of a candidate cell surface receptor.

BIOLOGICAL ACTIVITIES OF LPA

LPA is rapidly produced in thrombin-activated platelets[1] and growth factor-stimulated fibroblasts[2] and is a normal constituent of serum (Eichholtz et al., manuscript in preparation). This suggests that LPA may be secreted into the extracellular environment following its formation; our recent findings indeed indicate that newly produced LPA is rapidly released by activated platelets (Eichholtz et al., manuscript in preparation). The production of LPA in activated cells probably is secondary to the formation of PA through the action of a specific phospholipase A_2, as schematically illustrated in Fig. 1, but the

New Developments in Lipid-Protein Interactions and Receptor Function
Edited by K.W.A. Wirtz *et al.*, Plenum Press, 1993

existence of alternative metabolic routes leading to LPA formation cannot be excluded at present. Exogenous LPA is partially metabolized by cells to yield monoacylglycerol, but it appears that the biological activity of LPA is not attributable to one of its metabolites[3].

Exogenous LPA triggers rapid responses in many cell types (Table 1) and, in fibroblasts, stimulates long-term DNA synthesis and cell division[4-6]. Naturally occurring LPA (1-oleoyl) stimulates thymidine incorporation in fibroblasts with a half-maximal effect

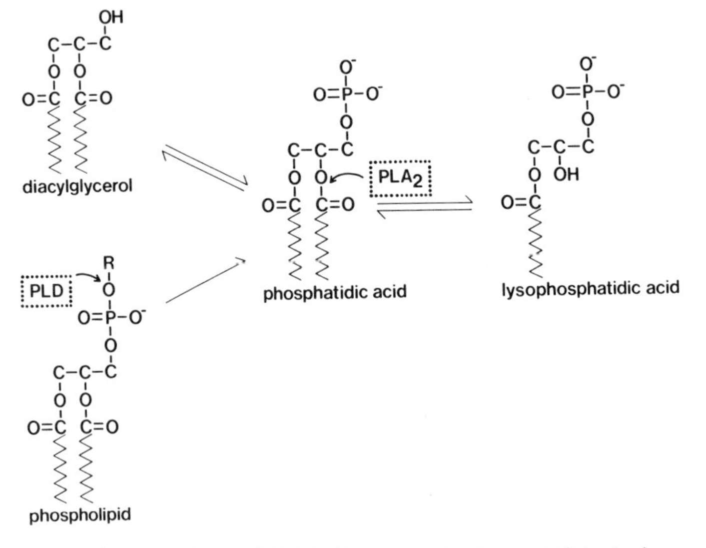

Figure 1. Proposed generation of LPA from newly formed PA during cell activation. Diacylglycerol is rapidly formed via receptor-linked activation of phospholipase C and is phosphorylated by diacylglycerol kinase to yield PA. PA may also be formed through activation of phospholipase D (PLD) acting on PC and PE (R denotes phospholipid headgroup). PA may be hydrolyzed by a phospholipase A_2 (PLA_2) to LPA. The contribution of other pathways to generating LPA remains to be explored.

observed at about 15 μM[5], a concentration that is significantly higher than those required for evoking the early events listed in Table 1, which are maximal at 0.5-1.0 μM. The discrepancy between these dose-response relationships remains unexplained at present. As with polypeptide growth factors, the mitogenic response to LPA requires long-term presence of the stimulus. When LPA is removed from the culture medium several hours after stimulation, the cells fail to undergo DNA synthesis. Furthermore, the mitogenic activity of LPA does not require the presence of peptide growth factors: neither insulin nor EGF were found to act in a synergistic fashion with LPA[5]. In this respect, LPA acts differently from certain mitogenic peptides which fail to stimulate cell proliferation unless synergizing growth factors are present[7].

Table 1. Rapid cellular responses to LPA [a]

Fibroblasts, epithelial cells,	phosphoinositide hydrolysis; arachidonate release; inhibition of adenylate cyclase
Neuroblastoma cells:	Ca^{2+} mobilization; morphological changes
MDCK epithelial cells:	Cl^- secretion
Smooth muscle:	Ca^{2+} mobilization; contraction
Platelets:	Ca^{2+} mobilization; aggregation
Xenopus oocytes:	Ca_i^{2+}-dependent Cl^- current
Dictyostelium disc.	chemotaxis

[a] References can be found in ref. 4.

SIGNAL TRANSDUCTION PATHWAYS

Several "classic" signal transduction pathways in the action of LPA have been identified[4,5]. These include: (i) GTP-dependent activation of phospholipase C, (ii) release of arachidonic acid presumably as a result of phospholipase A_2 activation, (iii) activation of phospholipase D[8], and (iv) pertussis toxin-sensitive inhibition of adenylate cyclase. These early events are characteristic for receptor-mediated processes; a candidate LPA receptor has recently been identified (a) as will be outlined below. A great variety of cell types respond to LPA by triggering the immediate breakdown of inositol phospholipids as measured by the formation of $InsP_3$[1,4,5] and diacylglycerol[5,10]. LPA-induced activation of phospholipase C results in rapid but transient rise in Ca^{2+}, which is primarily caused by the release of intracellularly stored Ca^{2+}. Concomitantly, protein kinase C is activated as shown by the phosphorylation of an endogenous 80 kDa protein substrate[5]. The kinetics and shape of the LPA-induced Ca^{2+} transient are virtually indistinguishable from those elicited by bradykinin and other hormones. Furthermore, LPA-induced Ca^{2+} mobilization is subject to homologous desensitization, a common feature of many signalling systems where agonist-induced attenuation of cellular responsiveness is thought to have an important regulatory role[11]. From studies on permeabilized cells it appears that the response is GTP-dependent, suggesting the involvement of a G protein (that is insensitive to pertussis toxin in human fibroblasts; ref. 5). Thus, the early phosphoinositide-breakdown response to LPA has all of the hallmarks of a G protein-coupled receptor event. It should be noted, however, that there is no correlation between early phosphoinositide hydrolysis and late initiation of DNA synthesis, supporting the notion that, contrary to common belief, activation of phospholipase C is neither required nor sufficient for induction of a mitogenic response[12].

In addition to activating phospholipase C, LPA inhibits cAMP accumulation in intact cells[5]. When adenylate cyclase is pre-stimulated by either forskolin or certain receptor agonists (isoproterenol, prostaglandin E_1, etc.), addition of LPA decreases cAMP accumulation by 60-70% within 10 min. This decrease in cAMP is dose-dependent and is completed blocked by pretreating the cells with pertussis toxin, indicating that LPA acts through the G_i protein that inhibits adenylate cyclase.

Furthermore, studies on N1E-115 and NG108-15 neuronal cells have revealed that LPA causes rapid changes in cytoskeletal organization (Jalink et al., manuscript submitted). These cells constitute a convenient model system to investigate early signaling events in the proliferative response. After serum starvation, these cells stop growing and subsequently begin to acquire various differentiated properties of mature neurons, including the formation of long neurites. Addition of 1-oleoyl-LPA (1μM) to growth factor-starved N1E-115 or NG108-15 cells causes rapid and dramatic changes in cell shape reminiscent of those observed during mitosis (Jalink et al., manuscript submitted). Virtually every flattened cell starts to round up as early as 5-10 sec after LPA addition, while rounding is complete within approx. 1 min. Almost simultaneously, growth cones begin to collapse and developing neurites retract. The effects of LPA on neuronal cell shape are dose-dependent, with some rounding detectable at doses as low as 10 nM and maximal responses at 0.5-1 μM. In the continuous presence of LPA, cells maintain their rounded shape for 5-10 min. At 10-20 min after the addition of LPA, however, most cells gradually resume a flattened morphology, but neurite outgrowth remains suppressed for at least a few hours. Remarkably, a second application of LPA to such respread cells leaves the shape unaltered. LPA-induced cell rounding is thus subject to homologous desensitization, as one would expect for a receptor-mediated process. Further evidence for the morphological response being receptor-mediated includes the finding that: (i) the drug suramin, which blocks LPA-receptor interaction (see below), inhibits LPA-induced shape changes; (ii) other, related lipids have no effect; (iii) microinjected LPA fails to mimic the action of extracellularly applied LPA; and (iv) the morphological response to LPA is fully mimicked by a synthetic peptide agonist of the cloned G protein-coupled thrombin receptor, whereas LPA and thrombin clearly act through separate receptors[13] (Jalink et al., manuscript submitted).

It appears that LPA- and thrombin-induced changes in neuronal cell shape are mediated by "contraction" of the cortical actin cytoskeleton with no direct involvement of microtubules. As for the signaling mechanism responsible for this novel action of LPA and thrombin: in N1E-115 cells, as in many other cells, LPA stimulates phosphoinositide hydrolysis leading to calcium mobilization and activation of protein kinase C. Yet, the following findings indicate that the PLC-Ca^{2+}-PKC pathway is not responsible for the observed changes in cell shape: (i), phosphoinositide-hydrolyzing neurotransmitters, Ca^{2+} ionophores and PKC-activating phorbol ester all fail to mimic LPA in inducing cell rounding; (ii), LPA-induced shape changes are not prevented by down-regulating PKC by long-term treatment of the cells with phorbol ester; and (iii), cells depleted of internal Ca^{2+} (by addition of ionophore in Ca^{2+}-free medium) show a normal morphological response to LPA despite the absence of a Ca^{2+} transient. Similarly, bacterial toxin-sensitive G proteins, adenylate cyclase or cyclic nucleotides have no apparent role in mediating LPA-induced shape changes, since LPA action is neither inhibited nor mimicked by treatment of the cells with pertussis or cholera toxin, cAMP and cGMP analogues or forskolin (10 μM).

Taken together, these results suggest that LPA- and thrombin-induced cell rounding is not mediated by known G protein-linked second messenger cascades. By analogy with the cell rounding induced by active protein tyrosine kinases[14,15] and microinjected cdc2 kinase[16], it seems plausible to assume that specific phosphorylations /dephosphorylations of certain actin-binding proteins[17,18] may be responsible for the morphological effects of LPA. In support of this, we observed that LPA-induced cell rounding is blocked by both microinjected vanadate and exogenously added pervanadate, which is a membrane-permeable form of vanadate[19], a widely-used inhibitor of protein tyrosine phosphatases. Furthermore, brief preincubation with such (non-specific) protein kinase inhibitors as genistein (50 μM), quercetin (50 μM) and staurosporine (1 μM) similarly inhibits cell rounding.

We have begun to analyze the possible role of the Src protein tyrosine kinase in the rapid neuronal shape changes. It was found that both LPA and thrombin induce a small but statistically significant increase in Src kinase activity in N1E-115 cells as measured in an immunocomplex kinase assay (Jalink et al., manuscript submitted). Although we cannot draw conclusions yet about cause-effect relationships, the results obtained are intriguing and deserve further study.

Activation of p21ras

The product of the ras proto-oncogene (p21ras) is a focal point of signal transduction by growth factor receptors, notably those with intrinsic tyrosine kinase activity. Given the mitogenic potency of LPA, we have examined the activation state of the Ras protein in quiescent fibroblasts following treatment with LPA. LPA causes the rapid accumulation of the GTP-bound, active form of Ras within one minute. The kinetics of activation is transient: the response subsides after about 10 minutes. This early response to LPA is similar to that elicited by EGF, although the extent of p21ras activation by LPA is somewhat weaker than observed with EGF. (van Corven et al., manuscript in preparation). Like the other early events in LPA action, p21.GTP accumulation conforms to all the criteria of a receptor-mediated process (van Corven et al., manuscript in preparation), including its dose-dependency (half-maximal effects close to 10 nM) and complete inhibition by suramin. Importantly, LPA-induced p21.GTP accumulation is blocked by prior treatment of the cells with pertussis toxin, whereas EGF-induced p21.GTP accumulation remains unaffected. Thus, a heterotrimeric G protein of the G_i subclass regulates p21ras through an as-yet-unknown effector pathway.

Future experiments should reveal whether activation of p21ras is uniquely observed with the putative LPA receptor, or else is a response to stimulation of other G protein-coupled receptors as well.

RECEPTOR IDENTIFICATION

As outlined above, the available evidence strongly suggests that LPA activates one or more specific G protein-coupled receptors at the cell surface. In a recent study, van der Bend et al.[9] succeeded in identifying a putative high-affinity LPA receptor. A ^{32}P-labeled LPA analogue containing a photoreactive fatty acid, diazirine-LPA, labels a membrane protein of apparent molecular mass 38-40 kD in various LPA-responsive cell types, with labeling being most prominent in neuronal cells and brain homogenates. Labeling is specific, in that only unlabeled LPA but not other phospholipids inhibit incorporation of diazirine-LPA into the 38-40 kD protein band; half-maximal inhibition is observed at approximately 10-20 nM unlabeled LPA. The 38-40 kD protein presumably represents a specific LPA cell surface receptor mediating at least part of the multiple cellular responses to LPA (for further details see ref. 9). The putative LPA receptor is not detectable in human neutrophils, which are biologically unresponsive to LPA[10]. Furthermore, binding of photoreactive LPA to its putative receptor is completely inhibited by the drug suramin. Our effort to search for more specific LPA receptor antagonist have been unsuccessful to date.

CONCLUDING REMARKS

The discovery that LPA exert profound effects on many cell types suggests that LPA may have a novel role as a lipid mediator. It appears that LPA is released by cells (notably activated platelets) perhaps in a manner similar to the secretion of platelet-activating factor, prostaglandins and other lipid agonists by activated cells. The released LPA may then activate target cells in a paracrine or autocrine fashion. At present it is premature to speculate further on the normal physiological (or perhaps pathophysiological) roles of LPA. Obviously, a major challenge for future studies is to establish the amino acid sequence of the putative LPA cell surface receptor, either through protein purification or, perhaps more straight-forward, via expression cloning in COS cells. Another challenge is to delineate the unique signal transduction events, involving activation of both the src and ras proto-oncogene products, that are responsible for reversal of the differentiated phenotype in neuroblastoma cells and initiation of DNA synthesis and cell division in fibroblasts and possibly other cell types.

ACKNOWLEDGEMENTS

Research related to this chapter was supported by the Netherlands Cancer Foundation and by the Netherlands Organization for Scientific Research (NWO). We thank José Overwater for preparing the camera-ready version of this manuscript.

REFERENCES

1 L.P. Watson, R.T. McConnell, and E.G. Lapetina, Decanoyl lysophosphatidic acid induces platelet aggregation through an extracellular site of action, Biochem J 232:61 (1985).
2 K. Fukami, and T. Takenawa, Formation of phosphatidic acid in PDGF-stimulated 3T3 cells is a potential mitogenic signal, J Biol Chem 267:10988 (1992).
3 R.L. van der Bend, J. de Widt, E.J. van Corven, W.H. Moolenaar, and W.J. van Blitterswijk, Metabolic conversion of the biologically active phospholipid, lysophosphatidic acid, in fibroblasts, Biochim Biophys Acta 1125:110 (1992).
4 W.H. Moolenaar, K. Jalink, and E.J. van Corven, Lysophosphatidic acid: a bioactive phospholipid with growth factor-like properties, Rev Physiol Biochem Pharmacol 119:47 (1992).
5 E.J. van Corven, A. Groenink, K. Jalink, T. Eichholtz, and W.H. Moolenaar, Lysophosphatidate-induced cell proliferation: identification and dissection of signaling pathways mediated by G proteins, Cell 59:45 (1989).
6 E.J. van Corven, A. van Rijswijk, K. Jalink, R. van der Bend, W.J. van Blitterswijk, and W.H. Moolenaar, Mitogenic action of lysophosphatidic acid and phosphatidic acid on fibroblasts, Biochem J 281:163 (1992).
7 E. Rozengurt, Early signals in the mitogenic response, Science 234:161 (1986).
8 R.L. van der Bend, J. de Widt, E.J. van Corven, W.H. Moolenaar, and W.J. van Blitterswijk, The biologically active phospholipid, lysophosphatidic acid, induces phosphatidylcholine breakdown via activation of phospholipase D, Biochem J 285:235 (1992).
9 R.L. van der Bend, J. Brunner, K. Jalink, E.J. van Corven, W.H. Moolenaar, and W.J. van Blitterswijk, Identification of a putative membrane receptor for the bioactive phospholipid, lysophosphatidic acid, EMBO J 11:2495 (1992c).
10 K. Jalink, E.J. van Corven, and W.H. Moolenaar, Lysophosphatidic acid, but not phosphatidic acid, is a potent Ca2+-mobilizing stimulus for fibroblasts, J Biol Chem 265:12232 (1990).
11 J.L. Benovic, M. Bouvier, H.G. Caron, and R.J. Lefkowitz, Regulation of adenylate cyclase-coupled β-adrenergic receptors, Ann Rev Cell Biol 4:405 (1988).
12 W.H. Moolenaar, G protein-coupled receptors, phosphoinositide hydrolysis and cell proliferation, Cell Growth & Differentiation 2:359 (1991).
13 K. Jalink, and W.H. Moolenaar, Thrombin receptor activation causes rapid and dramatic changes in neural cell shape independent of classic second messengers, J Cell Biol 118:411 (1992).
14 M. Chinkers, J.A. McKanna, and S. Cohen, Rapid rounding of human A431 cells by EGF, J Cell Biol 88:422 (1981).
15 R. Jove, and H. Hanafusa, Cell transformation by the viral src oncogene, Annu Rev Cell Biol 3:31 (1987).

16 N.J.C. Lamb, A. Fernandez, A. Watsin, J.-C. Labbé, and J.-C. Cavadore, Microinjection of cdc2 kinase induces marked changes in cell shape, cytoskeletal organization, and chromatin structure in mammalian fibroblasts, Cell 60:151 (1990).
17 T.D. Pollard, and J.A. Cooper, Actin and actin-binding proteins, Annu Rev Biochem 55:987 (1986).
18 T.P. Stossel, From signal to pseudopod. How cells control cytoplasmic actin assembly, J Biol Chem 264:18261 (1989).
19 I.G. Fantus, S. Kadota, G. Deragon, B. Foster, and B.I. Posner, Pervanadate mimics insulin action in rat adipocytes via activation of the insulin receptor tyrosine kinase, Biochemistry 28:8864 (1989).

THE ROLE OF LIPOPROTEIN RECEPTORS IN THE GROWTH OF CHICKEN OOCYTES

Wolfgang J. Schneider

Department of Molecular Genetics, The University of Vienna
Dr. Bohrgasse 9, A-1030 Vienna, Austria

INTRODUCTION

For several reasons, which I hope will become more apparent from the following discussion of results obtained over the past five years, my laboratory's interest started to center around a unique experimental system: the rapidly growing chicken oocyte. Commonly known as an egg yolk, this yellow sphere represents a single cell, the female zygote, and is the result of an enormous growth process that starts with a microscopically small cell within the ovary of a laying hen. During the last 7 days before being released as fully-grown oocyte (diameter, ~35 mm) from its follicle into the oviduct, this cell accumulates as much as 2 g of protein (mostly in the form of lipoproteins) each day. Thus, the oocytes of oviparous (egg-laying) animals not only provide a fascinating system to study lipoprotein transport, but also constitute a challenge in cell biological terms. How does such a giant cell function? How can it be assured to receive exactly those nutrient components which are required for sustenance of the developing embryo?

Up to 1986, there was a lack of biochemical information about the processes underlying the transport of yolk proteins into oocytes and concomitant oocyte growth of any vertebrate oviparous species. It was known that VLDL and VTG, as well as other yolk precursor proteins such as certain vitamin-binding proteins and transferrin, are synthesized in and secreted from the liver in estrogen-dependent fashion. Following their release from the plasma compartment within follicular cell layers, these macromolecules - in particular VLDL particles - had been shown to accumulate in coated pits on the oocytic plasma membrane. However, the modes of interaction between yolk precursor(s) and components of the surface membrane, as well as details on the actual uptake mechanism

New Developments in Lipid-Protein Interactions and Receptor Function
Edited by K.W.A. Wirtz *et al.*, Plenum Press, 1993

had not been investigated. Thus, my laboratory started to evaluate the possibility that such transport was receptor-mediated. In addition, we became interested in yolk formation *per se*, i.e., the events following uptake of yolk precursors into the oocyte that lead to their storage within the giant cell. The following sections summarize the results of our research along these lines.

RESULTS

The Chicken Oocyte Receptor for VLDL and Vitellogenin

To identify a protein that could interact with VLDL and VTG, detergent extracts of membranes prepared from growing ovarian follicles (from which most of the somatic tissue had been surgically removed) were subjected to SDS-PAGE, followed by electrophoretic transfer of the separated proteins to nitrocellulose membranes. These replicas were then incubated with radiolabeled ligands, i.e., either ^{125}I-VLDL or ^{125}I-VTG, the unbound ligand washed off, and bound material localized by autoradiography. Such "ligand blotting" (4) resulted in the first visualization of a membrane protein which migrates on SDS-PAGE, in the absence of sulfhydryl reducing agents, with an apparent M_r of 95,000 (5,21).

The function of this protein in receptor-mediated endocytosis of VLDL and VTG was demonstrated in several investigations (1,8,12,20). First, there was ample evidence that the 95-kDa protein is a homologue of the well-known mammalian LDL receptor (1,8,20). Antibodies directed against the bovine LDL receptor recognized the chicken protein; mammalian apo-B-containing lipoproteins bound to the avian receptor and *vice versa*; the chicken protein, like the LDL receptor, migrated much slower on SDS-PAGE under reducing than under non-reducing conditions, and was unable to bind ligands in which critical lysine residues had been charge-modified. More recently, the 95-kDa receptor has been purified to homogeneity (1), and information on its structure and localization has become available. Amino acid sequences of tryptic fragments of the chicken oocyte receptor were obtained, and its kinship to LDL receptors was confirmed through the demonstration of sequence conservation in three characteristic domains. In particular, the chicken receptor's internalization sequence, Phe-Asp-Asn-Pro-Val-Tyr, is identical with that in LDL receptors from mammals as well as *Xenopus laevis* (1). Thus, its involvement in receptor-mediated endocytosis was implied strongly. Second, ultrastructural immunolocalization demonstrated the receptor exclusively in the oocyte proper (mostly in coated pits and coated vesicular structures); somatic cells did not contain immunoreactive material (1). In agreement with these data was our finding in ligand- and Western blots that cultured somatic cells, such as chicken embryo fibroblasts or granulosa cells, did not appear to possess the 95-kDa receptor. However, in the course of the somatic cell culture studies we discovered a new protein with all the properties of a *bona fide* LDL receptor, described below.

The Somatic Cell-Specific Chicken LDL Receptor

When cultured chicken embryo fibroblasts are grown in the absence of lipoproteins and sterols, they subsist through synthesizing endogenous cholesterol. The biosynthetic pathway starts from acetyl-CoA and is energetically demanding; therefore, under normal circumstances (i.e., when sterols are present in the medium as components of LDL) the cells utilize them via receptor-mediated binding, uptake, and degradation of LDL particles. Under cholesterol-poor growth conditions, the receptor is synthesized at maximal rates so as to possibly overcome the deficit. In this "induced" state, the presence of a protein with an M_r of 130,000 on the surface of cultured somatic chicken cells could be demonstrated by ligand blotting (8). As expected from a sterol-regulated receptor, the expression of the 130-kDa protein was dramatically suppressed when the cells had been exposed to sterols and/or cholesterol-rich lipoproteins such as LDL (8). Also in concordance with the notion that this highly regulated protein represented the chicken LDL receptor, high density lipoprotein (HDL), which is known not to deliver cholesterol to cells, had no effect on the levels of the 130-kDa protein (7,8). Interestingly, despite these functional analogies as well as identity of M_r between the mammalian and chicken LDL receptors, key structural and ligand binding features set them further apart than the chicken oocyte receptor and the mammalian LDL receptor. First, as mentioned above, there is immunological crossreactivity between the classical LDL receptor and the 95-kDa protein, but not between avian and mammalian 130-kDa receptors. Second, the oocyte-specific receptor, but not the somatic chicken receptor, binds in addition to apo-B also apo-E (which is not found in birds!). Third, there is strong evidence for the ability of mammalian LDL receptors to interact with VTG (which is not synthesized by mammals!), a property shared with the oocyte receptor; again, the chicken somatic cell receptor appears to be different in that it cannot be shown to interact with VTG (19). In this context, VTG-recognition by mammalian LDL receptors is physiologically irrelevant, but in the avian system it would be detrimental, because oocyte-destined yolk precursor VTG would be subject to misrouting to somatic tissues. *Vice versa*, binding of the mammalian apo, apoE, to the chicken oocyte receptor might point to a common ancestral function of this apo and VTG, and raises the possiblity that modern members of the LDL receptor amily are recent descendants of the oocyte receptor (18). It appears, then, that the receptor dichotomy, supplying the oocyte with nutrients on one hand and regulating somatic lipid homeostasis on the other hand, has developed in response to the particular needs of the laying hen, and possibly all oviparous species. A synopsis of the results is presented in Fig. 1.

An important arising question from these studies was: are the two different receptor proteins the products of different genes as well? In order to gain insight into this aspect, we were fortunate to identify a strain of mutant hens that would provide us with valuable biochemical and genetic information.

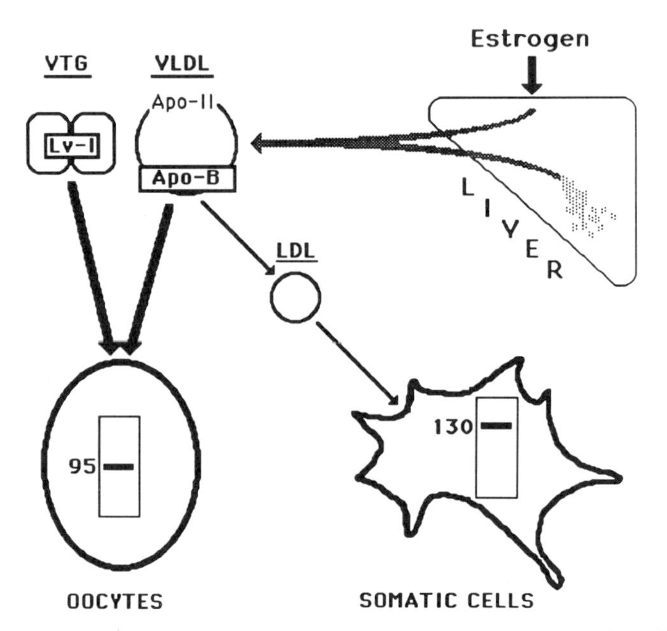

Figure 1. Lipoprotein Receptor Dichotomy in the Laying Hen.

In the laying hen, the synthesis of most, if not all, plasma components destined for uptake into growing oocytes takes place in the liver under the control of estrogen. Two major yolk components are lipoproteins: vitellogenin (VTG) and very low density lipoprotein (VLDL). The majority of these two yolk precursors is directed to the growign oocytes via a 95 kDa surface receptor. This receptor recognizes the lipovitellin-1 portion (Lv-1) of VTG and apoB of VLDL. The 95 kDa oocyte receptor is not expressed in somatic cells. Somatic cells are able to remove from the plasma small amounts of LDL generated from VLDL via lipolysis; in the laying hen, apo-VLDL-II (Apo-II) dramatically limits lipolysis of VLDL particles to LDL. Nevertheless, a 130 kDa receptor in somatic cells (which, in contrast to the 95 kDa oocyte receptor, is not expressed in zygotic cells) functions in systemic cholesterol homeostasis, analogous to the mammalian LDL receptor. In the mutant, non-laying R/O hen, the 95-kDa receptor is absent, leading to failure of oocyte growth, hyperlipidemia, and atherosclerotic lesions (see text).

The Restricted Ovulator (R/O) Hen

This animal model was first described in 1974 (10) as a mutant chicken strain with hereditary severe hyperlipidemia and absence of egg laying. Due to the endogenous hyperlipidemia, premature atherosclerosis is invariably associated with the abnormal pheontype; thus, the R/O animals may serve as a useful novel model for the study of atherogenesis. Importantly, breeding studies had established that the abnormal phenotype is the consequence of a single, sex-linked gene defect. Since ovarian follicles in R/O hens fail to reach mature size and oocyte growth stagnates at the 8-15 mm stage, the biochemical defect was thought to disrupt the transport of yolk precursors into oocytes. In order to directly test this hypothesis, we first studied the uptake of ^{125}I-VLDL into oocytes *in-vivo*, following the intravenous administration of radiotracer to normal laying hens and R/O females (12). There was indeed a dramatic difference in plasma disappearance of ^{125}I-VLDL: in normal layers, 95% of the injected dose was cleared within 24 hr, while in R/O hens, even after 96 hr greater than 50% remained in plasma. Concomitantly, over 60% of the VLDL had accumulated in normal oocytes, whereas there was negligible activity in the rudimentary R/O oocytes (12). These results prompted us to determine the ability of membranes prepared from mutant ovaries to bind ^{125}I-VLDL and/or ^{125}I-VTG in specific, high affinity fashion. Perhaps not surprising, we failed to detect such property in R/O samples, while normal ovarian membranes showed typical receptor sites for the two ligands. Finally, we subjected detergent extracts from both normal and mutant ovarian membranes to ligand- and immunoblotting (12). Both ^{125}I-VLDL and specific anti-receptor antibodies visualized a 95-kDa protein in normal tissue, but no reactivity was observed in the mutant. All of these results allowed the unambiguous conclusion that the biochemical defect underlying the non-laying R/O phenotype is indeed the lack of expression of a functional oocyte lipoprotein receptor.

Of particular importance were two additional observations. First, despite the absence of the 95-kDa receptor in the mutant, cultured R/O skin fibroblasts expressed the 130-kDa LDL receptor (9). Thus, the 95-kDa oocyte protein and the 130-kDa receptor of somatic cells are the products of different genes. Normal expression and function of the 130-kDa receptor in R/O animals further explains the absence of any abnormality in roosters of the mutant strain, who are phenotypically silent carriers of the oocyte-specific receptor defect.

Secondly, the results from the R/O animals laid to rest any doubts that the 95-kDa receptor, which initially was identified as VLDL-binding protein (5), might not be identical with the VTG-receptor. Namely, subsequent ligand blotting of R/O ovarian membrane extracts with ^{125}I-VTG was, as that with ^{125}I-VLDL, negative (20). Thus, on biochemical and genetic grounds, a single oocyte-specific protein is the receptor for both major yolk components, VLDL and VTG. Interestingly, as outlined below, there also seems to exist a common step in postendocytic processing of these ligands within the oocyte.

Cathepsin D: A Key Enzyme In Yolk Formation

Before we began our studies on intracellular routing of yolk lipids, a precursor-product relationship between plasma VLDL and VTG and certain isolatable yolk components had been established. For example, the VLDL fraction of yolk contains several polypeptides derived from apo-B, the major apo of circulating VLDL; plasma VTG is known to be converted into several distinct fragments termed lipovitellin(s), phosvitin(s), and phosvettes, respectively, upon uptake into the oocyte. We subsequently determined that an enzymatic activity localized in a particulate yolk fraction could convert intact plasma apo-B into yolk-specific fragments *in-vitro* (13). This conversion was blocked by the highly specific inhibitor of cathepsin D, pepstatin A. A similar effect of pepstatin A had been described for VTG-proteolysis in oocytes of *Xenopus laevis* (14). We purified cathepsin D from chicken oocytes by affinity chromatography on immobilized pepstatin A, and incubated plasma VLDL and VTG with the pure enzyme. Analysis by SDS-PAGE revealed that the enzyme catalyzed the *bona-fide* proteolytic processing of both plasma precursors into the known yolk products. The best match between the *in-vitro* breakdown products and yolk peptides was obtained under mildly acidic conditions, in agreement with the notion that such processing is likely to occur in an early endosomal (acidified) compartment.

In the course of the work, we also obtained a full-length cDNA for chicken cathepsin D from a follicle library (16). Sequence comparison of the 335-residue mature protein with other aspartyl proteinases suggests a high degree of known crucial features of this group of enzymes during evolution. A surprising finding was the absence of an inserted proteolytic "processing region", which in mammalian enzymes usually becomes excised during conversion from the single-chain to the two-chain form.

Finally, monospecific antibodies against chicken cathepsin D were raised and the enzyme localized ultrastructurally. The enzyme was found concentrated surrounding and overlapping with electron-dense yolk storage organelles, and was conspicuously absent from the cytoplasmic non-granular phase of the oocyte (16). These results suggest an intimate relationship between cathepsin D and yolk formation; it appears that as there is **one** receptor for the import of VLDL and VTG, there is also **one** enzyme for postendocytic proteolysis of both lipoproteins. From the point of view of cellular economy, it only makes sense to keep the number of components to a minimum; from an evolutionary point of view, the oocyte - a cell occupying a central position in survival of the species - might reveal a few more secrets about the essentials for cellular function. Another example of a multifunctional protein shall conclude the description of our work to date on this fascinating system.

The Role of Apolipoprotein VLDL-II

Besides Apo-B, which becomes cleaved upon import of VLDL into oocytes as described above, there is another VLDL-specific protein, termed apo-VLDL-II. Apo-VLDL-II is a small polypeptide (82 amino acids), whose hepatic synthesis is strictly estrogen-dependent. Thus, only the VLDL of laying hens, but not of roosters or immature hens, contains apo-VLDL-II; one laying hen VLDL particle contains 1 apo-B and as many as 25 apo-VLDL-II. It had long been thought that apo-VLDL-II, since it appears on VLDL at onset of laying, directs the VLDL particles to the oocyte receptor. However, we could clearly show that apo-VLDL-II is unable to interact with the 95-kDa oocyte receptor (11). Rather, we (11,13,17) and others (3.6) had observed that the VLDL of laying hens (which contains apo-VLDL-II) is significantly smaller (38 nm diameter on average) than apo-VLDL-II-free rooster VLDL (average diameter, 58 nm). To us this provided a clue to its role in oocyte-related lipoprotein metabolism, because other studies (15) had suggested that the extraoocytic follicular cell layers including a basement membrane (see below), through which yolk precursors must travel following their release from thecal capillaries, are impermeable to particles larger than ~40 nm. Thus, since apo-VLDL-II converts larger VLDL into smaller VLDL at onset of laying, it facilitates access of the lipoprotein to the 95-kDa receptor on the oocyte surface, a prerequisite for uptake into the zygote.

A second function of apo-VLDL-II was identified by us based on the previous finding (2) that extraoocytic somatic cells synthesize and secrete lipoprotein lipase (LPL). This enzyme, which hydrolyzes triacylglycerols, the energy-carrying component of lipoprotein particles such as VLDL, is expressed in particularly high levels in granulosa cells. These cuboidal cells form a monolayer, separated from the oocyte by the acellular perivitelline layer, and attached to a basement membrane on the viscerated aspect. Thus, as VLDL particles pass through intercellular gaps in the granulosa cell layer, they are subject to removal of triacylglycerols by LPL. However, VLDL isolated from yolk has exactly the same triacylglycerol content as that in plasma and therefore must have resisted the lipolytic action of LPL (13). In a series of experiments (17) we could indeed demonstrate that the presence of apo-VLDL-II on laying hen VLDL protects these particles from LPL attack. Whether or not this protective property is due merely to the fact that smaller lipoproteins are poorer substrates for LPL, or (also) is due to a direct inhibitory effect of apo-VLDL-II on LPL, must still be investigated. In any case, apo-

VLDL-II has two tightly linked functions which together assure that sufficient amounts of energy (in the form of triacylglycerols) can become stored inside the oocyte for subsequent use by the developing embryo.

FUTURE STUDIES

Many questions remain open about receptor-mediated control of oocyte growth. At the receptors' own level, their structures must be completely elucidated, and we must understand the molecular basis for their different ligand binding capacities. An important aspect is the failure of the 130-kDa receptor to recognize apo-E with concomitant ability of the oocyte-specific receptor to do so. Is this due to a gross or a subtle structural difference between the two receptors? An answer to this question could shed light on a long-standing unsolved problem in mammalian lipoprotein metabolism, that of a postulated separate receptor for apo-E containing lipoproteins of intestinal origin. One very recent candidate for this position, termed LDL receptor-related protein, has been found (i) to be identical to the receptor for a_2-macroglobulin (22), and (ii) to be present in chickens (19), which do not synthesize apo-E! Further insights into structural, physiologically relevant functional, and regulatory features of the chicken LDL receptor family members may reconcile some of these puzzling findings in viviparous species. Is cathepsin D the only postendocytically active proteinase? How is it regulated? Does it serve a function in procurement of nutrients to the embryo? These and many other questions concern the formation and utilization of yolk.

A third group of investigations await the cell biological aspects of oocyte growth. How can such a giant cell function? Is the nucleus in command at all times? How does the oocyte start to grow - what are the signals it must receive and how are they processed? How does the oocyte "know" when to cease growth and induce ovulation?

It has become clear from the many question marks in this final section that we have only begun to gain first insights into an intriguing biological system. Just the unravelling of lipoprotein transport and processing pathways in the laying hen alone has been a rewarding experience.

SUMMARY

Transport and removal of lipoproteins from the blood in the laying hen is of particular interest, because it is a system in which massive transport of lipid to one organ (the ovary) coexists with regulatory mechanisms for the control of lipid homeostasis in extraovarian tissues. In order to achieve this dual task, the laying hen expresses dichotomous receptor-mediated pathways. On one hand, very low density lipoproteins

(VLDL) and vitellogenin (VTG), which together form over 95% of the lipid in a fully-grown oocyte (i.e., an egg yolk), are transported into oocytes via a 95-kDa receptor protein. This receptor, termed oocyte VLDL/VTG receptor, is exclusively produced in growing oocytes, and is absent from somatic cells. It shows a high degree of structural similarity to other members of the so-called low-density-lipoprotein (LDL) receptor family, but in contrast to the LDL receptor, its expression is not suppressed by sterols. On the other hand, somatic cells, but not the oocytes, synthesize a 130-kDa receptor that recognizes VLDL-derived cholesterol-rich lipoproteins. This receptor is the functional analogue to the mammalian LDL receptor in that it mediates, at least in part, extraoocytic lipid homeostasis. The somatic LDL receptor of the chicken recognizes apolipoprotein (apo)B, but not VTG, in accordance with VTG's exclusive routing to growing oocytes. Within oocytes, both apoB of VLDL and VTG undergo limited specific postendocytic proteolytic processing. Recent studies have shown that this breakdown of macromolecular plasma precursor molecules is catalyzed by an endosomal form of cathepsin D, and is a key event in the formation of yolk, the major nutrient source for the developing embryo.

ACKNOWLEDGEMENTS

I have been fortunate to have such excellent collaborators as Drs. Ruedi Aebersold, Peter Bilous, Rajan George, Kozo Hayashi, Karl Kuchler, Johannes Nimpf, Esmond J. Sanders, Xinyi Shen, and Ernst Steyrer. Likewise, it was a pleasure to guide through their Ph.D. studies Mr. Dwayne L. Barber, Ian MacLachlan, Stefano Stifani, and Amandio Vieira. Last, but by no means least, all of us value the outstanding technical help of Rita Langford, Rita Lo, Grace Ozimek, Calla Shank-Hogue, and Barbara Steyrer. I appreciate the excellent secretarial help of Yolanda Gillam in preparation of this and many other manuscripts. I am grateful to the MRC Canada, the Heart and Stroke Foundation of Canada, the Alberta Heritage Foundation for Medical Research, and the Austrian Fonds zur Förderung der Wissenschaftlichen Forschung for Personnel and grant support of these studies.

REFERENCES

1. Barber, D.L., Sanders, E.J., Aebersold, R., and Schneider, W.J. (1991) *J. Biol. Chem.* **266**, 18761-18770.
2. Brannon, P.M., Cheung, A.H., and Bensadoun, A. (1978) *Biochim. Biophys. Acta* **531**, 96-108.
3. Burley, R.W., Sleigh, R.W., and Shenstone, F.S. (1984) *Eur. J. Biochem.* **142**, 171-176.

4. Daniel, T.O., Schneider, W.J., Goldstein, J.L., and Brown, M.S. (1983) *J. Biol. Chem.* **258**, 4606-4611.
5. George, R., Barber, D.L., and Schneider, W.J. (1987) *J. Biol. Chem.* **262**, 16838-16847.
6. Griffin, H.G., Grant, and Perry, M. (1982) *Biochem. J.* **206**, 647-654.
7. Hayashi, K., Ando, S., Stifani, S., and Schneider, W.J. (1989) *J. Lipid Res.* **30**, 1421-1428.
8. Hayashi, K., Nimpf, J., and Schneider, W.J. (1989) *J. Biol. Chem.* **264**, 3131-3139.
9. Hayashi, K., Nimpf, J., and Schneider, W.J., unpublished observations.
10. Ho, K.-J., Lawrence, W.D., Lewis, L.A., Liu, L.B., and Taylor, C.B. (1974) *Arch. Pathol.* **98**, 161-172.
11. Nimpf, J., George, R., and Schneider, W.J. (1988) *J. Lipid Res.* **29**, 657-667.
12. Nimpf, J., Radosavljevic, M., and Schneider, W.J. (1989) *J. Biol. Chem.* **264**, 1393-1398.
13. Nimpf, J., Radosavljevic, M., and Schneider, W.J. (1989) *Proc. Natl. Acad. Sci. USA* **86**, 906-910.
14. Opresko, L.K. and Karpf, R.A. (1987) *Cell* **51**, 557-568.
15. Perry, M.M. and Gilbert, A.B. (1979) *J. Cell Sci.* **39**, 257-262.
16. Retzek, H., Steyrer, E., Sanders, E.J., Nimpf, J., and Schneider, W.J., submitted.
17. Schneider, W.J., Carroll, R., Severson, D.L. and Nimpf, J. (1990) *J. Lipid Res.* **31**, 507-513.
18. Steyrer, E., Barber, D.L., and Schneider, W.J. (1990) *J. Biol. Chem.* **265**, 19575-19581.
19. Stifani, S., Barber, D.L., Aebersold, R., Steyrer, E., Shen, X., Nimpf, J., and Schneider, W.J. (1991) *J. Biol. Chem.* **266**. 19079-19087.
20. Stifani, S., Barber, D.L., Nimpf, J., and Schneider, W.J. (1990) *Proc. Natl. Acad. Sci. USA* **87**, 1955-1959.
21. Stifani, S., George, R., and Schneider, W.J. (1988) *Biochem. J.* **250**, 467-475.
22. Strickland, D., Ashcom, J.D., Williams, S., Burgess, W.H., Migliorini, M., and Argraves, W.S. (1990) *J. Biol. Chem.* **265**, 17401-17404.

LIST OF ABBREVIATIONS

DNA, desoxyribonucleic acid; R/O, restricted ovulator; SDS-PAGE, sodium dodecylsulfate-polyacrylamide gel electrophoresis

MODULATION OF OPIOID RECEPTOR FUNCTION BY MEMBRANE LIPIDS

Fedor Medzihradsky

Departments of Biological Chemistry and Pharmacology
University of Michigan Medical School
Ann Arbor, MI 48109-0606, U.S.A.

INTRODUCTION

Analogous to the dependence of soluble enzymes on the composition and characteristics of their aqueous milieu, the activity of functional membrane proteins is influenced by properties of the lipid bilayer. Specifically, in numerous studies the function of membrane-bound enzymes, receptors and transport carriers was correlated with membrane fluidity, a biophysical property largely determined by membrane composition at a given temperature (Shinitzky, 1987a; Aloia *et al.*, 1988; Viret *et al.*, 1990). The results of these studies have confirmed the significance of membrane microviscosity as a parameter that affects the dynamics and thereby, the biological activity of various membrane constituents. As shown in progressively fluidized membrane of turkey erythrocytes, the activation of adenylate cyclase following occupancy of the β-adrenergic receptor is a diffusion-controlled process (Hanski *et al.*, 1978). This model depicts the receptor and enzyme as mobile membrane components whose diffusion leads to collision-coupling and initiates signal generation. Thus, in this system, the fluidity of the membrane microenvironment has at least a modulating, if not a rate-limiting role. In contrast, in permanently coupled systems, such as the adenosine receptor, altered membrane microviscosity did not influence the coupling process (Braun and Levitzki, 1979). Both of these modes of coupling can be part of the overall mechanism of a receptor: approximately one-half of the α_2-adrenergic receptors that bound to GTP regulatory protein did so by collision-coupling, the others were precoupled (Neubig *et al.*, 1988).

On the other hand, modulation of function of membrane proteins also occurs by interaction with distinct membrane lipids. For example, a systematic study of the glucose carrier in erythrocytes has identified protein-lipid interaction as a major factor in determining transport activity, and has given specific significance to individual structural elements of the phospholipid molecule (Tefft *et al.*, 1986; Carruthers and Melchior, 1988). Furthermore, as shown with the nicotinic acetylcholine (McNamee and Fong, 1988) and β-adrenergic (McOsker *et al.*, 1983) receptors, their mechanisms involved interaction with specific membrane phospholipids.

SENSITIVITY OF OPIOID RECEPTORS TO LIPIDS

[Note: Since this overview is necessarily limited, the reader is referred to previous reviews of the field (Loh and Law, 1980; Medzihradsky and Carter, 1991).]

Key initial findings were the inhibition of opioid receptor binding in brain membranes treated with phospholipases, and the partial restoration of ligand binding by phosphatidylserine (Abood et al., 1978, 1980). In cultured neural cells, phospholipase C decreased the inhibition of PGE-stimulated adenylate cyclase by opioids (Law *et al.*, 1983), and unsaturated fatty acids reduced opioid receptor binding by lowering the Bmax (McGee and Kenimer, 1982) whereby both agonist and antagonist binding was affected (Ho and Cox, 1982). In addition to inhibiting ligand binding, exposure of synaptosomal membranes to phospholipase A_2 affected the coupling of δ opioid receptors to G protein (Lazar and Medzihradsky, 1990). The effects of lipids on the solubilized (Farahbakhsh *et al*, 1986) and partially purified (Hasegawa et al., 1987) opioid receptor have also been investigated. In the former work, binding of the opioid agonist [^{3}H]etorphine in reconstituted lipid vesicles was dependent on the lipid composition of the vesicles. In the second study, lipids with an acidic head group and polyunsaturated fatty acid backbone provided optimal conditions for ligand binding to the partially solubilized μ-receptor. With reference to the studies outlined above, it is of interest to note that treatment of brain membranes with phospholipase A_2 liberates polyunsaturated fatty acids (Butler and Abood, 1982).

MEMBRANE MODULATION OF OPIOID RECEPTOR MECHANISMS

Based on the structural characteristics of opioid ligands and on the physicochemical properties of the lipid bilayer, a model was forwarded depicting different localization of μ, δ, and κ opioid receptors in relation to the hydrophilic/hydrophobic regions of the membrane (Schwyzer, 1986). This model of membrane compartments for receptor binding sites (Figure 1) is supported by the distinct

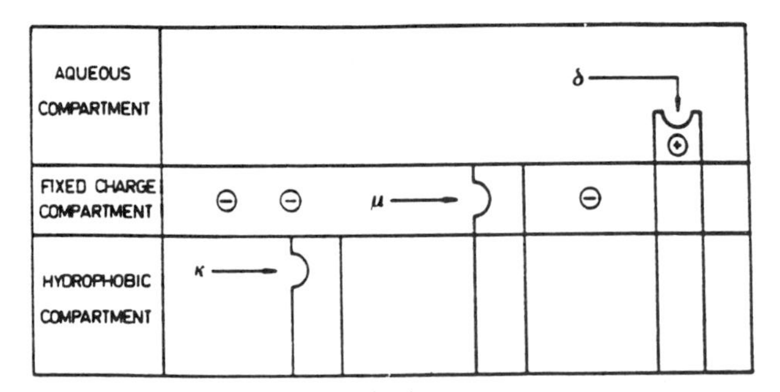

Figure 1. Proposed localization of opioid receptor binding sites in the membrane. (From Schwyzer, 1986. Reprinted with permission, American Chemical Society.)

ion dependence of opioid binding. While ligand binding to μ and δ receptors was subject to strong modulation by ions, in accordance with the proposed surface location of the respective binding sites, binding of κ opioids was much less susceptible to ion effects, again in agreement with the proposed hydrophobic receptor environment that is relatively inaccessible to charged species (Sargent *et al.*, 1988). Furthermore, increasing positive net charge on the molecule enhanced the selectivity of opioid peptides toward the μ-receptor (Schiller *et al.*, 1989). Earlier, a protein-lipid model for the opioid receptor was proposed depicting binding of opioid peptides and alkaloids to the protein and lipid part of the opioid receptor, respectively (Lee and Smith, 1980). It is now tempting to assume that the implicated lipid represents the boundary layer around the receptor, the essential environment for the function of many membrane proteins (see pertinent discussion later in the text).

To systematically assess the role of membrane environment in opioid action, the interaction of ligands with opioid receptors was studied in brain membranes modified by specific lipids. In view of the differential regulation of opioid agonist and antagonist binding by guanine nucleotides (Remmers and Medzihradsky, 1991a and 1991b), all experiments of this study were carried out in the presence of GTP-γ-S to keep the opioid receptor uncoupled from G protein. Considering the the results described in the previous section, free fatty acids were initially incorporated (Remmers *et al.*, 1990). The results revealed strong inhibition of receptor binding in membranes modified with unsaturated fatty acids, whereby the *cis*-isomers exhibited higher potency than their *trans* counterparts (Figure 2). At comparable membrane content, the *cis*-fatty acids also decreased membrane microviscosity to a larger extent (Figure 3).

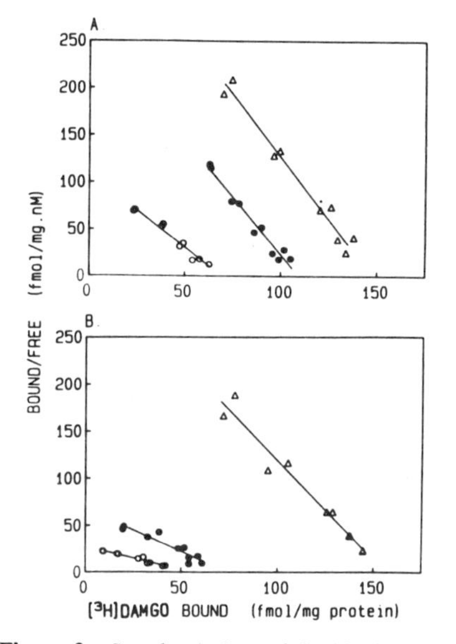

Figure 2. Scatchard plots of the binding of the μ-selective opioid peptide [^{3}H]DAMGO (Tyr-D-Ala-Gly-(Me)Phe-Gly-OH). Membranes were treated with fatty acids, added at a concentration of 0.87 μmol/mg of membrane protein. Subsequently, the equilibrium binding of [^{3}H]DAMGO in control membranes (Λ) and membranes modified with *cis*- (○) and *trans*- (●) vaccenic acid (A) or oleic (○) and elaidic (●) acids (B) was determined. (From Remmers *et al.*, 1990. Reprinted with permission, Raven Press, Ltd.)

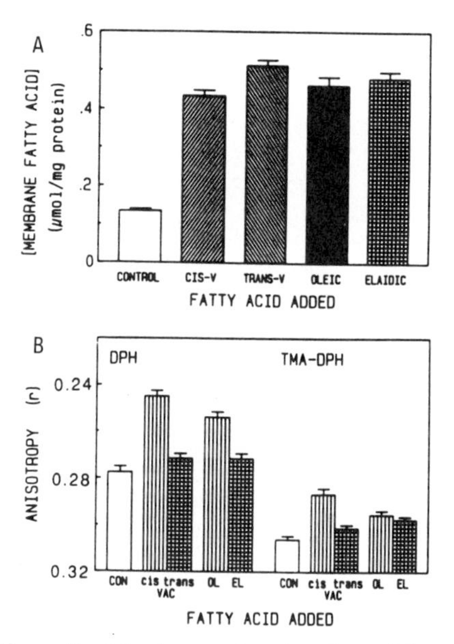

Figure 3. Membrane microviscosity and fatty acid content. Synaptosomal brain membranes were modified by the addition of 0.87 μmol fatty acid/mg protein. Following lipid extraction, methyl esters of fatty acids were formed and analyzed by GLC. Shown are the mean membrane content of fatty acids ($\pm$ SD) obtained in three to four experiments. Following incorporation of DPH and TMA-DPH into control and fatty acid-modified membranes, their fluorescence polarization was determined. The results are expressed as anisotropy values (r) whereby $r = (I_o - I_{90})/I_0 + 2I_{90})$, and I_0 and I_{90} represent the intensities of light when polarizers were in a parallel or perpendicular orientation, respectively. CON, control; EL, elaidic acid; OL, oleic acid; V or VAC, vaccenic acid. (From Remmer *et al.*, 1990. Reprinted with permission, Raven Press, Ltd.)

IMPLICATION OF MEMBRANE FLUIDITY

A. Modulation of Ligand Binding

The observed correlation between membrane microviscosity and modulation of opioid binding to receptor was further investigated by modifying synaptosomal brain membranes with either fluidizing (fatty acid) or rigidifying (cholesteryl hemisuccinate, CHS) lipids. Furthermore, the involvement of membrane surface or core was assessed by the use of both oleic acid (OA) and its methyl ester (MO). Accordingly, fluorescence anisotropy was determined with uncharged (diphenyhexatriene, DPH) and charged (trimethyl-DPH, TMA-DPH, and propionic acid-DPH, PA-DPH) fluidity probes (Figure 4).

Membrane modification by OA decreased microviscosity in the surface region of the membrane by 5% to 7%. In contrast, following the addition of MO, the fluidization of the membrane surface was marginal. This distinct pattern of fluidization caused by the charged and methylated fatty acids, respectively, was similar regardless of the position of unsaturation in the molecule (Fig. 5).

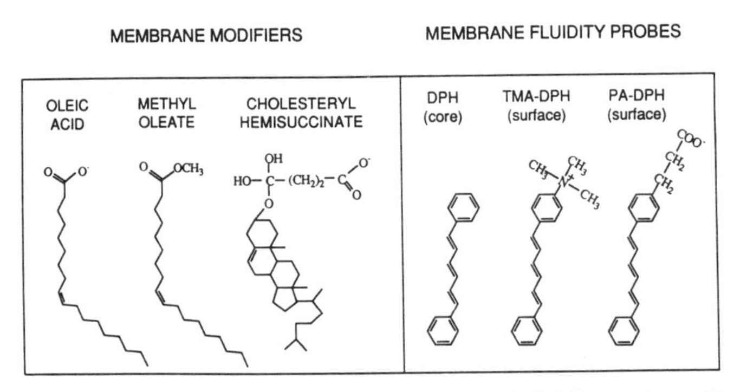

Figure 4. Structural features of membrane modifiers and of fluidity probes. (From Lazar and Medzihradsky, 1992. Reprinted with permission, Raven Press, Ltd.)

The reduction of membrane microviscosity by OA, both in the core and surface region of the membrane, was completely reversed following the addition of CHS. The sterol was particularly effective in restoring the microviscosity of membranes initially fluidized by OA (Fig. 6). Once the normal level of microviscosity at the membrane surface was restored, further rigidification progressed at a slower rate, and in control membranes low concentrations of CHS had no effect. In contrast, CHS strongly increased the microviscosity in the core of both OA-treated and control membranes (Fig. 6).

In the modified membranes, ligand binding to opioid receptors was inhibited in strict correlation with the degree of achieved fluidization at the membrane surface. Accordingly, opioid binding was largely unaffected in membranes treated with MO. On the other hand, free OA inhibited opioid binding by a distinct pattern of $\delta > \mu >> \kappa$, whereby the δ receptor, and to a lesser extent, the μ receptor, exhibited increasing sensitivity as the position of the double bond approached the carboxyl end of the molecule. Scatchard analysis of opioid binding revealed that OA, at 0.5 μmol/mg membrane protein, rendered δ receptor binding undetectable and considerably inhibited ligand interaction with the μ receptor (Table 1). On the other hand, to affect κ binding (studied in monkey brain membranes because of their adequate density of κ receptors) the concentration of OA had to be raised 2-fold relative to that modulating δ and μ

binding. However, even at that concentration, elaidic acid, the *trans* isomer of oleic acid with diminished fluidizing properties (Remmers *et al.*, 1990), was only marginally effective (Table 1). Methyl oleate did not significantly alter the parameters of ligand binding to any of the opioid receptors.

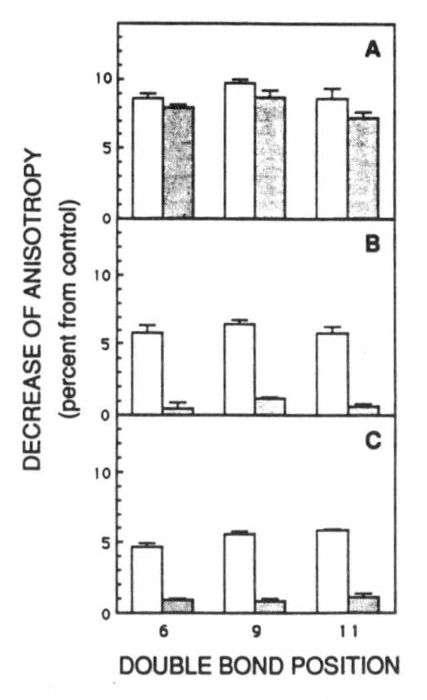

Figure 5. Microviscosity of rat brain membranes modified by fatty acids and their methyl esters. Synaptosomal membranes were modified by the addition of 0.5 μmol/mg membrane protein of either petroselenic (*cis*-Δ^6-octadecenoic) acid; oleic (*cis*-Δ^9-octadecenoic) acid; or *cis*-vaccenic (*cis*-Δ^{11}-octadecenoic) acid (□), or their corresponding methyl esters (■). After the incorporation of DPH (panel A), TMA-DPH (panel B) or PA-DPH (panel C) into control or modified membranes, fluorescence anisotropy was determined. (From Lazar and Medzihradsky, 1992. Reprinted with permission, Raven Press, Ltd.)

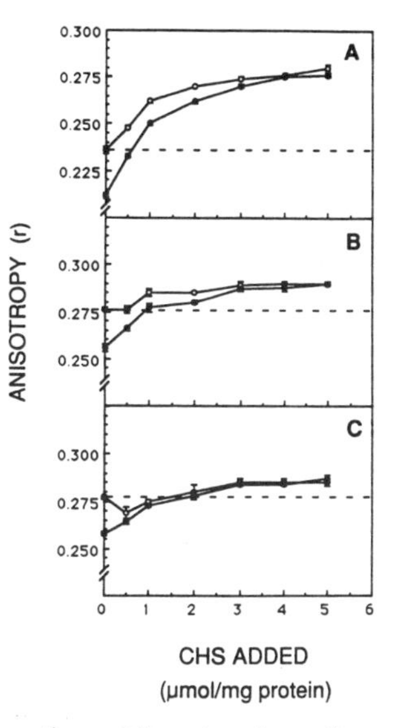

Figure 6. Microviscosity of rat brain membranes modified by cholesterol. Control synaptosomal membranes (○) and membranes initially treated with 0.5 μmol/mg membrane protein of oleic acid (●) were modified by the addition of increasing concentrations of cholesteryl hemisuccinate as shown. Following incorporation of DPH (panel A), TMA-DPH (panel B), or PA-DPH (panel C), fluorescence anisotropy (r) was determined. (From Lazar and Medzihradsky, 1992. Reprinted with permission, Raven Press, Ltd.)

In addition to restoring membrane microviscosity initially reduced by oleic acid (Figure 6), the subsequent addition of CHS reversed the inhibition of ligand binding by approximately 40 fmoles/mg protein at the μ and κ receptor (determined by the selective opioids [^{3}H]DAMGO and [^{3}H]U69,593, respectively), and by 90 fmoles/mg protein at the δ-opioid receptor (determined by the selective opioid [^{3}H]D-pen^2-D-Pen^5-enkephalin) to which binding was initially abolished by OA (Table 1). After treatment with CHS only the K_D of [^{3}H]DPDPE remained higher relative to the control value. In untreated membranes the addition of CHS increased the B_{max} of δ-, μ-, and κ-opioids, by 38%, 25% and 23%, respectively. In these membranes, as in those treated with fatty acids, the K_D at the δ site was affected most (Table 1).

B. Modulation of Receptor Conformation

Changes in lipid composition and membrane fluidity were shown to affect the function of membrane proteins by influencing their conformational states (Fong and McNamee, 1987; Aloia *et al.*, 1988). For example, membrane fluidization by either unsaturated fatty acids or ethanol inhibited the transition to the K^+-sensitive conformation of Na, K-ATPase (Swann, 1984), and a decrease in membrane fluidity induced conformational changes in the lipid-buried subunit of mitochondrial H^+-ATPase that were transmitted to the aqueous subunit of the enzyme (Zhang and Yang, 1989). In view of these findings, it is conceivable that the effects of OA and CHS described in the previous section could reflect the distortion and restoration, respectively, of opioid receptor conformation that is optimal for ligand binding. Furthermore, the observed strong inhibition of opioid receptor binding by CHS may be the consequence of displacement by the sterol of essential lipids in the boundary layer surrounding the receptor.

TABLE 1. Parameters of Opioid Binding in Modified and Control Membranes

Modifiers added	Amount added (μmol/mg of protein)	K_D (nM)	B_{max} (fmol/mg of protein)	n
Binding of [^{3}H]DPDPE				
None		3.52 (0.33)	97.7 (8.1)	45
OA	0.5	ND	ND	42
MO	0.5	5.99 (1.09)	90.0 (17.0)	44
CHS	3.0	11.70 (1.90)	135.0 (12.0)	35
OA/CHS	0.5/3.0	16.80 (5.40)	89.5 (16.8)	38
Binding of [^{3}H]DAMGO				
None		0.36 (0.02)	109 (3.0)	46
OA	0.5	0.92 (0.13)	45.0 (4.7)	46
MO	0.5	0.42 (0.03)	117 (6.0)	47
CHS	3.0	0.35 (0.02)	136 (4.0)	42
OA/CHS	0.5/3.0	0.40 (0.02)	84.7 (2.2)	40
Binding of [^{3}H]U69,593				
None		0.91 (0.08)	60.2 (3.3)	41
OA	1.0	0.84 (0.15)	17.5 (1.9)	38
MO	1.0	1.07 (0.06)	72.0 (2.9)	42
CHS	3.0	0.49 (0.06)	73.4 (4.4)	40
OA/CHS	1.0/3.0	0.82 (0.08)	60.9 (3.7)	39
EA	1.0	1.20 (0.08)	51.2 (2.4)	40

The binding parameters at equilibrium were obtained from nonlinear regression analysis, fitting the data to a one-site model for ligand binding. Shown are parameter means and SEM values (in parentheses) computed from the total number of data points (n) obtained in three experiments. OA, oleic acid; MO, methyl oleate; CHS, cholesteryl hemisuccinate; EA, elaidic acid. ND, not detectable. (From Lazar and Medzihradsky, 1992. Reprinted with permission, Raven Press, Ltd.)

Considering the evidence for a role of membrane fluidity in the conformational transition of proteins and in the modulation of opioid receptor binding, the conformational change of opioid receptor in membranes with altered microviscosity was investigated. Based on the ability of sodium to enhance antagonist binding while reducing that of agonist (Pert and Snyder, 1974) and to protect the opioid receptor from inactivation by the sulfhydryl-reactive agent N-ethylmaleimide (NEM; Simon and Groth, 1975), a two-state model for the opioid receptor was forwarded in which sodium induces the conversion of an "agonist" to an "antagonist" conformation (Figure 7). In the latter state, the model depicts obstructed interaction of sodium with functional sulfhydryl residues on the receptor (Simon and Hiller, 1981).

While in unmodified synaptosomal membranes 150 mM NaCl enhanced and attenuated, respectively, the receptor binding of the opioid antagonist [^{3}H]naltrexone and agonist [^{3}H]etorphine, this modulating effect of sodium was strongly diminished in membranes rigidified by the incorporation of CHS (Figure 8). Furthermore, incorporation of OA into membranes initially treated with CHS restored the ability of sodium to distinctly affect the binding of opioid agonists and antagonists (Figure 9). As described earlier in this chapter, the incoporation of CHS and OA markedly altered the microviscosity of brain membranes (Figures 5 and 6). Additional evidence for the involvement of membrane microviscosity in the modulation of opioid receptor conformation was provided by the complete restoration of the enhancement by sodium of [^{3}H]naltrexone binding in CHS-treated membranes that were subsequently fluidized by temperature elevation (Lazar and Medzihradsky, 1993).

Incubation of untreated membranes with NEM caused a progressive inhibition of [^{3}H]naltrexone binding, whereby inactivation of the μ receptor exhibited pseudo first-order kinetics (Figure 10). In these membranes, sodium potently attenuated the rate of receptor inactivation, increasing the $t_{1/2}$ by 185%. In contrast, in membranes rigidified by CHS the enhancement of the $t_{1/2}$ by sodium was limited to 42%. Subsequent addition of OA to the CHS-modified membranes restored sodium protection of the receptor toward that observed in untreated, control membranes (Fig. 10).

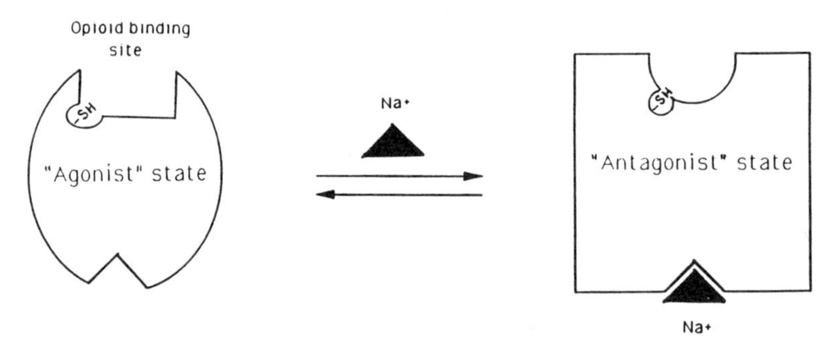

Figure 7. Model for the sodium-induced conformational change of the opioid receptor. (Modified from Simon and Hiller, 1981. Reprinted with permission, Little, Brown & Co.)

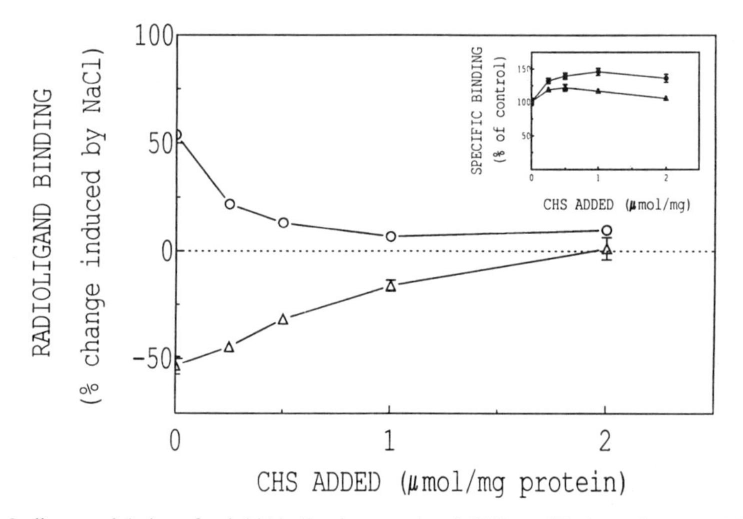

Figure 8. Sodium modulation of opioid binding in control and CHS-modified membranes. After treating membranes with increasing concentrations of CHS, the specific binding of 0.5 nM [^{3}H]etorphine (∧, ▲) and 0.5 nM [^{3}H]naltrexone (○, ●) was determined in the absence (▲, ●) and presence (∧, ○) of 150 mM NaCl. Ligand binding was carried out in the presence of 50 nM oxymorphindole to block ligand binding to δ receptors. Plotted are the sodium effects on ligand binding relative to binding performed in the absence of sodium (depicted in the inset). (From Lazar and Medzihradsky, 1993. Reprinted with permission, Raven Press, Ltd.)

These results are consistent with an inhibition of the sodium-induced conformational transition of the receptor in the rigidified membranes as compared to the conversion between the "agonist" and "antagonist" states occurring in control membranes. The involvement of membrane microviscosity in the latter process was strongly supported by restoration of the sodium effect on ligand binding and on receptor protection following fluidization of the CHS-treated membranes with oleic acid or by temperature.

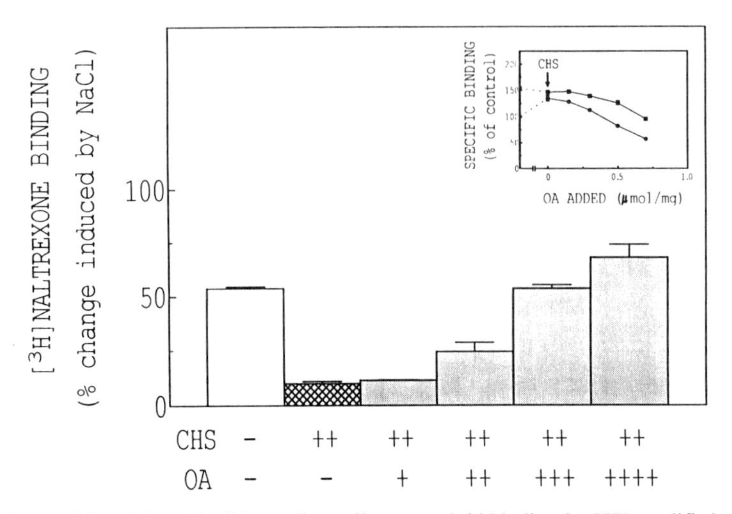

Figure 9. Reversal by oleic acid of the sodium effect on opioid binding in CHS-modified membranes. Membranes treated with 2 μmol CHS/mg protein (++) were subsequently modified by the addition of increasing concentrations (per mg membrane protein) of OA: 0.15 μmol (+); 0.3 μmol (++); 0.5 μmol (+++); and 0.7 μmol (++++). The specific binding of 0.5 nM [^{3}H]naltrexone in control, CHS- and OA-modified membranes was then determined in the absence (●) and presence (■) of 150 mM NaCl and in the presence of 50 nM oxymorphindole (inset). Plotted is the enhancement of ligand binding by sodium relative to binding determined in its absence. (From Lazar and Medzihradsky, 1993. Reprinted with permission, Raven Press, Ltd.)

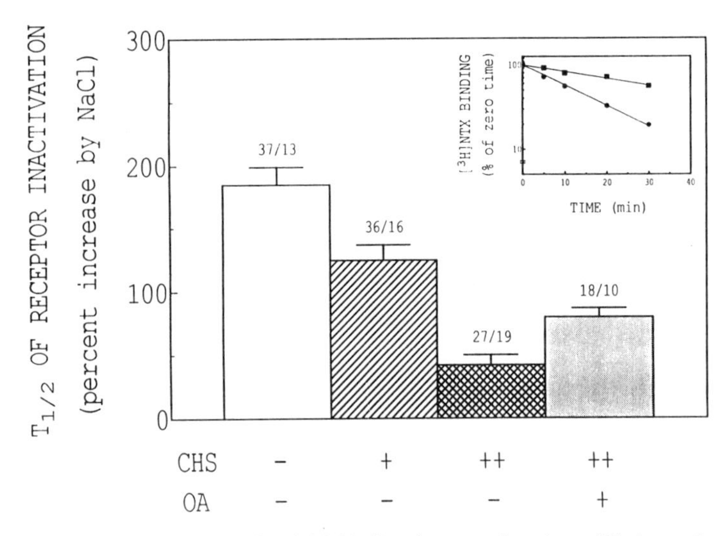

Figure 10. Inactivation by NEM of opioid binding in control and modified membranes. After pretreatment with 0 (-), 1 (+), or 2 (++) μmol CHS and with 0(-) or 0.2 μmol OA/mg membrane protein (+), brain membranes were incubated with 3 mM NEM for various periods of time in the absence (●) or presence (■) of 150 mM NaCl. Subsequently, binding of 0.5 nM [^{3}H]naltrexone in the presence of 150 mM NaCl and 50 nM oxymorphindole was determined. Plotted is the effect of Na^+ on the $t_{1/2}$ of receptor inactivation determined by ligand binding: the numbers above the depicted bars represent the $t_{1/2}$ values in the presence/absence of sodium.The inset depicts the time-dependent inactivation of receptor by NEM in untreated membranes, observed in a representative experiment. (From Lazar and Medzihradsky, 1993. Reprinted with permission, Raven Press, Ltd.)

CONCLUSION AND OUTLOOK

In spite of the strong evidence for modulation of opioid receptor function in the membrane environment, the molecular mechanisms by which membrane lipids influence ligand-receptor-effector interactions are not yet understood. Is the modulation caused by the lipid bulk order or by the interaction of the receptor with a specific lipid in the membrane microenvironment? Can these two modes of lipid action be resolved and quantified? Which of the phospholipids is primarily interacting with the receptor in the boundary layer, and which structural element in the molecule of a given lipid is responsible for bringing about the observed gross modulations? As shown recently, intricate structural details of the lipid bilayer contribute specifically to modulation of functional membrane proteins such as the glucose carrier in erythrocytes (Carruthers and Melchior, 1988). As outlined in the Introduction, lipid modulation of membrane composition and biophysical property has strong significance in influencing opioid signal transduction involving lateral mobility of receptor-effector components. In this respect, opioid receptors were shown to couple to G proteins (Clark *et al.*, 1989; Remmers and Medzihradsky, 1991a; Carter and Medzihradsky, 1993), and the myristoylation and prenylation of these transducers, likely to mediate their interaction with both lipids and proteins (Marshall, 1993), was recently described (Hepler and Gilman, 1992).

In elucidating the molecular basis of the effect of lipids on receptor function, it is of primary interest to focus on the lipids localized in the boundary (annular) layer in the immediate vicinity of the receptor (*e.g.*, Lentz, 1988). If the receptor and membrane phospholipids are both carrying a fluorescent tag, than their interaction can be studied by resonance energy transfer. Since the quantitative relationship in the energy transfer between two fluorophores is determined by the distance between them, the results of such experiments provides information on the proximity of two different molecular species, each labeled by a fluorescent tag with overlapping spectral characteristics. Thereby, the efficiency of the energy transfer is measured as the extent of reduction (quenching) of the donor quantum yield. A particularly interesting feature of this approach is its potential to characterize the lipid microenvironment of a membrane-bound functional protein (Devaux and Seigneuret, 1985). By measuring energy transfer between the fluorescent receptor and different fluorescent phospholipids (or cholesterol) incorporated into the bilayer, the lipid molecule that distributes into the boundary layer can be identified. In addition to using fluorescence spectroscopy at a steady-state (*e.g.*, Verbist *et al.*, 1991), the interaction between fluorescently-labeled phospholipids and receptor can be studied at a much higher resolution and sensitivity using the technique of time-resolved polarized fluorescence (Van Hoek *et al.*, 1987; Van Paridon *et al.*, 1988). Significantly, the latter method can also provide information on lipid order and mobility in the boundary layer.

A crucial aspect in interpreting the interaction between a membrane-bound functional protein and incorporated lipid is the mode of insertion of the latter into the membrane. Accurate interpretation of the fluorescence spectra, assumed to reflect the interaction between a membrane phospholipid and receptor, requires a complete, unperturbed incorporation of the added lipid into the phospholipid bilayer, rather than other, nonspecific membrane associations such as adherence. Membrane modification by lipid transfer proteins fulfills the above requirements (Wirtz, 1991). Recent advances in the use of these proteins further increase the specificity of lipid incorporation by eliminating nonspecific transfer, previously contributed to by the employed lipid vesicles (Van Paridon *et al.*, 1988). With this methodological approach the incorporated lipid becomes an integral part of the lipid bilayer whereby either the surface or membrane core can be preferentially modified (Figure 11). As shown, the target membrane in Experiment 2 will have the same surface composition, but a different core due to the specific acyl chains attached to the PC head group. On the other hand, the membrane core in Experiment 3 will resemble that of the control membrane in Experiment 1, but the surface will be different. This two-tier chemical modification can then be complemented by the differential determination of microviscosity at the membrane surface and core using hydrophilic and lipophilic probes, respectively.

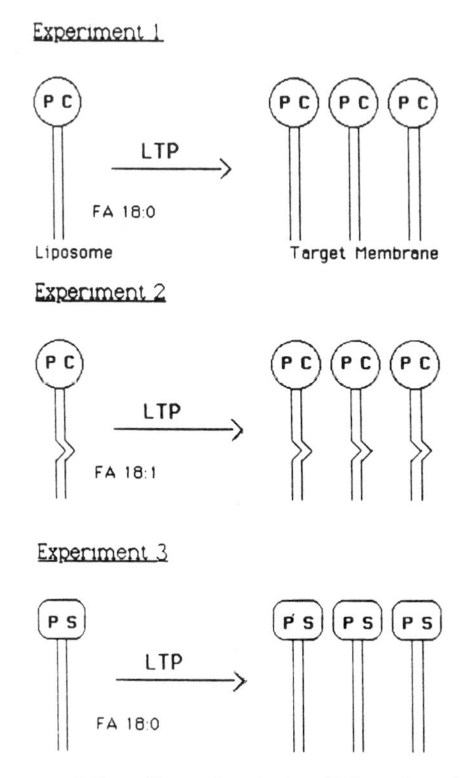

Figure 11. Schematic representation of experiments in which preferentially the membrane surface (experiments 1 and 3) or membrane core (experiments 1 and 2) is modified by phospholipid transfer proteins (LTP). The twisted line in the acyl chain of Experiment 2 indicates unsaturation. PC, phosphatidylcholine; PS, phosphatidylserine; FA, fatty acyl chain of phospholipids. (From Medzihradsky, 1988. Reprinted with permission, Pergamon Press, Ltd.)

Within the goal to study opioid receptor mechanisms in the membrane environment by the methodological approach described above, fluorescent opioids containing the nitrobenzo-oxa-diazole (NBD) residue were synthesized (Archer *et al.*, 1992). A common characteristic of the NBD-labeled fluorescent ligands appears to be preservation of high-affinity binding to the respective receptors. The molecular feature that contributes to this favorable characteristics of the NBD-labeled compounds is the small size of the fluorophore, particularly when compared to fluorescein or rhodamine. The limited bulk of the NBD residue also offers considerable advantages in the use of such ligands to study receptor dynamics in the plasma membrane. As observed with the labeled opioids, the NBD fluorescence has a strong dependence on solvent polarity, a property that aids in discriminating between the fluorescence of receptor-bound, membrane-sequestered, and free ligand in the assay medium. In addition, the high fluorescence intensity of the two synthesized opioid alkaloids enables their detection at subnanomolar concentrations. Recently, a new generation of fluorescent opioids was introduced carrying the fluorophore BODIPY, a nonpolar probe exhibiting high affinity for membranes (Archer and Medzihradsky, manuscript in preparation).

Energy transfer between opioid receptor (labeled by fluorescent ligands) and fluorescent phospholipids and cholesterol (incorporated into neural membranes by transfer proteins) has the potential to disclose the protein-lipid interactions that contribute to the observed modulation of receptor mechanisms by membrane lipids. Such experiments should also reveal the distribution of phospholipid molecules from the bulk pool of membrane lipids into the boundary layer. The understanding of the specific lipid requirements of opioid receptors has particular significance for the eventual reconstitution of the purified protein. As shown with other receptor systems, the functional properties of purified receptors incorporated into lipid vesicles of random composition are quite different, and frequently vastly inferior, than those determined in the native membranes.

Numerous physiological conditions have been shown to influence the lipid composition and fluidity of biological membranes. These include diet (McMurchie, 1988), development (Hitzemann and Harris, 1984), and aging (Shinitzky, 1987b). In addition, drug administration (Heron *et al.*, 1982) and various pathological conditions have resulted in altered membrane content of lipids and cholesterol (*e.g.*, Aloia and Boggs, 1985). For example, in experimentally induced cerebral ischemia (Yoshida *et al.*, 1983) significant increases in the concentration of free fatty acids in brain were described. Therefore, the potent effects of membrane lipids on opioid receptor function, summarized in this chapter, may have significant implications for the modulation of opioid action *in vivo,* thus emphasizing the need to understand the underlying molecular mechanisms.

ACKNOWLEDGEMENTS

The secretarial assistance of Ms. Becky J. McLaughlin in preparing the manuscript is gratefully acknowledged. The author is a recipient of USPHS grant DA 04087 and of a Fogarty Senior International Fellowship for 1992-93.

REFERENCES

Abood, L.G., Butler, M., and Reynolds, D., 1980, Effect of calcium and physical state of neural membranes on phosphatidylserine requirement for opiate binding. *Mol. Pharmacol.* 17:290.

Abood, L.G., Salem, N., MacNiel, M., and Butler, M., 1978, Phospholipid changes in synaptic membranes by lipolytic enzymes and subsequent restoration of opiate binding with phosphatidylserine. *Biochim. Biophys. Acta* 530:35.

Aloia, R.C. and Boggs J.M., 1985, "Membrane Fluidity in Biology, Vol. 3: Disease Processes", Academic Press, Inc., New York.

Aloia, R.C., Curtain, C.C., and Gordon, L.M., 1988, Advances in Membrane Fluidity, Vol. 2: Lipid Domains and the Relationship to Membrane Function. Alan R. Liss, Inc., New York.

Archer, S., Medzihradsky, F., Seyed-Mozaffari, A., and Emmerson, P.J., 1992, Synthesis and characterization of 7-nitrobenzo-2-oxa-1,3-diazole (NBD)-labeled fluorescent opioids. *Biochem. Pharmacol.* 43:301.

Braun, S. and Levitzki, A., 1979, Adenosine receptor permanently coupled to turkey erythrocyte adenylate cyclase. *Biochemistry* 18:2134.

Butler, M. and Abood, L.G., 1982, Use of phospholipase A_2 to compare phospholipid organization in synaptic membrane, myelin and liposomes. *J. Membr. Biol.* 66:1.

Carruthers, A. and Melchior, D.L., 1988, Role of the bilayer lipids in governing membrane transport processes, in "Lipid Domains and the Relationship to Membrane Function", R.C. Aloia, C.C. Curtain, and L.M. Gordon, eds., Alan R. Liss, Inc., New York.

Carter, B.D. and Medzihradsky, F., 1993, G_o mediates the coupling of μ opioid receptor to adenylyl cyclase in cloned neural cells and brain. *Proc. Natl. Acad. Sci. USA* (in press).

Clark, M.J., Nordby, G.L., and Medzihradsky, 1989, Relationship between opioid receptor occupancy and stimulation of low-Km GTPase in brain membranes. *J. Neurochem.* 52:1162.

Devaux, P.F. and Seigneuret, M., 1985, Specificity of lipid-protein interactions as determined by spectroscopic techniques. *Biochim. Biophys. Acta* 822:63.

Farahbakhsh, Z.T., Deamer, D.W., Lee, N.M., and Loh, H.H., 1986, Enzymatic reconstitution of brain membrane and membrane opiate receptors. *J. Neurochem* 46:953.

Fong, T.M. and McNamee, M.G., 1987, Stabilization of acetylcholine receptor secondary structure by cholesterol and negatively charged phospholipids in membranes. *Biochemistry* 26:3871.

Hanski, E., Rimon, G., and Levitzki, A., 1978, Adenylate cyclase activation by the β-adrenergic receptors is a diffusion-controlled process. *Biochemistry* 18:846.

Hasegawa, J.-I., Loh, H.H., and Lee, N.M., 1987, Lipid requirement for μ opioid receptor binding. *J. Neurochem.* 49:1007.

Hepler, J.R. and Gilman, A., 1992, G proteins. *Trends Biochem. Sci.* 17:383.

Heron, D., Shinitzky, M., Zamir, N., and Samuel, D., 1982, Adaptive modulations of brain membrane fluidity in drug addiction and denervation supersensitivity. *Biochem. Pharmacol.* 31:2435.

Hitzemann, R.J. and Harris, R.A., 1984, Developmental changes in synaptic membrane fluidity: a comparison of 1,6-diphenyl-1,3,5-hexatriene (DPH) and 1-[4-(trimethyl-amino)phenyl]-6-phenyl-1,3,5-hexatriene (TMA-DPH). *Brain Res.* 316:113.

Ho, W.K.K. and Cox, B.M., 1982, Reduction of opioid binding in neuroblastoma x glioma cells grown in medium containing unsaturated fatty acids. *Biochim. Biophys. Acta* 688:211.

Law, P.Y., Griffin, M.T., Koehler, J.E., and Loh, H.H., 1983, Attenuation of enkephalin activity in neuroblastoma x glioma NG108-15 hybrid cells by phospholipases. *J. Neurochem.* 40:267.

Lazar, D.F. and Medzihradsky, F., 1990, Differential inhibition of δ-opiate binding and low-Km GTPase stimulation by phospholipase A_2 treatment. *Progr. in Clin. Biol. Res.* 328:113.

Lazar, D.F. and Medzihradsky, F., 1992, Altered microviscosity at brain membrane surface induces distinct and reversible inhibition of opioid receptor binding. *J. Neurochem.* (in press).

Lazar, D.F. and Medzihradsky, F., 1993, Altered transition between agonist- and antagonist-favoring states of μ opioid receptor in brain membranes with modified microviscosity. *J. Neurochem.* (in press).

Lee, N.M. and Smith, A.D., 1980, A protein-lipid model of the opiate receptor. *Life Sci.* 26:1459.

Lentz, B.R., 1988, Organization of membrane proteins, *in* "Lipid Domains and the Relationship to Membrane Function", R.C. Aloia, C.C. Curtain, and L.M. Gordon, eds., Alan R. Liss, Inc., New York.

Loh, H.H. and Law, P.-Y., 1980, The role of membrane lipids in receptor mechanisms. *Ann. Rev. Pharmacol. Toxicol.* 20:201.

Marshall, C.J., 1993, Protein prenylation: a mediator of protein-protein interactions. *Science* 259:1865.

McGee, R., Jr. and Kenimer, J.G., 1982, The effects of exposure to fatty acids on opiate receptors, prostaglandin E_1 receptors, and adenylate cyclase activity of neuroblastoma x glioma hybrid cells. *Mol Pharmacol.* 22:360.

McMurchie, E.J., 1988, Dietary lipids and the regulation of membrane fluidity and function, in Physiological Regulation of Membrane Fluidity", R.C. Aloia, C.C. Curtain, and L.M. Gordon, eds., Alan R. Liss, New York.

McNamee, M.G. and Fong, M., 1988, Effects of membrane lipids and fluidity on acetylcholine receptor function, *in* "Lipid Domains and the Relationship to Membrane Function", R.C. Aloia, C.C. Curtain, and L.M. Gordon, eds. Alan R. Liss, Inc., New York (1988).

McOsker, C.C., Weiland, G.A., and Zilversmit, D.B., 1983, Inhibition of hormone-stimulated adenylate cyclase activity after altering turkey erythrocyte phospholipid composition with a nonspecific lipid transfer protein. *J. Biol. Chem.* 258:13017.

Medzihradsky, F., 1988, Membrane modulation of opioid receptor mechanisms by membrane lipids: an investigated approach. *Adv. in Biosci.* 75:41.

Medzihradsky, F. and Carter, B.D., 1991, Opioid receptor mechanisms, in "Biochemistry and Physiology of Substance Abuse", R.R. Watson, ed. CRC Press, Boca Raton, FL.

Neubig, R.R., Gantzos, R.D., and Thomsen, W.J., 1988, Mechanism of agonist and antagonist binding to α_2-adrenergic receptors: evidence for a precoupled receptor-guanine nucleotide protein complex. *Biochemistry* 27:2374.

Pert, C.B. and Snyder, S.H., 1974, Opiate receptor binding of agonists and antagonists affected differentially by sodium. *Mol. Pharmacol.* 10:868.

Remmers, A.E. and Medzihradsky, F., 1991a, Reconstitution of high-affinity opioid agonist binding in brain membranes. *Proc. Natl. Acad. Sci. U.S.A.* 88:2171.

Remmers, A.E. and Medzihradsky, F., 1991b, Resolution of biphasic binding of the opioid antagonist naltrexone in brain membranes. *J. Neurochem.* 57:1265.

Remmers, A.E., Nordby, G.L., and Medzihradsky, F., 1990, Modulation of opioid receptor binding by *cis* and *trans* fatty acid. *J. Neurochem.* 55:1993.

Sargent, D.F., Bean, J.W., Kosterlitz, H.W., and Schwyzer, R., 1988, Cation dependence of opioid receptor binding supports theory on membrane-mediated receptor selectivity. *Biochemistry* 27:4974.

Schiller, P.W., Nguyen, T.M.-D., Chung, N.N., and Lemieux, C., 1989, Dermorphan analogues carrying an increased positive net charge in their "messages" domain display extremely high μ opioid receptor selectivity. *J. Med. Chem.* 32:698.

Schwyzer, R., 1986, Molecular mechanisms of opioid receptor selection. *Biochemistry* 25:6335.

Shinitzky, M., 1987a, "Physiology of Membrane Fluidity", Vol. 1, CRC Press, Boca Raton, FL.

Shinitzky, M., 1987b, Patterns of lipid changes in the membranes of the aged brain. *Gerontology* 33:149.

Simon, E.J. and Groth, J., 1975, Kinetics of opiate receptor inactivation by sulfhydryl reagents: evidence for conformational change in presence of sodium ions. *Proc. Natl. Acad. Sci. USA* 72:2404.

Simon, E.J. and Hiller, J.M., 1981, Opioid peptides and opiate receptors, *in* "Basic Neurochemistry", G.J. Siegel, R.W. Albers, B.W. Agranoff, and R. Katzman, eds. Little, Brown and Co., Boston.

Swann, A.C., 1984, Free fatty acids and (Na^+,K^+)-ATPase: effects on cation regulation, enzyme conformation, and interactions with ethanol. *Arch. Biochem. Biophys.* 233:354.

Tefft, R.E., Carruthers, A., and Melchior, D.L., 1986, Reconstituted human erythrocyte sugar transporter activity is determined by bilayer lipid head groups. *Biochemistry* 25:3709.

Van Hoek, A., Vos, K.,m and Visser, A.J.W.G., 1987, Ultrasensitive time-resolved polarized fluorescence spectroscopy as a tool in biology and medicine. *J. Quantum Electr.* 23:1812.

Van Paridon, P.A., Shute, J.K., Wirtz, K.W.A., and Visser A.J.W.G., 1988, A fluorescence decay study of parinaroyl-phosphatidylinositol incorporated into artificial and natural membranes. *Eur. Biophys. J.* 16:53.

Verbist, J., Gadella, T.W.J., Raeymaekers, L., Wuytack, F., Wirtz, K.W.A., and Casteels, R., 1991, Phosphoinositide-protein interactions of the plasma-membrane Ca^{2+} transport ATPase as revealed by fluorescence energy transfer. *Biochim. Biophys. Acta* 1063:1.

Viret, J., Daveloose, D., and Leterrier, F., 1990, Modulation of the activity of functional membrane proteins by the lipid bilayer fluidity, *in* "Membrane Transport and Information Storage", R.C. Aloia, C.C. Curtain, and L.M. Gordon, eds. Alan R. Liss, Inc., New York.

Wirtz, K.W.A., 1991, Phospholipid transfer proteins. *Ann. Rev. Biochem.* 60:73.

Yoshida, S., Inoh, S., Asano, T., Sano, K., Shimasaki, H., and Ueta, N., 1983, Brain free fatty acids, edema, and mortality in gerbils subjected to transient, bilateral ischemia, and effect of barbiturate anesthesia. *J. Neurochem.* 40:1278.

Zhang, X.F. and Yang, F.Y., 1989, Further study on the role of Mg^{2+} in lipid-protein interaction in reconstituted porcine heart mitochondrial H^+-ATPase. *Biochim. Biophys. Acta* 976:53.

A PHARMACOLOGICAL APPROACH TO IDENTIFY HORMONE SIGNALING PATHWAYS CONTROLLING GENE REGULATION IN *DICTYOSTELIUM*

Ron D.M. Soede[1], Dorien J.M. Peters[1], Bernd Jastorff[2], Peter J.M. Van Haastert[3] and Pauline Schaap[1]

[1]Cell Biology and Genetics, Department of Biology, University of Leiden, Kaiserstraat 63, 2311 GP Leiden, The Netherlands
[2]Institute for Organic Chemistry, University of Bremen, NW2 Leobener Strasse, D-2800 Bremen, Federal Republic of Germany
[3]Department of Biochemistry, University of Groningen, Nijenborgh 16, 9747 AG Groningen, The Netherlands

SUMMARY

In *Dictyostelium discoideum*, extracellular cAMP functions as a hormone-like signal and induces chemotaxis and the expression of several classes of developmentally regulated genes. We describe a novel set of non-hydrolysable cAMP derivatives, which can selectively activate cell surface cAMP receptors (CAR) or the intracellular cAMP-dependent kinase (PKA), two putative target proteins for the effects of cAMP. Comparison of the affinities of these derivatives for CAR and PKA, to their efficacy as agonists for induction of three classes of cAMP-regulated genes, shows that CAR is the first target for cAMP regulation of gene expression.

To unravel involvement of specific cAMP signal transduction pathways in control of gene regulation, we used the partial antagonist 8-*p*-chlorophenylthioadenosine 3',5'-monophosphate (8-CPT-cAMP). This cAMP derivative selectively induces expression of aggregative genes but not of the postaggregative prestalk and prespore genes. 8-CPT-cAMP mimicks cAMP induced activation of adenylyl cyclase and guanylyl cyclase, but inhibits instead of activates phospholipase C (PLC). Our data suggest specific involvement of PLC activation in postaggregative but not preaggregative gene regulation.

INTRODUCTION

In cell-cell communication secreted signaling molecules play an important role. Individual cells integrate the diverse stimuli that reach them to respond adequately. The

New Developments in Lipid-Protein Interactions and Receptor Function
Edited by K.W.A. Wirtz *et al.*, Plenum Press, 1993

response depends on the properties of the cell such as the type of receptors that are expressed, the signal transduction pathway to which these receptors are linked and the cellular functions that are activated. To understand signal transduction pathways and their control in gene regulation, intensive research has been done in the lower eukaryote *Dictyostelium discoideum*. Signal transduction in this organism is very similar to that in higher eukaryotes.

In the unicellular amoebal stage of development, *Dictyostelium discoideum* cells live in the soil and feed by phagocytosis of bacteria. When their food source is exhausted a specific developmental program is initiated (figure 1). Cells start to synthesize and secrete pulses of cAMP, which serve as chemoattractant for surrounding cells. Excited cells move towards the cAMP source and start to synthesize and secrete cAMP themselves. In this way a population of cells is able to relay cAMP signals over large distances. This process results in the formation of multicellular aggregates, involving up to 10^5 cells.

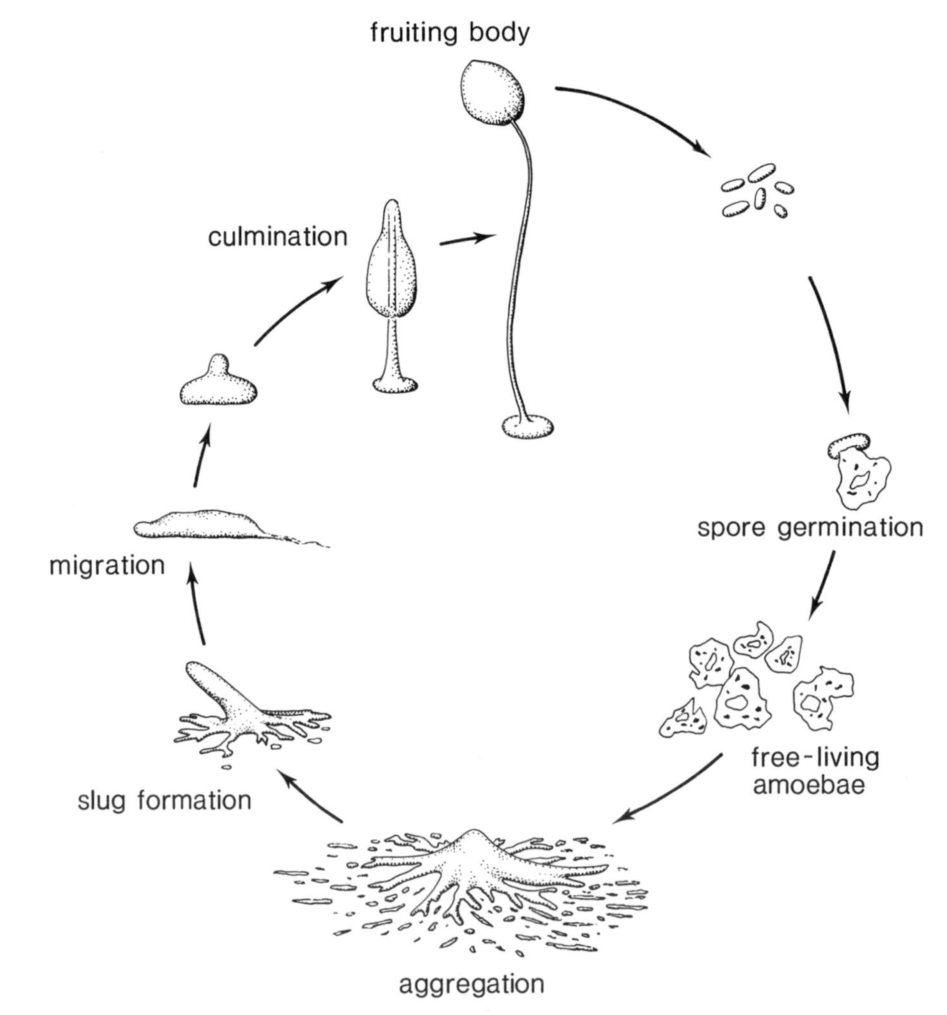

Figure 1. Life cycle of *Dictyostelium discoideum*.

At the slug stage, cells differentiate into two major cell types, prestalk cells and prespore cells. A specific cell pattern is formed and maintained in the slug: the prestalk cells are located in the anterior region, the prespore cells in the posterior region. The cell aggregate undergoes several morphogenetic changes and finally forms a fruiting body, consisting of a spherical head of spores, held aloft by a slender stalk. This stalk is formed by highly vacuolized, rigid stalk cells. If sufficient food is available, spores can germinate and give rise to new amoeboid cells.

During *Dictyostelium discoideum* development, extracellular cAMP not only functions as a chemoattractant but also as a hormone-like signal; it induces the expression of several classes of genes (see Williams, 1988). In the present study, we shall focus our attention on signal transduction pathways involved in cell type specific gene regulation. We describe pharmacological approaches used to identify specific signal transduction components involved in gene regulation.

REGULATION OF GENE EXPRESSION DURING DEVELOPMENT

Stage- and cell type specific gene expression in *Dictyostelium discoideum* has been intensively investigated; many developmentally regulated genes have been cloned and regulation of their expression has been analysed. Several signal molecules regulating gene expression have been characterized and possible functions have been suggested for other molecules, which are in some cases partially purified. From these data a critical role for cAMP and other hormones in regulation of gene expression becomes evident.

In *Dictyostelium discoideum*, developmental regulation of gene expression is started upon exhaustion of their food source. Expression of genes specific for the growth phase of development decreases (Kopachik *et al.*, 1985; Singleton *et al.*, 1987,1988) and a class of "early" genes are transcribed. The expression of these genes is considered to be induced by amino acid deprivation and by secreted factors, that are used to sense cell density (Grabel and Loomis, 1978; Clarke *et al.*, 1987; Gomer *et al.*, 1991). Early gene expression is transient, repression is in many cases induced by cAMP (Singleton *et al.*, 1988; Mann and Firtel, 1989).

A second class of gene products, required for the aggregation process, accumulates after about four hours into starvation. Their genes code for cAMP receptors, cAMP-phosphodiesterase (PDE), adhesive contact sites A (csA) and a G-protein α-subunit (Gα2) (Darmon *et al.*, 1975; Gerisch *et al.*, 1975; Lacombe *et al.*, 1986; Noegel *et al.*, 1986; Klein *et al.*, 1988; Kumagai *et al.*, 1989). The chemotactic signal, nanomolar cAMP pulses, accelerates the expression of these aggregative genes. During aggregation a third class of genes are transcribed, which are later preferentially expressed at the anterior prestalk region. These prestalk-related genes are positively regulated by both nanomolar cAMP pulses or continous stimulation with micromolar cAMP concentrations. This class includes genes coding for cysteine proteases as well as a number of other proteins of unknown function (Barklis and Lodish, 1983; Mehdy *et al.*, 1983; Mehdy and Firtel, 1985).

After aggregates have formed, expression of aggregative genes declines. At this stage, a major class of spore-specific genes are expressed (Barklis and Lodish, 1983; Mehdy *et al.*, 1983; Morrissey *et al.*, 1984). Expression of spore-specific genes is induced by micromolar concentrations of cAMP. Two stalk specific genes, *ecmA* and *ecmB*, coding for extracellular stalk matrix proteins are also transcribed during the formation of tight aggregates (Jermyn *et al.*, 1987). Expression of these genes is induced by a Differentiation Inducing Factor (DIF) (Williams *et al.*, 1987).

TRANSDUCTION OF EXTRACELLULAR cAMP SIGNALS

A role for cAMP as morphogen motivated further research to identify signaling pathways controlling gene regulation. Extracellular cAMP binds to highly specific cell surface cAMP receptors (Henderson, 1975). The genes encoding four homologous cAMP receptors (CARs) have been cloned and the deduced amino acid sequences predict proteins with seven transmembrane spanning domains, which are characteristic of surface receptors interacting with G-proteins (Klein *et al.*, 1988; Saxe *et al.*, 1991).
This makes cAMP transmembrane processing similar to transduction of a large number of

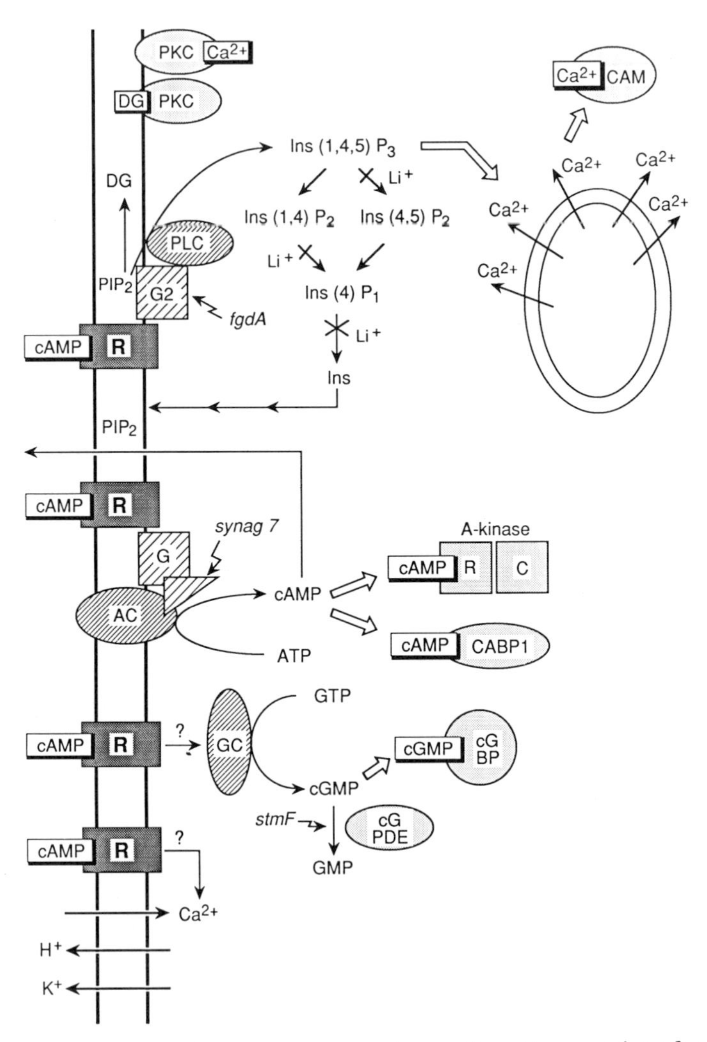

Figure 2. cAMP signal transduction pathways in *Dictyostelium*. Representation of currently identified components of cAMP transduction pathways. R: cAMP surface receptor; AC: adenylyl cyclase; GC: guanylyl cyclase; PLC: phospholipase C; G: guanine nucleotide regulatory protein; CABP: cAMP binding protein; A-kinase: cAMP-dependent protein kinase; CAM: calmodulin; PKC: Ca^{2+}-dependent protein kinase; DG: diacylglycerol; PIP_2: phosphatidylinositolbisphosphate; Ins(1,4,5)P_3: inositol 1,4,5-trisphosphate.

well-known hydrophilic hormones in higher vertebrates as epinephrine, dopamine, serotonin, luteinizing hormone, vasopressin and many others (Birnbaumer, 1990). The different CARs show similar, but not identical cyclic nucleotide specificity (Johnson *et al.*, 1992), and are expressed at different stages of development. This suggests that the four CARs may have different functions during development.

Interaction of cAMP with cell surface receptors activates a number of intracellular target enzymes (figure 2), such as adenylyl cyclase (Roos *et al.*, 1975), guanylyl cyclase (Mato *et al.*, 1977; Wurster *et al.*, 1977) and phospholipase C (Europe-Finner and Newell, 1987; Van Haastert et al., 1989).
This activation results in rapid increases in the levels of the intracellular messengers $InsP_3$, cGMP and cAMP, which peak at respectively 6, 10 and 60 seconds. $InsP_3$ induces release of Ca^{2+} from non-mitochondrial stores (Europe-Finner and Newell 1986; Abe *et al.*, 1988). Extracellular cAMP also induces an influx of Ca^{2+} (Wick *et al.*, 1978; Milne and Coukell, 1991) and efflux of K^+-ions and H^+-ions (Malchow *et al.*, 1978; Aeckerle *et al.*, 1985).

The activation of adenylyl cyclase and phospholipase C is mediated by different G-proteins (Theibert and Devreotes, 1986; Van Haastert *et al.*, 1987; Van Haastert *et al.*, 1989). It is not yet known if G-proteins are involved in the activation of guanylyl cyclase. At present eight distinct G-protein α-subunits and one ß-subunit have been cloned (Lilly *et al.*, 1988; Kumagai *et al.*, 1989; Pupillo *et al.*, 1989; Hadwiger *et al.*, 1991; Wu and Devreotes, 1991). The function of only one of these G-proteins is clear: G2, a large hetero-trimeric G-protein, most likely couples the cAMP receptor to PLC (Coukell *et al.*, 1983; Kesbeke *et al.*, 1988; Kumagai *et al.*, 1989).

Several target proteins for the potential intracellular messengers Ca^{2+}, cAMP and cGMP have been identified and the genes for some have been cloned. Ca^{2+}-ions can interact with calmodulin (Clarke *et al.*, 1980; Marshak *et al.*, 1984) or activate a protein kinase C (PKC), which is stimulated by Ca^{2+}, phospholipids or phorbol esters (Luderus *et al.*, 1989). cGMP can interact with a cGMP-dependent protein kinase (PKG) (Wanner and Wurster, 1990) and an intracellular cGMP binding protein, which changes its kinetics upon interaction with oligonucleotides (Parissenti and Coukell, 1990). The binding of this protein is most effectively altered by GC-rich DNA sequences and may therefore interact with regulatory DNA sequences, which are frequently GC-rich in *Dictyostelium*. Intracellular cAMP can activate a cAMP-dependent protein kinase (De Gunzburg and Veron, 1982; Mutzel *et al.*, 1987), which consists of only one regulatory and one catalytic subunit. This in contrast to the PKA holoenzyme of higher eukaryotes, which consist of two regulatory and two catalytic subunits.

It is evident that cAMP induces a large repertory of intracellular responses, which are each potential intermediates for its regulation of gene expression. Of particular interest is the strong homology between transmembrane signal transduction of cAMP in this lower eukaryote and the processing of many hormone-like signals in mammalian systems. In the following paragraphs we will describe pharmacological approaches used in our laboratory, to identify signal transduction pathways involved in developmental regulation of gene expression.

SELECTIVE ACTIVATION OF SIGNAL TRANSDUCTION COMPONENTS BY NON-HYDROLYSABLE cAMP DERIVATIVES

An initial step in analyzing transduction pathways involved in gene regulation, is to identify the first target for the effects of cAMP. *Dictyostelium* cells exhibit at least four cAMP cell surface receptors, an intracellular cAMP-dependent protein kinase and a cell surface cAMP-phosphodiesterase, which are all possible candidates to transduce the cAMP signal. To elucidate which of these cAMP binding proteins is involved in the regulation of a specific class of cAMP induced genes, we used a set of systematically modified non-hydrolysable cAMP derivatives (figure 3, table 1). In these derivatives an axial (Sp) exocyclic oxygen atom is replaced by a sulphur atom, which renders the cyclic phosphate ring resistent to degradation by PDE. Further modifications were introduced to induce specificity for binding to either CAR (2'deoxy and 5'amino derivatives) or to PKA (substitutions at 8 and 6 positions and replacement of adenine by benzimidazole) or to increase lipophilicity and render the derivatives membrane-permeable (6-thioethyl, 8-bromo and benzimidazole derivatives).

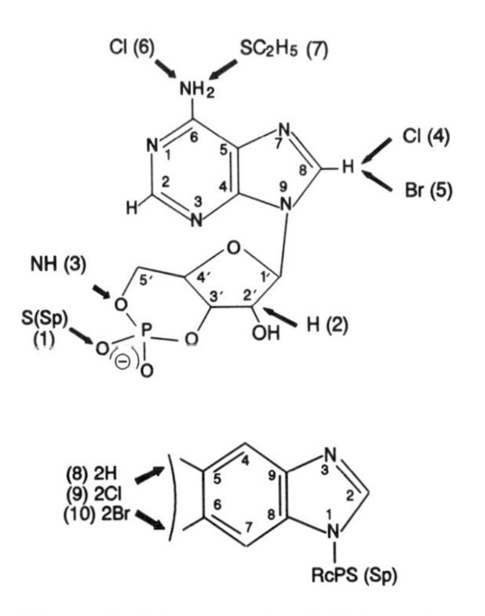

Figure 3. Structures of cAMP derivatives.

Table 1. Nomenclature of cAMP derivatives

No.	Abbreviation	Name Sp-isomer
1	Sp-cAMPS	adenosine 3':5'-monophosphorothioate
2	Sp-2'HcAMPS	2'-deoxyadenosine 3':5'-monophosphorothioate
3	Sp-5'NHcAMPS	5'-deoxy-5'aminoadenosine 3':5'-monophosphorothioate
4	Sp-8ClcAMPS	8-chloroadenosine 3':5'-monophosphorothioate
5	Sp-8BrcAMPS	8-bromoadenosine 3':5'-monophosphorothioate
6	Sp-6ClcPUMPS	6-chloropurineriboside 3':5'-monophosphorothioate
7	Sp-6SEtcPUMPS	6-thioethylpurineriboside 3':5'-monophosphorothioate
8	Sp-cBIMPS	benzimidazoleriboside 3':5'-monophosphorothioate
9	Sp-DClcBIMPS	5,6-dichlorobenzimidazoleriboside 3':5'-monophosphorothioate
10	Sp-DBrcBIMPS	5,6-dibromobenzimidazoleriboside 3':5'-monophosphorothioate

The affinity of all cAMP derivatives for CAR, PKA, and PDE was determined by measuring their potency to displace ^{3}H-cAMP from binding to the three cAMP binding proteins. An example of the dose response curve for displacement of ^{3}H-cAMP from binding to CAR by five derivatives is presented in figure 4. IC_{50} values (concentrations

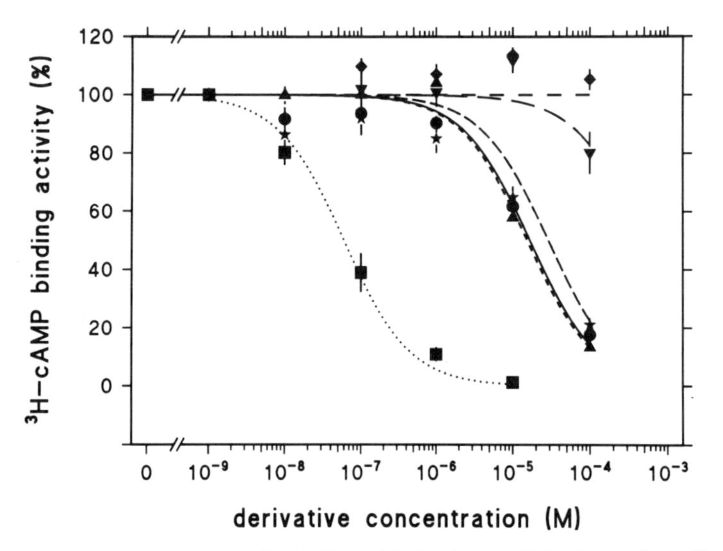

Figure 4. Dose response curve for binding of derivatives to CAR. Vegetative cells of *D. discoideum* strain Ax3 were suspended in PB (10 mM Na/K phosphate buffer pH 6.5) and incubated for 1 min with ^{3}H-cAMP and different concentrations of cAMP (■), Sp-cAMPS (●), Sp-5'NHcAMPS (▲), Sp-2'HcAMPS (★), Sp-8BrcAMPS (▼), Sp-6SEtcPUMPS (◆). Cells were seperated from unbound ligand by centrifugation through silicon oil. Data are presented as percentage of the amount of ^{3}H-cAMP bound to cells in the absence of derivative. Means and SE of three experiments performed in triplicate are presented.

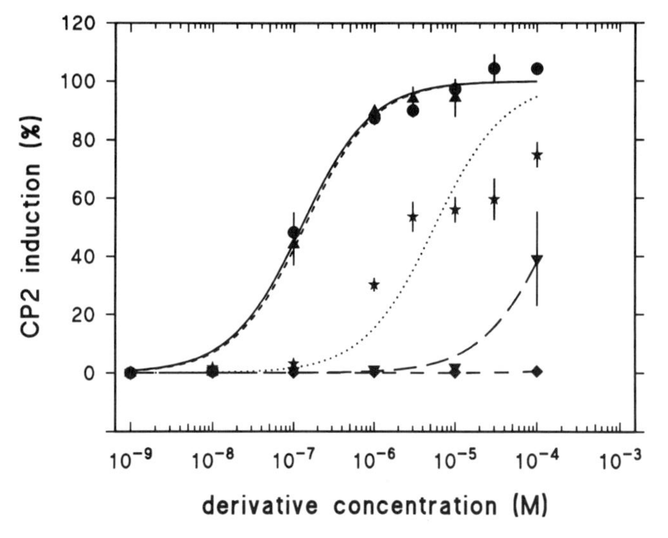

Figure 5. Dose response curve for induction of CP2 gene expression by cAMP derivatives. Vegetative CP2-*lacZ* cells were starved for 4 hr on PB agar at 3 x 10^6 cells/cm^2, subsequently resuspended to 3 x 10^6 cells/ml PB and incubated for 6 hr as 100 μl aliquots in microtiter plate wells, containing different concentrations of Sp-cAMPS (●), Sp-5'NHcAMPS (▲), Sp-2'HcAMPS (★), Sp-8BrcAMPS (▼), and Sp-6SEtcPUMPS (◆). Cells in microtiter plates were lysed by freeze-thawing and ß-galactosidase activities were measured using a Biorad Elisa reader (OD_{415}). Means and SE of three experiments performed in triplicate are presented.

inducing halfmaximal inhibition of ^{3}H-cAMP binding) calculated from dose response curves are presented in figure 6.

Replacement of the axial exocyclic oxygen by sulphur (Sp-cAMPS) reduces affinity for

CAR about 80 fold, for PKA 20 fold and for PDE 200 fold. The additional 2'H and 5'NH modifications, which interfere with hydrogen bond formation at these positions, have little effect on binding to CAR, but reduce binding to PKA respectively 1800 and 300 fold compared to Sp-cAMPS. The 8Cl and 8Br modifications, which change the distribution of the molecule to preferentially *syn* conformation, reduce affinity to CAR respectively 20 and 40 fold but do not effect binding to PKA. The derivatives substituted at the 6-position, as well as the three Sp-benzimidazoles are very poor CAR agonists, but show high affinity to PKA. Except for Sp-5'NHcAMPS, all derivatives show very low affinity for PDE. To conclude, among the tested derivatives, compounds 2 and 3 are highly selective for binding to CAR, while compounds 4-10 are selective PKA agonists.

All cAMP derivatives were also tested for induction of three classes of cAMP

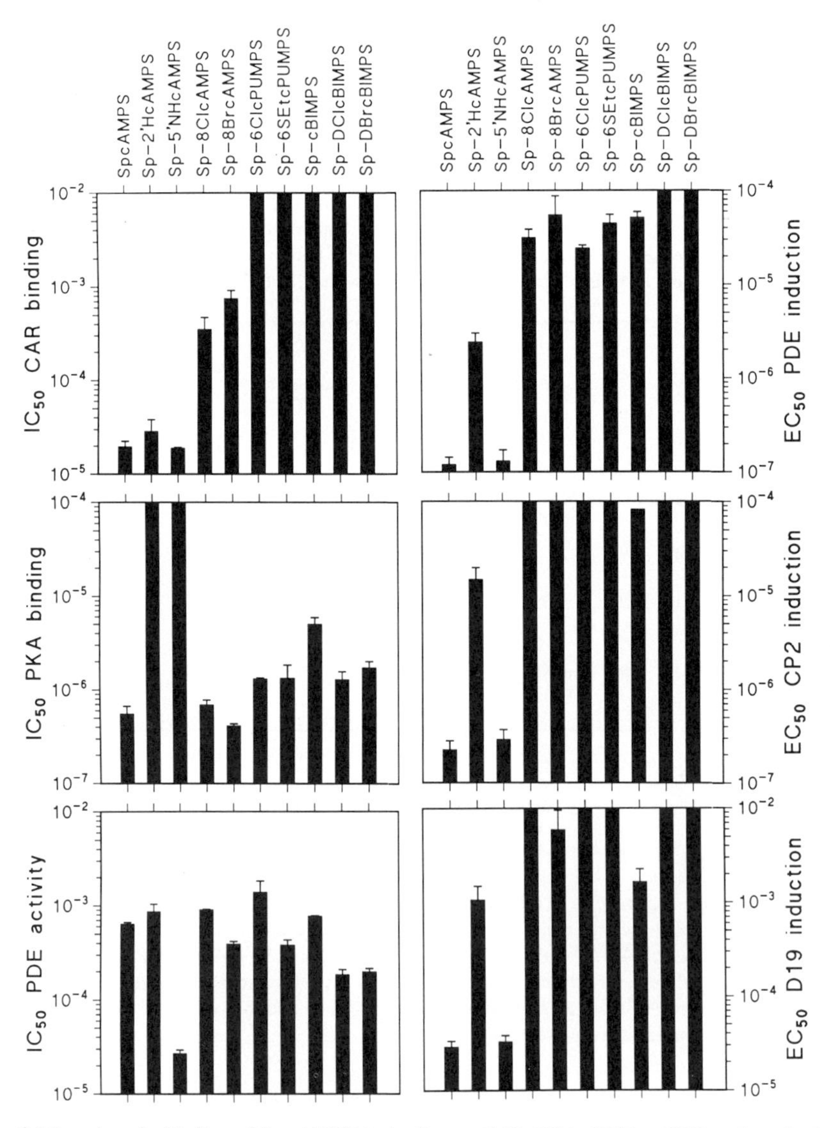

Figure 6. IC_{50} values for binding of Sp-cAMPS derivatives to CAR, PKA, PDE and EC_{50} values for induction of PDE, CP2, D19 gene expression by Sp-cAMPS derivatives.

regulated genes. To measure effects of different concentrations of cAMP derivatives on gene expression, we used cell lines transformed with vectors carrying fusions of the *E.coli lacZ* gene with respectively the promoter of the prespore gene D19 (Dingermann *et al.*, 1989), the prestalk gene CP2 (Pears and Williams, 1987; Datta and Firtel, 1988) and the promoter of the PDE gene (Faure *et al.*, 1990). Promoter activity results in synthesis of the ß-galactosidase enzyme, which can be detected with a highly sensitive and quantitative spectrophotometric assay. Induction of CP2 gene expression by five different derivatives is shown in figure 5. EC_{50} values (concentrations required for halfmaximal induction), calculated from the effects of all derivatives on the three different classes of genes are presented in figure 6.

When comparing the affinities of the cAMP derivatives for CAR, PKA, and PDE to their efficacy as agonists for induction of three classes of cAMP-regulated genes, it appears that for all the three genes the highly selective PKA agonists Sp-6ClcPUMPS, Sp-6SEtcPUMPS, Sp-DClcBIMPS and Sp-DBrcBIMPS are virtually ineffective to induce expression. The other PKA agonists Sp-8BrcAMPS and Sp-8ClcAMPS are more effective, and can induce low levels of gene expression at very high concentrations, but these compounds also bind moderately well to CAR. The CAR agonist Sp-5'NHcAMPS induces expression of PDE, CP2 and D19 genes very efficiently. This specificity profiles suggests that the first target for cAMP regulation of gene expression is a surface cAMP receptor.

Of further interest are the observations that the concentration requirements for gene induction by Sp-cAMPS differ considerably for the three classes of genes. This suggests that multiple receptors with different affinities may mediate cAMP induced expression of PDE, CP2 or D19 gene expression. Binding of derivatives to surface cAMP receptors probably represents binding to the chemotactic receptor CAR1, which is the most abundant binding activity at this stage. The observation that Sp-2'HcAMPS binds to CAR1 almost as effectively as Sp-cAMPS, but requires a 10 to 50 fold higher concentration than Sp-cAMPS to induce gene expression, suggests that other CARs than CAR1 control gene expression.

SELECTIVE INDUCTION OF GENE EXPRESSION BY THE PARTIAL ANTAGONIST 8-CPT-cAMP

The data presented above indicate that surface cAMP receptors and not protein kinase A are the first target for effects of cAMP on gene regulation. We now come to the question whether distinct receptors regulate the expression of a class of genes, through activation of a specific signaling pathway. As shown in figure 2, binding of cAMP to CARs activates the intracellular target enzymes, adenylyl cyclase, guanylyl cyclase and phospholipase C, resulting in rapid increases in the levels of the second messengers $InsP_3$, cGMP and cAMP.

We used the partial chemotactic antagonist 8-*p*-chlorophenylthioadenosine 3':5'-monophosphate (8-CPT-cAMP) to dissect signaling pathways involved in gene regulation. In *Dictyostelium*, partial chemotactic antagonists typically inhibit chemotaxis to cAMP at concentrations where they bind to receptors (Van Haastert and Kien, 1983), but induce some chemotaxis at supersaturating concentrations. This in contrast to full antagonists, such as adenosine 3':5'-monophosphorothioate, Rp-isomer (Rp-cAMPS), which inhibit chemotaxis at all concentrations. Full antagonists cannot activate any second messenger pathway in *Dictyostelium* (Van Haastert, 1987). This is however, not the case for partial antagonists. 8-CPT-cAMP shows 150 fold lower affinity for CARs than cAMP (Van Ments-Cohen *et al.*, 1991) and induces normal accumulation of cAMP and cGMP at about 50 fold higher concentrations than cAMP (Peters *et al.*, 1991). However, 8-CPT-cAMP cannot induce an $InsP_3$ response at any concentration and very surprisingly, induces a decrease rather than an increase of $InsP_3$ levels (Figure 7).

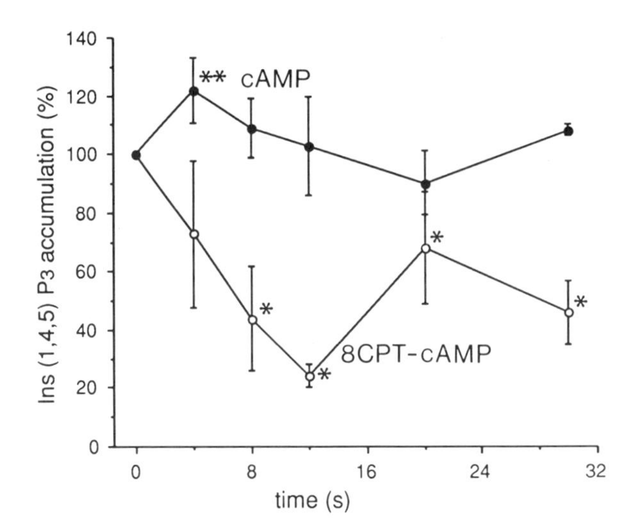

Figure 7. Induction of $InsP_3$ accumulation. Vegetative cells starved for 4 hr in PB were resuspended in 40 mM Hepes, pH 6.5. Three batches of cells were simultaneously stimulated with 10^{-6} M cAMP (•) or 10^{-4} M 8-CPT-cAMP (▴). At the indicated time intervals, aliquots were transferred to perchloric acid, and $InsP_3$ levels were measured. Data are expressed as percentage of basal level (stimulation with water). Means and SEMs of three experiments done in triplicate are presented. **, Data are significant below basal level; *, data are significant above basal level (Student's test, $P \leq 0.05$).

We measured effects of 8-CPT-cAMP on expression of three classes of cAMP regulated genes; the aggregative genes, which are expressed in response to nanomolar cAMP pulses, and the prestalk and prespore genes which are expressed in response to constant levels of cAMP. Figure 8 shows effects of different concentrations of pulses of cAMP and 8-CPT-cAMP on expression of the aggregative genes, coding for the chemotactic cAMP receptor CAR1 (Klein *et al.*, 1988) and adhesive contact sites A (csA) (Noegel *et al.*, 1986). Despite its lower affinity for surface cAMP receptors, pulses of 8-CPT-cAMP induces transcription of the csA and CAR1 genes almost as effectively as cAMP. This strongly suggest that $InsP_3$ accumulation is not involved in inducing expression of these genes.

We also measured effects of cAMP, 8-CPT-cAMP and two other cAMP analogs on expression of the prestalk gene CP2 and the prespore gene D19. The other analogs 8-bromoadenosine 3': 5'-monophosphate (8-Br-cAMP) and 6-chloropurineriboside 3':5'-

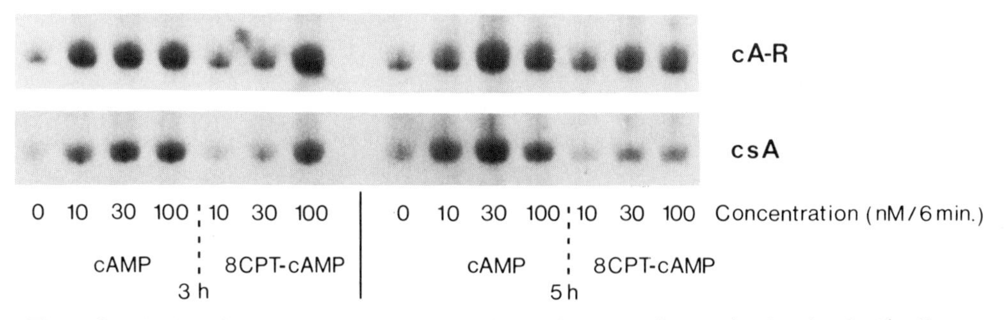

Figure 8. Induction of aggregative gene expression. Vegetative NC4 cells were incubated at 5×10^6 cells per ml in PB and stimulated with the indicated concentrations of cAMP or 8-CPT-cAMP at 6-min intervals. mRNA was isolated after 3 hr of incubation. Nothern (RNA) blots were probed with csA and CAR1 cDNAs.

monophosphate (6-Cl-cPUMP) bind to CARs with respectively 220 fold and 2500 fold lower affinity than cAMP and were used as controls. Figure 9 shows that 6-Cl-cPUMP and 8-Br-cAMP can induce expression of the D19 and CP2 genes at concentrations which are somewhat lower than expected from their relative affinity for surface receptors. This is a consequence of the fact that cAMP and cAMP derivatives are rapidly degraded by cAMP-phosphodiesterases during incubation with cells. Derivatives which are active at concentrations where cAMP-PDE activity is more or less saturated are less affected, and therefore appear to have increased efficacy. In contrast to 6-Cl-cPUMP and 8-Br-cAMP, which display even lower affinity for CAR than 8-CPT-cAMP, the latter derivative cannot induce expression of either the prespore gene D19 or the prestalk gene CP2. This suggest that InsP3 accumulation may be required for activation of expression of these genes.

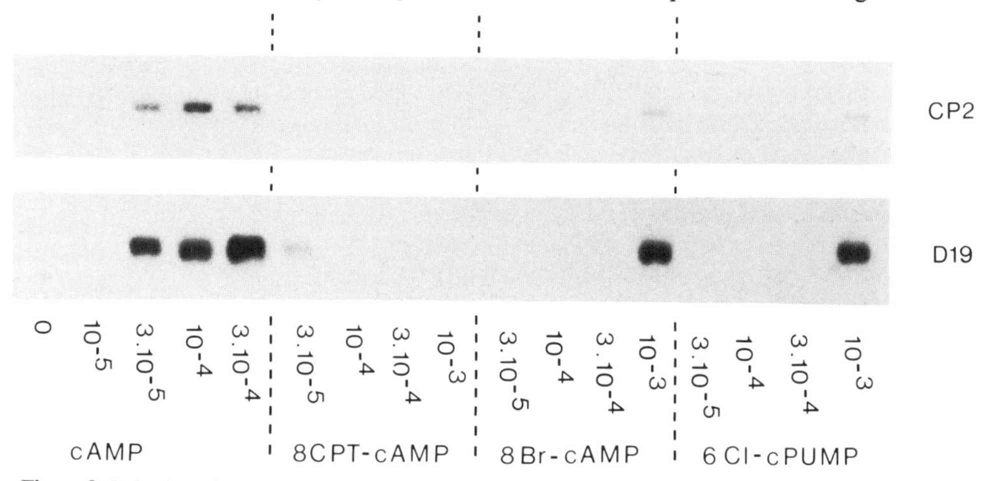

Figure 9. Induction of prespore and prestalk gene expression. Aggregation-competent cells were resuspended to 10^7 cells per ml in PB and stimulated with the indicated concentrations of cAMP, 8-Br-cAMP, 6-Cl-cPUMP, or 8-CPT-cAMP at 60-min intervals. mRNA was isolated after 3 or 5 hr of incubation, and Northern blots were probed with, respectively, CP2 and D19 cDNAs.

DISCUSSION

During *Dictyostelium* development extracellular cAMP regulates expression of several classes of genes. In this study we have used pharmacological approaches to unravel cAMP signal transduction pathways involved in cAMP-induced gene regulation. Binding and activation of a receptor protein is the first step in transduction of any hormone-like signal. *Dictyostelium* cells contain several putative targets for the effects of cAMP: four highly homologous surface cAMP receptors (CARs) (Saxe *et al.*, 1991), a cAMP-dependent protein kinase (PKA) (Mutzel *et al.*, 1987) and a cAMP-phosphodiesterase (PDE) (Lacombe *et al.*, 1986). Early studies comparing the nucleotide specificity of different cAMP-induced responses with the specificity of *Dictyostelium* cAMP binding proteins showed that cAMP-induced accumulation of second messengers and chemotaxis are mediated by CARs (De Wit *et al.*, 1982; Van Haastert and Kien, 1983; Theibert and Devreotes, 1986) and not by PKA or PDE. Using a limited number of cAMP derivatives, it was demonstrated that nucleotide specificity of induction of gene expression is similar to that of CAR and dissimilar from PKA and PDE (Schaap and Van Driel, 1985; Oyama and Blumberg, 1986). However, recent molecular genetic evidence strongly suggests requirement of PKA activation for gene regulation. Inactivation of PKA by overexpression of native R-subunit or R-subunits lacking the cAMP binding sites, blocks development and inhibits expression of the cAMP-induced prestalk and prespore genes (Simon *et al.*, 1989; Firtel and Chapman, 1990; Harwood *et al.*, 1992a,b).

A major drawback of earlier pharmacological analysis of gene regulation by cAMP derivatives is inherent to the fact that gene expression requires several hours of exposure to cAMP (derivative), during which the agonist is rapidly degraded by extracellular PDE. cAMP degradation products as adenosine are furthermore known to exert both positive and negative effects on expression of cAMP-regulated genes (Weijer and Durston, 1985; Schaap and Wang, 1986; Spek *et al.*, 1988). Since not all cAMP derivatives are equally sensitive to degradation (Van Haastert *et al.*, 1983), and the effect of their degradation products are not known, this could result in unpredictable interference with gene expression. In general, the efficacy of poor CAR agonists will be overestimated, since at the high concentrations, where these compounds are active, their degradation is more or less saturated.

In order to overcome these difficulties, we have used a novel set of non-hydrolysable cAMP derivatives. Some derivatives were additionally modified to reduce polarity and enhance membrane permeability. Two of these derivatives, Sp-8BrcAMPS and Sp-6SEtcAMPS were shown to enter the cell and activate PKA (Schaap *et al.*, in press). We show here that the nucleotide specificity profile of cAMP-regulated aggregative genes, prespore and prestalk genes is highly similar to that of CAR, and dissimilar to PKA and PDE. Furthermore, even the highly lipophilic PKA agonists Sp-8BrcAMPS and Sp-6SEtcAMPS, which enter the cell and can activate PKA, cannot induce expression of any of the tested cAMP regulated genes. These data yield definite evidence that CARs are the first targets for cAMP-induced gene expression and also show that activation of PKA is not sufficient to induce gene expression. A role for cAMP as intracellular messenger controlling gene regulation was also contradicted by findings that mutants defective in activation of adenylyl cyclase show virtually normal induction of expression of cAMP regulated genes (Schaap *et al.*, 1986; Bozzaro *et al.*, 1987; Mann *et al.*, 1988).

cAMP receptors can activate three different target enzymes in *Dictyostelium*, adenylyl cyclase, guanylyl cyclase and phospholipase C. The partial cAMP antagonist 8-CPT-cAMP can activate adenylyl and guanylyl cyclase activation, but cannot activate phospholipase C. We show here that 8-CPT-cAMP induces normal activation of aggregative genes, but cannot induce expression of prestalk and prespore genes. This suggests that second messengers $InsP_3$ and diacylglycerol (DAG), which are produced during phospholipase C activation may mediate prestalk and prespore gene expression. In case of prespore gene expression, this suggestion is corroborated by findings that inhibitors of Ca^{2+} mobilization effectively inhibit prespore gene expression (Schaap *et al.*, 1986; Blumberg *et al.*, 1989). Furthermore prespore gene expression can be induced under special conditions by $InsP_3$/DAG pulses (Ginsburg and Kimmel, 1989), and is inhibited by LiCl, which inhibits cAMP-induced $InsP_3$ accumulation (Peters *et al.*, 1989). Remarkably, expression of prestalk genes is counteracted by $InsP_3$/DAG pulses (Ginsburg and Kimmel, 1989) and is neither inhibited by LiCl nor by inhibitors of Ca^{2+} mobilization. It seems therefore likely that prestalk gene expression is mediated by a presently unknown signaling response which cannot be activated by 8-CPT-cAMP. The observation that 8-CPT-cAMP induces normal expression of aggregative genes indicates that the second messengers $InsP_3$ and DAG do not mediate aggregative gene expression. Since aggregative gene expression is also normally induced in mutants defective in adenylate cyclase activation, this leaves cGMP or a presently unknown intracellular messenger as a putative mediator of this response.

The present study shows that specifically modified cAMP derivatives are useful tools to unravel involvement of specific cAMP signal transduction pathways in gene regulation. Using both pharmacological and molecular genetic approaches we will continue to substantiate the evidence for involvement of specific second messengers in gene regulation and to identify signaling components which are activated by specific second messengers.

REFERENCES

Abe, T., Maeda, Y., Iijima, T., 1988, Transient increase of the intracellular Ca^{2+} concentration during chemotactic signal transduction in *Dictyostelium discoideum* cells, Differentiation 39:90-96.

Aeckerle, S., Wurster, B. and Malchow, D., 1985, Oscillations and cyclic AMP-induced changes of the K^+ concentration in *Dictyostelium discoideum*, EMBO J., 4:39-43.

Barklis, E. and Lodish, H.F., 1983, Regulation of *Dictyostelium discoideum* mRNAs specific for prespore or prestalk cells, Cell 32:1139-1148.

Birnbaumer, L., Abramwitz, J. and Brown, A.M., 1990, Receptor-effector coupling by G-proteins, Biochim.Biophys.Acta 1031:163-224.

Blumberg, D.D., Comer, J.F. and Walton, E.M., 1989, Ca^{2+} antagonists distinguish different requirements for cAMP-mediated gene expression in the cellular slime mold, *Dictyostelium discoideum*, Differentiation 41:14-21.

Bozzaro, S., Hagmann, J., Noegel, A., Westphal, M., Calautti, E. and Bogliolo, E., 1987, Cell differentiation in the absence of intracellular and extracellular cyclic AMP pulses in *Dictyostelium discoideum*, Dev.Biol., 123:540-548.

Clarke, M., Bazari, W.L. and Kayman, S.C., 1980, Isolation and properties of calmodulin from *Dictyostelium discoideum*, J.Bacteriol. 141:397-400.

Clarke, M., Kayman, S.C. and Riley, K., 1987, Density-dependent induction of discoidin-I synthesis in exponentially growing cells of *Dictyostelium discoideum*, Differentiation 34:79-87.

Coukell, M.B., Lappano, S. and Cameron, A.M., 1983, Isolation and characterization of cAMP unresponsive (frigid) aggregation-deficient mutants of *Dictyostelium discoideum*, Dev.Genet. 3:283-297.

Darmon, M., Brachet, P., and Pereira da Silva, L.H., 1975, Chemotactic signals induce cell differentiation in *Dictyostelium discoideum*, Proc.Natl.Acad.Sci.USA 72:3163-3166.

Datta, S. and Firtel, R.A., 1988, An 80-bp cis-acting regulatory region controls cAMP and developmental regulation of a prestalk gene in *Dictyostelium*, Genes Dev. 2:294-304.

De Gunzburg, J. and Veron, M., 1982, A cAMP-dependent protein kinase is present in differentiating *Dictyostelium discoideum* cells, EMBO J. 1:1063-1068.

De Wit, R.J.W., Arents, J.C. and Van Driel, R., 1982, Ligand binding properties of the cytoplasmic cAMP-binding protein of *Dictyostelium discoideum*, FEMS Lett., 145:150-154.

Dingermann, T., Reindl, N., Werner, H., Hildebrandt, M., Nellen, W., Harwood, A., Williams, J. and Nerke, K., 1989, Optimalization and *in situ* detection of *E.coli* ß-galactosidase gene expression in *Dictyostelium discoideum*, Gene 85:353-362.

Europe-Finner, G.N. and Newell, P.C., 1986, Inositol1,4,5-trisphosphate induces calcium release from a non-mitochondrial pool in amoebae of *Dictyostelium*, Biochim. Biophys.Acta 887:335-340.

Europe-Finner, G.N. and Newell, P.C., 1987, Cyclic AMP stimulates accumulation of inositol trisphosphate in *Dictyostelium*, J.Cell.Sci. 87:221-229.

Faure, M., Franke, J., Hall, A.L., Podgorski, G.J. and Kessin, R.H., 1990, The cyclic nucleotide phosphodiesterase gene of *Dictyostelium discoideum* contains three promoters specific for growth, aggregation, and late development, Mol.Cell.Biol. 10:1921-1930.

Firtel, R.A. and Chapman, A.L., 1990, A role for cAMP-dependent protein kinase A in early *Dictyostelium* development, Genes Dev., 4:18-28.

Gerish, G., Fromm, H., Huesgen, A. and Wick, U., 1975, Control of cell-contact sites by cyclic AMP pulses in differentiating *Dictyostelium* cells, Nature 255:547-549.

Ginsburg, G. and Kimmel, A.R., 1989, Inositol trisphosphate and diacylglycerol can differentially modulate gene expression in *Dictyostelium*, Proc.Natl.Acad.Sci.USA 86:9332-9336.

Gomer, R.H., Yuen, I.S. and Firtel, R.A., 1991, A secreted $80x10^3$ M_r protein mediates sensing of cell density and the onset of development in *Dictyostelium*, Development 112:269-278.

Grabel, L. and Loomis, W.F., 1978, Effector controlling accumulation of N-acetylglucos-aminidase during development of *Dictyostelium discoideum*, Dev.Biol. 64:203-209.

Gross, J.D., Bradbury, J., Kay, R.R. and Peacey, M.J., 1983, Intracellular pH and the control of cell differentiation in *Dictyostelium discoideum*, Nature 303:244-245.

Hadwiger, J.A., Wilkie, T.M., Strathmann, M. and Firtel, R.A., 1991, Identification of *Dictyostelium* Gα genes expressed during multicellular development, Proc.Natl.Acad.Sci.USA 88:8213-8217.

Harwood, A.J., Hopper, N.A., Simon, M-N., Bouzid, S., Veron, M. and Williams, J.G., 1992a, Multiple roles for cAMP-dependent protein kinase during *Dictyostelium* development, Dev.Biol., 149:90-99.

Harwood, A.J., Hopper, N.A., Simon, M-N., Driscoll, D.M., Veron, M. and Williams, J.G., 1992b, Culmination in *Dictyostelium* is regulated by the cAMP-dependent protein kinase, Cell, 69:615-624.

Henderson, E.J., 1975, The cyclic adenosine 3':5'-monophosphate receptor of *Dictyostelium discoideum*, J.Biol.Chem. 250:4730-4736.

Janssens, P.M.W., De Jong, C.C.C., Vink, A.A. and Van Haastert, P.J.M., 1989, J.Biol.Chem. 264:4329-4335.

Jermyn, K.A., Berks, M., Kay, R.R. and Williams, J.G., 1987, Two distinct classes of prestalk-enriched mRNA sequences in *Dictyostelium discoideum*, Development 100:745-755.

Johnson, R.L., Van Haastert, P.J.M., Kimmel, A.R., Saxe III, C.L., Jastorff, B. and Devreotes, P.N., 1992, The cyclic nucleotide specificity of three cAMP receptors in *Dictyostelium*, J.Biol.Chem. in press.

Kesbeke, F., Snaar-Jagalska, B.E. and Van Haastert, P.J.M., 1988, Signal transduction in *Dictyostelium* fgd A mutants with a defective interaction between surface cAMP receptors and a GTP-binding regulatory protein, J.Cell Biol. 107:521-528.

Klein, P.S., Sun, T.J., Saxe III, C.L., Kimmel, A.R., Johnson, R.L. and Devreotes, P.N., 1988, A chemoattractant receptor controls development in *Dictyostelium discoideum*, Science 241:1467-1472.

Kopachik, W., Bergen, L.G. and Barclay, S.L., 1985, Genes selectively expressed in proliferating *Dictyostelium* amoebae, Proc.Natl.Acad.Sci.USA 82:8540-8544.

Lacombe, M-L., Podgorski, G.J., Franke, J. and Kessin, R.H., 1986, Molecular cloning and developmental expression of the cyclic nucleotide phosphodiesterase gene of *Dictyostelium discoideum*, J.Biol.Chem. 261:16811-16817.

Lilly, P., Klein, P., Theibert, A., Vaughan, R., Pupillo, M., Saxe, C., Kimmel, A. and Devreotes, P., 1988, Receptor G-protein interactions in the development of *Dictyostelium*, Bot.Acta 101:123-127.

Luderus, M.E.E., Van der Most, R.G., Otte, A.P. and Van Driel, R., 1989, A protein kinase C-related enzyme activity in *Dictyostelium discoideum*, FEBS Lett. 253:71-75.

Malchow, D., Nanjundiah, V., Wurster, B., Eckstein, F. and Gerisch, G., 1978, Cyclic AMP-induced pH changes in *Dictyostelium discoideum* and their control by calcium, Biochim.Biophys.Acta 538:473-480.

Mann, S.K.O. and Firtel, R.A., 1989, Two-phase regulatory pathway controls cAMP receptor-mediated expression of early genes in *Dictyostelium*, Proc.Natl.Acad.Sci.USA 86:1924-1928.

Mann, S.K.O., Pinko, C. and Firtel, R.A., 1988, cAMP regulation of early gene expression in signal transduction mutants of *Dictyostelium*, Dev.Biol., 130:294-303.

Marshak, D.R., Clarke, M., Roberts, D.M. and Watterson, D.M., 1984, Structural and functional properties of calmodulin from the eukaryotic microorganism *Dictyostelium discoideum*, Biochemistry 23:2891-2899.

Mato, J.M., Krens, F.A., Van Haastert, J.P.M. and Konijn, T.M., 1977, 3':5'-Cyclic AMP depedent 3':5'-cyclic GMP accumulation in *Dictyostelium discoideum*, Proc.Natl.Acad.Sci. USA 74:2348-2351.

Mehdy, M.C., Ratner, D. and Firtel, R.A., 1983, Induction and modulation of cell-type-specific gene expression in *Dictyostelium*, Cell 32:763-771.

Mehdy, M.C. and Firtel, R.A., 1985, A secreted factor and cyclic AMP jointly regulate cell-type-specific gene expression in *Dictyostelium discoideum*, Mol.Cell.Biol. 5:705-713.

Milne, J.L. and Coukell, M.B., 1991, A Ca^{2+} transport system associated with the plasma membrane of *Dictyostelium discoideum* is activated by different chemoattractant receptors, J.Cell.Biol. 112:103-110.

Morrissey, J.H., Devine, K.M. and Loomis, W.F., 1984, The timing of cell-type-specific differentiation in *Dictyostelium discoideum*, Dev.Biol. 103:414-424.

Mutzel, R., Lacombe, M-L., Simon, M-N., De Gunzburg, J. and Veron, M., 1987, Cloning and cDNA sequence of the regulatory subunit of cAMP-dependent protein kinase from *Dictyostelium discoideum*, Proc.Natl.Acad.Sci.USA 84:6-10.

Noegel, A., Gerish, G., Stadler, J. and Westphal, M., 1986, Complete sequence and transcript regulation of a cell adhesion protein from aggregating *Dictyostelium* cells, EMBO J. 5:1473-1476.

Oyama, M. and Blumberg, D.D., 1986, Interaction of cAMP with the cell-surface receptor induces cell-type specific mRNA accumulation in *Dictyostelium discoideum*, Proc.Natl.Acad.Sci.USA 83:4819-4823.

Parissenti, A.M. and Coukell, M.B., 1990, Effects of DNA and synthetic oligodeoxyribonucleotides on the binding properties of a cGMP-binding protein from *Dictyostelium discoideum*, Biochim.Biophys.Acta 1040:294-300.

Pears, C.J. and Williams, J.G., 1987, Identification of a DNA sequence element required for efficient expression of a developmental regulated and cAMP-inducible gene of *Dictyostelium discoideum*, EMBO J. 6:195-200.

Peters, D.J.M., Van Lookeren Campagne, M.M., Van Haastert, P.J.M., Spek, W. and Schaap, P., 1989, Lithium ions induce prestalk-associated gene expression and inhibit prespore gene expression in *Dictyostelium discoideum*, J.Cell Sci. 93:205-210.

Pupillo, M., Kumagai, A., Pitt, G.S., Firtel, R.A. and Devreotes, P.N., 1989, Multiple α subunits of guanine nucleotide-binding proteins in *Dictyostelium*, Proc.Natl.Acad.Sci.USA 86:4892-4896.

Roos, W., Nanjundiah, V., Malchow, D. and Gerish, G., 1975, Amplification of cyclic AMP signals in aggregating cells of *Dictyostelium discoideum*, FEBS Lett. 53:139-142.

Saxe III, C.L., Johnson, R.L., Devreotes, P.N. and Kimmel, A.R., 1991, Expression of a cAMP receptor gene of *Dictyostelium* and evidence for a multigene family, Genes Dev. 5:1-8.

Schaap, P., Van Ments-Cohen, M., Soede, R.D.M., Brandt, R., Firtel, R.A., Dorstmann, W., Genieser, H-G., Jastorff, B. and Van Haastert, P.J.M., Cell-permeable non-hydrolysable cAMP derivatives as tools for analysis of signaling pathways controlling gene regulation in *Dictyostelium*, J.Biol.Chem., in press.

Schaap, P. and Van Driel, R., 1985, Induction of post-aggregative differentiation in *Dictyostelium discoideum* by cAMP. Evidence of involvement of the cell surface cAMP receptor, Exp.Cell Res. 159:388-398.

Schaap, P., Van Lookeren-Campagne, M.M., Van Driel, R., Spek, W., Van Haastert, P.J.M. and Pinas, J., 1986, Postaggregative differentiation induction by cyclic AMP in *Dictyostelium*: Intracellular transduction pathway and requirement for additional stimuli, Dev.Biol. 118:52-63.

Schaap, P. and Wang, M., 1986, Interactions between adenosine and oscillatory cAMP signaling regulate size and pattern in *Dictyostelium*, Cell 45:137-144.

Simon, M-N., Driscoll, D., Mutzel, R., Part, D., Williams, J. and Veron, M., 1989, Overproduction of the regulatory subunit of the cAMP-dependent protein kinase blocks the differentiation of *Dictyostelium discoideum*, EMBO J., 8:2039-2043.

Singleton, C.K., Delude, R.L., McPherson, C.E., 1987, Characterization of genes which are deactivated upon the onset of development in *Dictyostelium discoideum, Dev.Biol.* 119:433-441.

Singleton, C.K., Gregoli, P.A., Manning, S.S. and Northington, S.J., 1988, Characterization of genes which are transiently expressed during the preaggregative phase of development of *Dictyostelium discoideum*, Dev.Biol. 129:140-146.

Spek, W., Van Drunen, K., Van Eijk, R. and Schaap, P., 1988, Opposite effects of adenosine on two types of cAMP-induced gene expression in *Dictyostelium* indicate the involvement of at least two different intracellular pathways for the transduction of cAMP signals, FEBS Lett., 228:231-234.

Theibert, A. and Devreotes, P.N., 1986, Surface receptor-mediated activation of adenylate cyclase in *Dictyostelium*. Regulation by guanine nucleotides in wild-type cells and aggregation deficient mutants, J.Biol.Chem. 261:15121-15125.

Van Haastert, P.J.M., 1987, Down-regulation of cell surface cyclic AMP receptors and desensitization of cyclic AMP-stimulated adenylate cyclase by cyclic AMP in *Dictyostelium discoideum*, J.Biol.Chem. 262:7700-7704.

Van Haastert, P.J.M., De Vries, M.J., Penning, L.C., Roovers, E., Van der Kaay, J., Erneux, C. and Van Lookeren Campagne, M.M., 1989, Chemoattractant and guanosine5'-[γ-thio]triphosphate induce the accumulation of inositol1,4,5-trisphosphate in *Dictyostelium* cells that are labeled with [^{3}H]inositol by electroporation, Biochem.J. 258:577-586.

Van Haastert, P.J.M., Kesbeke, F., Konijn, T.M., Baranaik, J., Stec, W., Jastorff, B., 1987, "Biophosphates and their analogues - Synthesis, Structure, Metabolism and Activity," Elsevier Science Publishers BV, Amsterdam.

Van Haastert, P.J.M. and Kien, E., 1983, Binding of cAMP derivatives to *Dictyostelium discoideum* cells. Activation mechanism of the cell surface cAMP receptor, J.Biol.Chem., 258:9636-9642.

Van Ments-Cohen, M., Genieser, H-G., Jastorff, B., Van Haastert, P.J.M. and Schaap, P., 1991, Kinetics and nucleotide specificity of a surface cAMP binding site in *Dictyostelium discoideum*, which is not down-regulated by cAMP, FEMS Lett., 82:9-14.

Wang, M., Van Haastert, P.J.M. and Schaap, P., 1986, Multiple effects of differentiation-inducing factor on prespore differentiation and cyclic-AMP signal transduction in *Dictyostelium*, Differentiation 33:24-28.

Wang, M. and Schaap, P., 1989, Ammonia depletion and DIF trigger stalk cell differentiation in intact *Dictyostelium discoideum* slugs, Development 105:569-574.

Weijer, C.J. and Durston, A.J., 1985, Influence of cAMP and hydrolysis products on cell type regulation in *Dictyostelium discoideum*, J.Embryol.Exp.Morphol. 86:19-37.

Williams, J.G., 1988, The role of diffusable molecules in regulating the cellular differentiation of *Dictyostelium discoideum*, Development 103:1-16.

Williams, J.G., Ceccarelli, A., McRobbie, S., Mahbubani, H., Kay, R.R., Earley, A., Berks, M. and Jermyn, K.A., 1987, Direct induction of *Dictyostelium* prestalk induction by DIF provides evidenc that DIF is a morphogen, Cell 49:185-192.

Wu, L. and Devreotes, P.N., 1991, *Dictyostelium* transiently expresses eight distinct G-protein α-subunits during its developmental program, Biochem.Biophys.Res.Commun. 179:1141-1147.

Wurster, B., Schubiger, K., Wick, U. and Gerisch, G., 1977, Cyclic GMP in *Dictyostelium discoideum*. Oscillations and pulses in response to folic acid and cyclic AMP signals, FEBS Lett. 76:141-144.

GLYCOSYL-PHOSPHATIDYLINOSITOL: ROLE IN NEUROTROPHIC FACTORS SIGNALLING

Matias A. Avila, Yolanda León, Beatriz Gil,
and Isabel Varela-Nieto

Instituto de Investigaciones Biomédicas
Consejo Superior de Investigaciones Científicas
Arturo Duperier 4, 28029 Madrid, Spain

INTRODUCTION

In vertebrates, the formation and maintenance of neuronal connections are subject to regulation by multiple target-derived, diffusible neurotrophic factors. Nerve growth factor (NGF), brain-derived neurotrophic factor (BDNF), and neurotrophin 3 (NT-3) are members of a family of structurally related proteins termed neurotrophins that promote the growth and survival of neurons in the central and peripheral nervous systems (for a review see Barde, 1990). Each of these proteins bind to at least two membrane receptors. One is the low affinity NGF receptor (p75), which binds each member of the neurotrophin family with similar affinity, but different rate constants (Rodriguez-Tebar et al., 1992). The other is one of a family of tyrosine kinase receptors: *trk*A binds only NGF, the related *trk*B receptor binds BDNF and NT-3, and *trk*C binds NT-3 alone (Hempstead et al., 1991; Klein et al., 1991a; Klein et al., 1991b; Lamballe et al., 1991). Following neurotrophin binding the members of the *trk* family are rapidly autophosphorylated at tyrosine residues (for a review see Yancopoulos et al., 1990). How receptor autophosphorylation is coupled to the modulation of intracellular processes is not well understood. One hypothesis suggest that the receptor kinase activity catalyses the phosphorylation of cellular protein substrates. A second hypothesis suggests that autophosphorylation of the receptor would lead to variations in their interactions with other membrane components. These two pathways need not to be mutually exclusive, and in fact they may operate synergistically to coordinate the complex series of cellular responses to neurotrophic factors.

Glycosyl-phosphatidylinositol (GPI) has been found in several cell membranes and bears a remarkable resemblance with the GPI anchor of membrane proteins (Ferguson and Williams, 1988; Udenfriend et al., 1991). The hydrolysis of GPI produces a rapid intracellular accumulation of its polar head group, an inositol-phosphoglycan (IPG) which has been shown to mimic a variety of the biological effects of insulin (Saltiel and

New Developments in Lipid-Protein Interactions and Receptor Function
Edited by K.W.A. Wirtz *et al.*, Plenum Press, 1993

Cuatrecasas, 1986; Mato et al., 1987; Kelly et al., 1987; Machicao et al., 1990). Insulin stimulates the hydrolysis of a membrane GPI containing inositol, sugars and saturated fatty acids in myocytes, hepatoma, hepatocytes, adypocytes, T-lymphocytes and CHO cells (for a review see Larner et al., 1989). Moreover, a variety of growth factors with tyrosine kinase receptors such as insulin-like growth factor, epidermal growth factor and interleukin-2 have also been found to stimulate GPI hydrolysis in target cells, implicating the GPI system in the intracellular transmission of a variety of signals (for a review see Mato et al., 1991). NGF also promotes the hydrolysis of a membrane GPI in PC-12 cells where NGF is known to have profound biological effects (Chan et al., 1989). Recent experiments have demonstrated that a rat liver derived IPG is able to copy the effects of insulin on the early developing inner ear of the chick embryo (Varela-Nieto et al., 1991). Insulin regulates cell division in the otic vesicle whilst NGF stimulates cell proliferation in the associated cochleo-vestibular ganglion (CVG), an effect mediated trough low affinity receptors (Represa et al., 1989; Bernd and Represa, 1989). An interesting possibility is, therefore, that IPG would be conserved for some of the developmental actions of insulin and NGF, which could use a common signalling pathway, shared perhaps with other related neurotrophic factors, to regulate cell growth.

The work discussed below provides further support for the involvement of this GPI/IPG pathway in transducing the mitogenic effects of NGF on the early developing inner ear by showing: 1) The presence of endogenous GPI and IPG, the later with strong mitogenic activity, 2) The ability of NGF to stimulate GPI hydrolysis in parallel with its biological activity, and 3) The ability of anti-IPG antibodies to block the biological effects of NGF.

EXPERIMENTAL PROCEDURES

Preparation of Explant Cultures

Cochleo-vestibular ganglia (CVG; statoacoustic ganglion) were aseptically isolated from 3-day-old chick embryos, as previously described (Bernd and Represa, 1989) and staged according to Hamburger & Hamilton, 1951. The standard culture medium consisted of serum-free M-199 medium with Hank's salts and glutamine (Flow laboratories), supplemented with 15 mM $NaHCO_3$. Incubations were carried out at 37º C in a water-saturated atmosphere containing 5% CO_2.

Histology

CVG were fixed in 4% para-formaldehyde and processed for histology. Determinations of tissue volume were made using morphometrical methods on serial sections as described in detail in (Giraldez et al., 1987).

CVG Labelling and GPI Purification

CVG were labelled for 24 hr in the presence of 100 μCi/ml of the various labels (New England Nuclear). At the end of the incubation period, CVG were collected, placed in 0.2 ml of phosphate-saline buffer and 0.2 ml of ice-cold 10% trichloroacetic acid were added to each sample. After standing for 15 min at 4ºC, cellular lipids were extracted and GPI purified as indicated in (Mato et al., 1987).

Purification of the IPG

IPG was prepared by treating GPI purified from either rat liver or three days old chicken embryos with bacterial phosphatidylinositol-specific phospholipase C, as previously described (Varela et al., 1990). Phospholipase C from Bacillus thuringiensis was a generous gift of Dr S. Udenfriend (Roche Institute of Molecular Biology, New Jersey. USA). The concentration of IPG was calculated by measuring free amino groups, considering that each molecule of IPG contains one amino group. The biological activity of IPG from both sources was assessed in vitro by testing its capacity to inhibit the phosphorylation of histone IIA by the cyclic AMP-dependent protein kinase (Villalba et al., 1988).

Immunodetection of chick embryo-purified IPG by anti-IPG antibodies

Polyclonal rabbit anti-IPG antibodies were raised and tested as described in (Romero et al., 1990). IgG was purified from immune sera by ammonium sulfate fractionation followed by affinity chromatography with protein A-agarose. The specifity of the recognition was assessed as described in (Represa et al., 1991). Briefly, IPG purified from either rat liver or chick embryos and several simple sugars and inositols were spotted on a silica G60 t.l.c.-plate (Merck). Antigens were detected by autoradiography as previously described (Rougon et al., 1986).

RESULTS AND DISCUSSION

The early development of the vertebrate inner ear involves the thickening and invagination of the ectoderm and the formation of the otic vesicle. At developmental stage 18, it consists of a fluid filled cavity lined by a transporting epithelia and an attached ganglion (Giraldez et al., 1988). Within 48 hr the otocyst goes through a distinct period of cell proliferation evolving towards a more complex structure with signs of growth and morphogenesis. The CVG also proliferates pari pasu with the otic vesicle. NGF receptors have been described in CVG where NGF shows a distinct mitogenic effect (Bernd & Represa, 1989). We investigated whether IPG also stimulates growth in CVG. Figure 1A (upper panels) compares the effect of culturing CVG for 24 hr with NGF (50 ng/ml) or IPG (10 μM). Endogenous IPG was purified by phosphatidylinositol-specific-phospholipase C hydrolysis of a GPI purified from chick embryos and it was assayed for biological effects. The activity of chick-derived IPG was evaluated in vitro by determining its ability to inhibit the catalytic subunit of cyclic AMP-dependent protein kinase (Villalba et al., 1988).

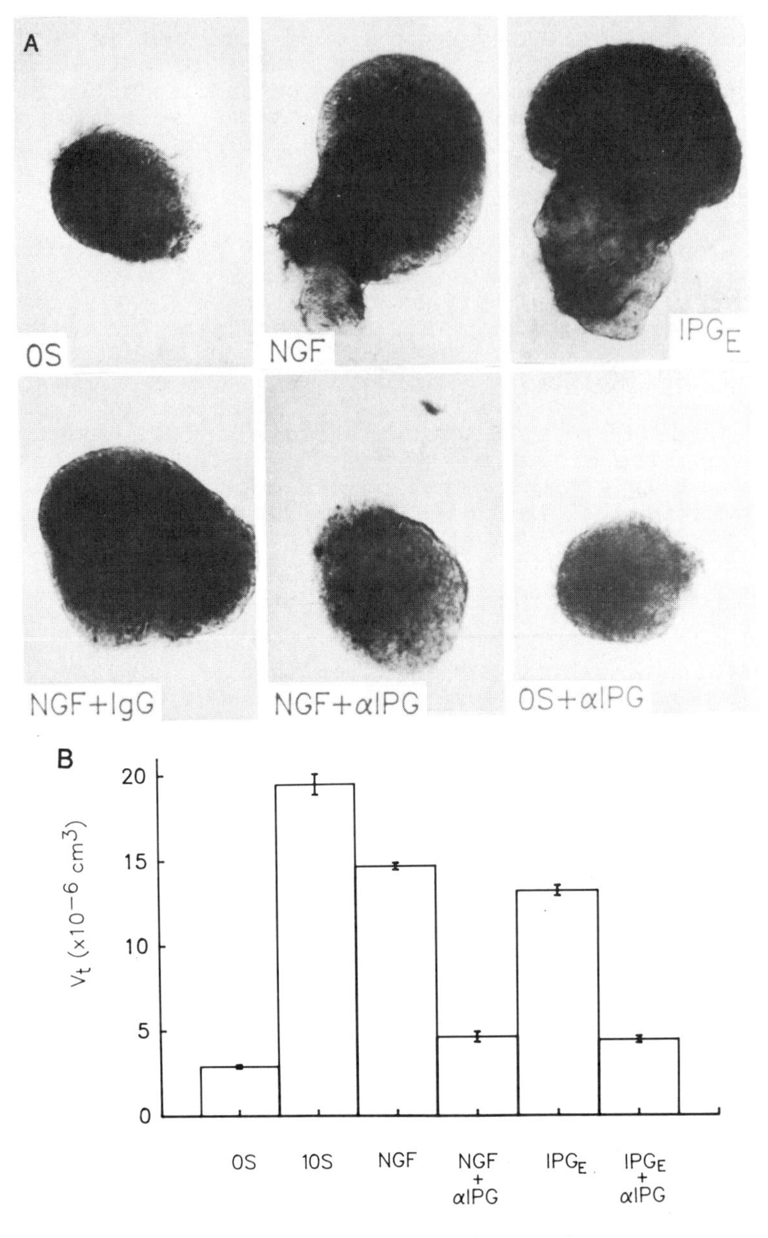

Figure 1. Effect of IPG on CVG proliferation and blockade of the mitogenic effect of NGF by anti-IPG antibodies

A. Isolated CVG were grown in culture for 24 hr in M-199 alone (OS), or in M-199 containing NGF (NGF), embryonic derived IPG (IPG_E), NGF plus IgG from pre-immune rabbit (NGF + IgG), NGF plus anti-IPG antibody (NGF + anti-IPG), embryonic derived IPG plus anti-IPG antibody (IPG_E + anti-IPG). Concentrations used were: NGF=50 ng ml^{-1}; IPG=5 μg ml^{-1}; anti-IPG antibody=20 μg ml^{-1} and pre-immune IgG=20 μg ml^{-1}.
B. Effect of anti-IPG antibody on tissue volume. Measurements of tissue volume were done on ganglia from experiments performed as in A that were processed for histology. Symbols as in A. 10S indicates ganglia grown in M-199 plus 10% FCS. Results are mean ± S.E. of three experiments.

Chicken-derived IPG induced cell proliferation of isolated CVG to the same extent of that previously observed with NGF and liver-derived IPG (Varela-Nieto et al., 1991; and Fig. 1A). The increase in size produced by the addition of IPG was very similar to that of NGF at saturating concentrations (see Bernd and Represa, 1989). The relative contribution of the vestibular and cochlear ganglia to the final mass increase was also comparable

Blockade of NGF-induced proliferation by anti-IPG antibodies

An anti-IPG antibody has been reported to selectively block a variety of insulin actions (Romero et al., 1990). This antibody (20 μg/ml) was able to abolish both IPG and NGF stimulation of CVG growth, as estimated from the size of the CVG (Fig. 1A, lower panels) and morphometrical measurements of tissue volume (Fig. 1B). A similar result was observed with IPG that was purified either from chick embryo or rat liver. Anti-IPG antibodies on the other hand, had no effect on the proliferation of either control CVG kept in culture without additions (0S), or CVG stimulated to grow with 10% serum (10S). The IgG fraction purified from serum of a non immunized rabbit was unable to block the effects of IPG or NGF.

Hydrolysis of GPI by NGF

The glycolipid precursor of IPG was characterized by incubating isolated CVG overnight with radiolabelled precursors of GPI: [^{3}H]glucosamine, [^{3}H]galactose, [^{3}H]myristic acid, or [^{3}H]palmitic acid. Polar lipids were then extracted and resolved by sequential acid-base t.l.c. (Mato et al, 1987). After these treatments, only a labelled fraction of Rf 0.5 was recovered, which was further purified by double dimension t.l.c.. As shown in figure 2A a single spot was observed by fluorography of the plate. The chromatographic profile obtained for CVG-isolated GPI was identical to that observed for rat liver GPI. Further studies were performed with the single fraction obtained after t.l.c. in a basic solvent system. CVG-isolated GPI incorporated [^{3}H]glucosamine, [^{3}H]galactose, [^{3}H]myristic acid, and [^{3}H]palmitic acid. This pattern is similar to that previously reported for the insulin modulated GPI (Mato et al., 1987). The presence of GPI was confirmed by demonstrating that the glucosamine molecule was covalently linked to phosphatidylinositol. The Rf 0.5 glycolipid fraction from ganglia metabolically labelled with [^{3}H]glucosamine was treated with either sodium nitrite (pH 3.75), or phospholipase C from <u>B. thuringiensis</u>. Both treatments produced the hydrolysis of the CVG-isolated GPI, measured as radioactivity released in the aqueous phase. The percentage of GPI hydrolysis was 61.1% for the deamination treatment, and 34.65% for the phosphodiesteric cleavage (data determined in two experiments performed in duplicate). These results indicate that CVG cells contain a GPI molecule that could serve as the endogenous source of IPG.

The ability of NGF to hydrolyse GPI was determined by incubating isolated CVG prelabelled with [^{3}H]glucosamine, with NGF (25 ng/ml) (Figures 2B & 2C).

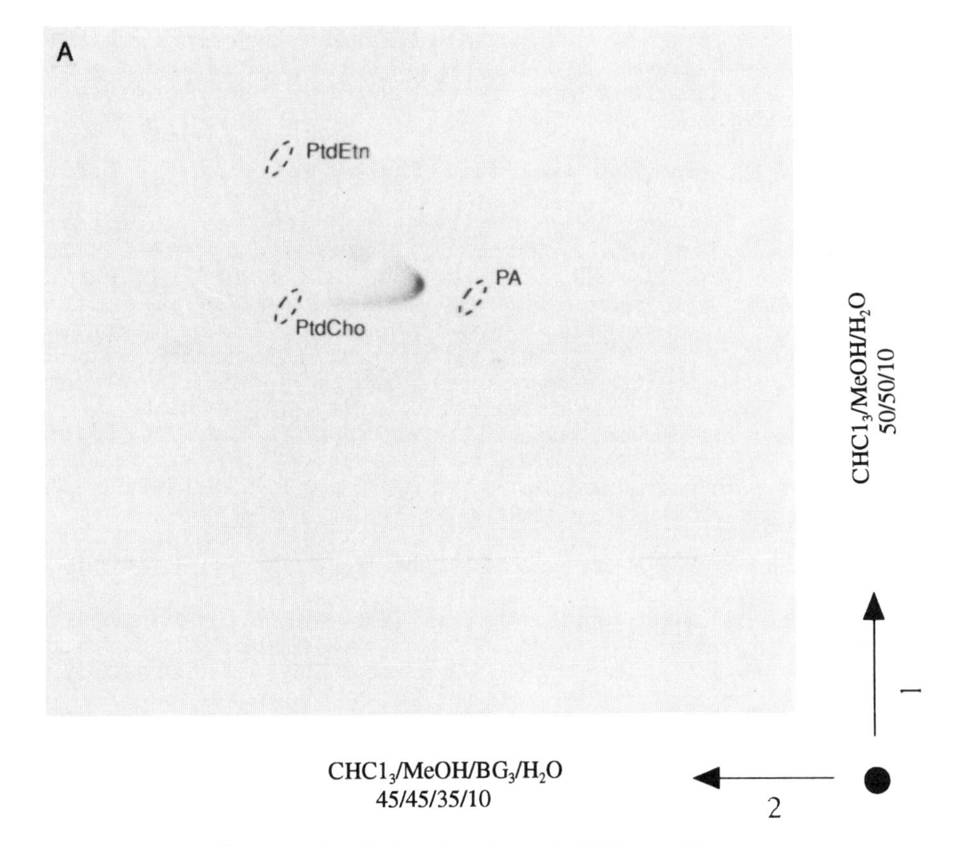

Figure 2. Hydrolysis of GPI by NGF.

A. Two dimensional t.l.c. of ^{3}H-GPI isolated from chick embryos. ^{3}H-GPI was spotted at one corner of the plate (Or.) and the chromatogram developed first in chloroform /methanol /water (50/50/10 vol/vol), dried briefly, turned through 90° and further developed in chloroform /methanol /ammonia /water (45/45/3.5/10 vol/vol). Finally the plate was dried and labelled-GPI detected by fluorography of the plate. Phospholipid standards were visualized with I_2 and are indicated by dotted lines.
B. Time course of hydrolysis of GPI. Cochlear ganglia were incubated with ^{3}H-glucosamine and exposed to NGF, 25 ng ml^{-1}. The fraction of GPI hydrolysed was plotted in ordinates against the time of incubation. dpm of [^{3}H]glucosamine incorporated per ganglion at zero time was 981 ± 155. Results are mean ± S.E. of 4 different experiments.
C. Dose response effect of NGF on GPI hydrolysis. Experiments were done as described in B. GPI hydrolysed was measured after incubation for 1 min with the concentrations of NGF indicated in abscissa. Results are from one representative experiment out of three. The inset shows the cell proliferation rate of CVG as a function of NGF in the incubation medium.

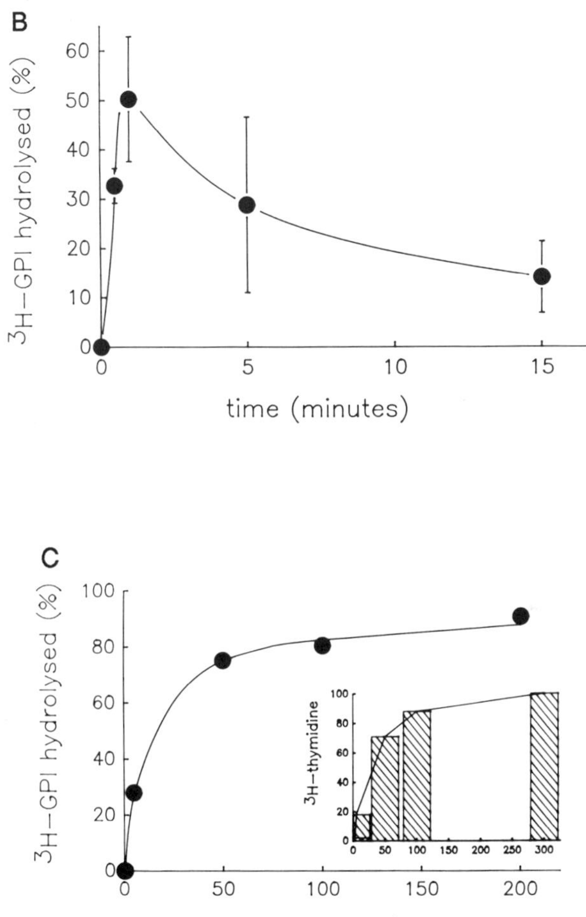
B
3H−GPI hydrolysed (%)
0
10
20
30
40
50
60
0
5
10
15
time (minutes)
C
3H−GPI hydrolysed (%)
0
20
40
60
80
100
0
50
100
150
200
NGF (ng.ml−1)
3H−thymidine
0
20
40
60
80
100
0
50
100
150
200
250
300

GPI hydrolysis was measured at the times indicated and analyzed by t.l.c. (Fig. 2B). NGF induced a loss of GPI of about 30% within 0.5 min and 50% within 1 min. Recovery started immediately thereafter and the new steady-state value was reached about 10 min after the addition of the growth factor. The dose-dependency of GPI hydrolysis was studied following a similar experimental approach (Fig. 2C). Hydrolysis was almost complete at concentrations above 100 ng/ml NGF, the half maximal effect occurring at 25 ng/ml NGF. These results indicate that the NGF is able to regulate the rate of hydrolysis of a plasma membrane GPI. This hydrolysis may represent an early event in the transduction of the NGF signal in CVG.

It is interesting to note that the dose-response profile of GPI hydrolysis overlaps with that of cell proliferation induced by NGF (Fig. 2C, inset, data from Bernd and Represa, 1989). This would indicate that the mitogenic effect of NGF could be linked to the hydrolysis of GPI.

CVG from 72 hr chick embryos show only low affinity NGF receptors (Bernd and Represa, 1989), thus the activation of the GPI/IPG system would be linked to specific NGF binding to this kind of receptors. The low-affinity NGF receptor has been associated with the p75 membrane bound protein and it is also shared by other neurotrophic factors such as BDNF (Rodriguez-Tebar et al., 1990) and NT3 (Ernfors et al., 1990). An interesting possibility would be that these other neurotrophic factors were also active to induce cell proliferation in the CVG and that they use GPI hydrolysis as a common signalling pathway.

Recent experiments have shown that both BDNF and NT3 modulate cell proliferation of CVG (Avila, M.A., Represa, J., Romero, G., Miner, C., Mato, J.M., Giraldez, F. and Varela-Nieto, I., manuscript in preparation). BDNF and NT3 are able to replicate the proliferative effect of NGF but at much lower concentrations (1-5 ng/ml). These doses are consistent with BDNF and NT3 acting trough their high affinity receptors. In parallel, BDNF and NT3 promote the rapid cleavage of GPI in a dose-dependent manner. The resulting IPG release is a crucial step for the transduction of the mitogenic signal, since anti-IPG antibodies are able to impair the effect of BDNF and NT3 on ganglion growth. These data suggests that the mitogenic effect of BDNF and NT3 is coupled to the hydrolysis of GPI. For insulin it has been shown that GPI hydrolysis requires an intact tyrosine kinase activity of its membrane receptor (Villalba et al., 1990). If neurotrophins were acting in the CVG trough the trk receptor family, then GPI hydrolysis may well be an early consequence of the stimulation of the receptor tyrosine kinase activity.

Acknowledgements

We wish to thank Dr. S. Udenfriend for the phosphatidylinositol specific phospholipase C, Dr. G. Romero for the anti-IPG antibody and Dr. Y.A. Barde for kindly providing us with NT3 and BDNF. Special thanks are due to Dr. A. Rodriguez-Tebar for many useful discussions and criticisms. M.A.A. and B. Gil are fellows of Ministerio de Educación y Ciencia, Spain. Y.L. is fellow of Comunidad de Madrid, Spain. This work was supported in part by grants from Dirección General de

Investigación, Ciencia y Tecnología (PM90-0001), Europharma, Comunidad Autonoma de Madrid. This work has been done in collaboration with the team working at Dr. F. Giraldez laboratory. Dep. Bioquímica, Biología Molecular y Fisiología. Facultad de Medicina. Universidad de Valladolid. 47005 Valladolid. Spain.

Figures 1 & 2 were originally published in Proc. Natl. Acad. Sci. USA. (Represa et al., 1991).

Footnotes

Abbreviations used: Nerve growth factor (NGF); Brain-derived neurotrophic factor (BDNF); Neurotrophin-3 (NT-3); inositol phosphoglycan (IPG); cochleo-vestibular ganglion (CVG); glycosyl-phosphatidylinositol (GPI).

REFERENCES

Barde, Y.A. The nerve growth factor family. 1990. Prog. Growth Factor Res. 2: 237.

Bernd, P. and Represa, J. 1989. Characterization of nerve growth factor receptors in the embryonic otic vesicle and cochleovestibular ganglion. Dev. Biol. 134:11.

Chan, B.L., Chao, M.V. and Saltiel, A.R. 1989. Nerve growth factor stimulates the hydrolysis of glycosyl-phosphatidylinositol in PC-12 cells: A mechanism of protein kinase C regulation. Proc. Natl. Acad. Sci. USA 82:1756.

Ernfors, P., Ibañez, C.F., Ebendal, T., Olson, L. and Persson, H. 1990. Molecular cloning and neurotrophic activities of a protein with structural similarities to ß-nerve growth factor: developmental and topographical expression in the brain. Proc. Natl. Acad. Sci. USA. 87:5454.

Ferguson, M.A.J. and Williams A.F. 1988. Cell-surface anchoring of proteins via glycosyl-phosphatidylinositol structures. Ann. Rev. Biochem. 57: 285.

Giraldez, F., Represa, J.J., Borondo, L. and Barbosa, E. 1987. Polarization and density of Na-pumps in the inner ear of the chick embryo during early stages of development. Development 100:271.

Hamburger, V. and Hamilton, H.L. 1951. A series of normal stages in the development of the chick embryo. J. Morphol. 88:49.

Hempstead, B.L., Martin-Zanca, D., Kaplan, D.R., Parada, L.F., and Chao, M.V. 1991. High affinity NGF binding requires coexpression of the trk proto-oncogene product and the low affinity NGF receptor. Nature 350:678.

Kelly, K., Mato, J.M. and Jarret, L. 1987. Glucose transport and antilipolysis are differentially regulated by the polar head group of an insulin-sensitive glycophospholipid. Proc. Natl. Acad. Sci. 84:6404.

Klein, R., Jing, S., Nanduri, V., O'Rourke, E., and Barbacid, M. 1991a. The trk proto-oncogene encodes a receptor for nerve growth factor. Cell 65:189.

Klein, R., Nanduri, V., Jing, S., Lamballe, F., Tapley, P., Bryand, S., Cordon-Cardo, C., Jones, K.R., Reichardt, L.F. and Barbacid, M. 1991b. The trkB tyrosine protein kinase is a receptor for brain derived neurotrophic factor and neurotrophin 3. Cell. 66:395.

Lamballe, F., Klein, R. and Barbacid, M. 1991. trkC, a new member of the trk family of tyrosine protein kinases, is a receptor for neurotrophin 3. Cell. 66:967

Machicao, F., Mushack, J., Seffer, E., Ermel, B. and Haring, H.U. 1990. Mannose, glucosamine and inositol monophosphate inhibit the effects of insulin on lipogenesis. Further evidence for a role for inositol phosphate-oligosaccharides in insulin action. Biochem. J. 266:909.

Mato, J.M., Kelly, L.L., Abler, A. and Jarret, L. 1987. Identification of a novel insulin-sensitive glycophospholipid from H35 hepatoma cells. J. Biol. Chem. 262: 2131.

Mato, J.M., Alvarez, L., Clemente, R., Guadaño, A., Avila, M.A., Ochoa, P and Varela-Nieto, I. 1991. Glycosyl-phosphatidylinositol: role in signal transduction. in Diabetes 1991. Rifkin, H., Colwell, J.A. and Taylor, S.I. eds. Elsevier Science Publishers.

Larner, J., Romero, G., Kennington, A.S., Lilley, K., Kilgour, E., Zhang, C., Heimark-Gamez, C., Houston, D.B., and Huang, L.C. 1990. Duality in the mechanism of action of insulin. Adv. Second Messenger Phosphoprotein Res. 573:297.

Represa, J., Miner, C., Barbosa, E. and Giraldez, F. 1988. Bombesin and other growth factors activate cell proliferation in chick embryo otic vesicles in culture. Development. 102:87.

Represa, J., Avila, M.A., Miner, C., Giraldez, F., Romero, G., Clemente, R., Mato, J.M. and Varela-Nieto, I. 1991. Glycosyl-phosphatidylinositol/ inositol phosphoglycan: A signalling system for the low-affinity nerve growth factor receptor. Proc. Natl. Acad. Sci. USA. 88:8016.

Rodriguez-Tebar, A., Dechant, G., and Barde, Y.A. 1990. Binding of Brain-derived neurotrophic factor to the nerve growth factor receptor. Neuron 4:487.

Rodriguez-Tebar, A., Dechant, G., Gotz, R. and Barde, Y.A. 1992. Binding of neurotrophin-3 to its neuronal receptors and interactions with nerve growth factor and brain-derived neurotrophic factor. EMBO J. 11:917.

Romero, G., Gámez, G., Huang, L.C., Lilley, K. and Luttrell, L. 1990. Anti-inositolglycan antibodies selectively block some of the actions of insulin in intact BC_3H1 cells. Proc. Natl. Acad. Sci. USA. 87:1476.

Rougon, G., Dubois, C., Buckley, N., Magnani, J.L. and Zollinger, W. 1986. A monoclonal antibody against Meningococcus group B polysaccharides distinguishes embryonic from adult N-CAM. J. Cell. Biol. 103:2429.

Saltiel, A.R. and Cuatrecasas, P. 1986. Insulin stimulates the generation from hepatic plasma membranes of modulators derived from an inositol glycolipid. Proc. Natl. Acad. Sci. 83:5793.

Udenfriend, S., Micanovic, R. and Kodula, K. 1991. Structural requeriments of a nascent protein for processing to a PI-G anchored form: studies in intact cells and cell-free systems. Cell Biol. International Reports. 15:739.

Varela, I., Avila, M., Mato, JM and Hue, L. 1990. Insulin-induced phospho-oligosaccharide stimulates amino acid transport in isolated rat hepatocytes. Biochem. J. 267: 541.

Varela-Nieto, I., Represa, J., Avila, M.A., Miner, C., Mato, J.M. and Giraldez, F. 1991. Inositol phospho-oligosaccharide stimulates cell proliferation in the early developing iner ear. Dev. Biol. 143:432.

Villalba, M., Kelly, K. and Mato, J.M. 1988. Inhibition of cAMP-dependent protein kinase by the polar head group of an insulin sensitive glycophospholipid. Biochim. Biophys. Acta 968:69.

Villaba, M., Alvarez, J.F., Russell, D.S., Mato, J.M. and Rosen, O.M. 1990. Hydrolysis of glycosyl-phosphatidylinositol in response to insulin is reduced in cells bearing kinase-deficient insulin receptors. Growth Factors. 2:91.

PHARMACOLOGICAL, ELECTROPHYSIOLOGICAL AND MOLECULAR ANALYSIS OF A DOPAMINE RECEPTOR IN A SPECIFIC SUBSET OF NEURONS IN THE CENTRAL NERVOUS SYSTEM OF THE POND SNAIL *LYMNAEA STAGNALIS*

Cindy C. Gerhardt, Hans J. Lodder*, Ellen van Kesteren, Rudi J. Planta, Karel S. Kits*, Harm van Heerikhuizen, and Erno Vreugdenhil

Department of Biochemistry and Molecular Biology
*Department of Neurophysiology
Vrije Universiteit
De Boelelaan 1083
1081 HV Amsterdam
The Netherlands

INTRODUCTION

Dopamine (DA) is an important neurotransmitter in both vertebrates and invertebrates, where it is acting primarily in the brain but also in the perifery. In the vertebrate brain, DA is involved in initiation and execution of movement, maintenance of emotional stability, and regulation of pituitary function. DA exerts its physiological effects through interaction with specific cell surface receptors. Because this interaction is implicated to play a role in the pathophysiology of Parkinsons disease (1), in schizophrenia (2), and in drug addiction (3), vertebrate DA receptors have been extensively studied for many years.

Vertebrate DA receptors belong to the superfamily of G-protein coupled receptors. They are classically divided into two classes, the D1-receptor (D1-R) and the D2-receptor (D2-R) These classes differ in their pharmacological properties (several D1-R and D2-R selective agonists and antagonists are known) and in their intracellular responses (4). The D1-R, after binding DA, stimulates adenylyl cyclase via G_s, leading to an increase in cAMP and thereby to activation of cAMP dependent protein kinases (7,8). The D2-R on the other hand, interacts with G_i or G_o, resulting in an inhibition of adenylyl cyclase (leading to a decrease in cAMP levels) or an inhibition of phosphoinositol hydrolysis and a subsequent decrease in intracellular calcium concentration (reviewed in 6). However, both biochemical and pharmacological data suggested that the D1-R and D2-R classification is inadequate (7). Molecular biological studies have demonstrated that DA-receptors possess the typical architecture common to most G-protein coupled receptors. Members of this superfamily can be structurally recognized by the fact that they contain seven highly conserved, hydrophobic regions that putatively form seven transmembrane α-helices. Cloning of DA-receptors revealed that many of the observed discrepancies concerning the D1/D2-R classification can be accounted for by different receptor subtypes encoded by distinct genes. Two D1-R subtypes are now known, D1 and D5, and four D2-R subtypes, D2A, D2B, D3 and D4 (reviewed in 8). The D2A-R and the D2B-R are translation products of two mRNA's derived from a common pre-mRNA by alternative splicing. The D2B-R protein has a 29 amino acid insertion in the third cytoplasmatic loop as compared to the D2A-R. All five different DA-R subtypes exhibit a specific pharmacological profile and are expressed in a tissue-specific manner. Also on the level of their genomic organization, a distinction between D1-like and D2-like subtypes can be made. The D1 and D5 receptor are encoded on single exons, while all D2-like subtypes are encoded by split genes.

In invertebrate species DA occurs widely and is usually the most abundant catecholamine present (9). Especially in molluscs convincing evidence for DA acting as a neurotransmitter has been obtained (see 10 for a review). DA synthesis and metabolism qualitatively resembles that of vertebrates and also (re)uptake mechanisms for DA exist in molluscs.

In the periphery, DA can act as a neuromuscular transmitter directly effecting smooth muscle contraction. In *Aplysia*, DA is shown to be present in large quantities in the gill. The effect of DA on isolated gill muscle fibers is an immediate contraction (11). In the opaline gland of *Aplysia*, DA also induces contraction. In addition, it enhances the size of subsequent neurally evoked gland contractions, and also the size of the excitatory junction potential. This suggests a modulatory function for DA in opaline release (12). Furthermore, DA, together with acetylcholine and serotonin, functions as a neurotransmitter in the heart. In *Lymnaea*, DA evokes a fast and short excitatory effect on the auricle (13).

In the molluscan CNS, it has been shown that neuronal networks regulating defensive behaviour (14), respiratory rythms (15), and feeding patterns (16) are under dopaminergic control. Most research, however, has focused on the cellular effects of DA receptor stimulation. For instance, many reports have been published on the effects of DA on Na^+, K^+, and Ca^{++}- currents, resulting in depolarization, hyperpolarization, or a biphasic reponse of the neuronal membrane. Also in molluscs DA can, but does not always, induce changes in cAMP production (reviewed in 10 and 17). The molluscan CNS provides a good model system to study central actions of DA and DA receptors. It is easily accessible for experimental manipulation, it is simply organized in a number of ganglia and it has large neurons which often can be identified by shape, colour and position.

The CNS of the pond snail *Lymnaea stagnalis* is composed of 11 ganglia situated in a ring around the oesophagus (see Fig. 1). In total, it consists of no more than 15,000 neurons some of which contain extremely large cell bodies (up to 200 μm). In addition, for a number of these neurons the function is known. They can form defined networks, some of which can be reproduced *in vitro* (15). As the size of the neurons is large, they are especially suited for electrophysiological analysis and micro-injection of DNA or proteins.

DA is present in considerable amounts in the CNS of *Lymnaea stagnalis* (18,19). DA immunoreactivity is predominantly present in the cerebral and pedal ganglia. Some DA-positive cellbodies are found, but the majority of immunostaining for DA is located within the axons. One large cell in the right pedal ganglion (RPeD1) is strongly immunopositive. This large cell is therefore called the Giant Dopamine Cell. RPeD1 itself responds to DA, suggesting the presence of DA autoreceptors in this cell (18,20). DA also affects the Light Green Cells (LGCs), large neuroendocrine cells located in the cerebral ganglia which synthesize and secrete four different proteins strongly related to human insulin, the molluscan insulin related peptides (MIPs) (21). Application of DA on LGCs induces a membrane hyperpolarization, mainly mediated by an increase in potassium conductance (22). In the presence of the D2-R receptor antagonist (-)-sulpiride, this hyperpolarization is blocked, and DA then causes an increase in excitability of the LGCs. SKF 38393, a selective D1-R agonist, also increases the excitability of the LGCs. It has therefore been suggested that both D1- and D2-like dopamine receptors are present in the LGCs (23).

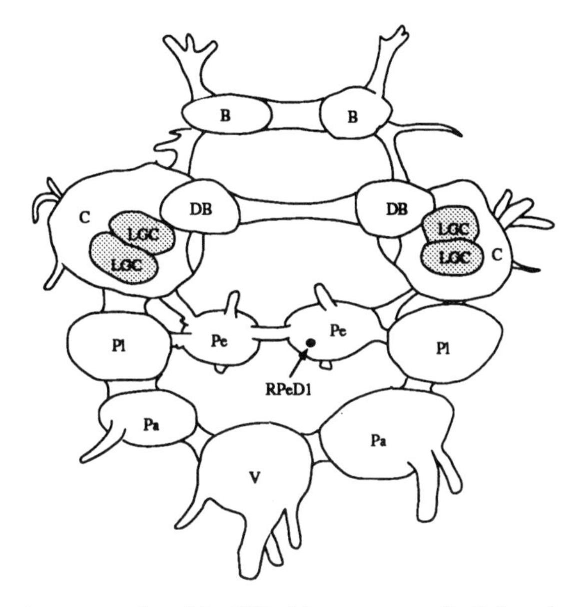

Figure 1. Schematic representation of the CNS of *Lymnaea stagnalis*. B=buccal ganglia, C=cerebral ganglia, DB=dorsal bodies, LGC=light green cells, Pe=pedal ganglia, RPeD1=giant dopamine cell, Pl=pleural ganglia, Pa=partietal ganglia, V=visceral ganglion.

In this chapter, a detailed description of the DA-receptors present in the LGCs is given. Based on pharmacological data, it is concluded that these DA receptors differ considerably from vertebrate DA receptors, making it difficult to classify them as a D1-R or a D2-R subtype. A further electrophysiological analysis indicates the presence of a DA-inhibited Ca^{++}-current, and shows the role of cAMP in the modulation of DA-induced K^{+}- and Ca^{++}-currents. Finally, the molecular cloning of G-protein coupled receptors expressed in LGCs has elucidated the (partial) sequences of a serotonin receptor and a putative dopamine receptor.

PHARMACOLOGICAL PROFILE OF THE LGC DA-RECEPTOR

DA induces a biphasic response in the LGCs, thus suggesting the presence of both D1- and D2-like receptors. Stimulation of the D2-like receptor induces an instantaneous hyperpolarization, while a small depolarization is manifested after about 5 min. To find out whether the fast hyperpolarizing response is mediated via a D2-like receptor, several DA-R agonists and antagonists were tested for their ability to induce or inhibit LGC membrane hyperpolarization (24).

The effect on the membrane potential of the LGCs was determined for DA and eleven DA receptor agonists, in a concentration of 10 μM (see Table 1). DA itself, 6,7-diOH-aminotetralin (6,7-diOH-AT), 6,7-diOH dipropyl-aminotetralin (6,7-diOH-DPAT), 5,6-diOH-AT and 5,6-diOH-DPAT all exerted hyperpolarizing effect on the LGCs. However, selective D2-R agonists like LY 141865, LY 171555, and N 0437 did not hyperpolarize the LGCs. The D1-R selective agonist SKF 38393 showed a small depolarizing effect. DA and 6,7-diOH-AT were half-maximally effective at a concentration of approximately 1 μM.

Table 1. Hyperpolarizing effects of DA and DA receptor agonists on LGC membrane potentials. After dissection of the CNS an LGC was penetrated with a microelectrode to record the membrane potential. DA was applied, and the induced hyperpolarization was measured (left column). Then DA was washed out, the agonist was applied, and the effect was measured (middle column). This procedure was performed in 6 different CNS for each DA receptor agonist. In the right column agonist-induced hyperpolarization is expressed as % of DA-induced hyperpolarization. Data represent mean values ± S.E.M. [DA]=[agonist] = 10 μM.

Agonist	Hyperpolarization (mV)		% effect
	DA	Agonist	
6,7-diOH-AT	-15.2 ± 1.0	-10.5 ± 1.3	68.7 ± 4.0
6,7-diOH-DPAT	-16.8 ± 0.5	-2.6 ± 0.4	15.3 ± 2.1
5,6-diOH-AT	-14.9 ± 0.9	-1.8 ± 0.3	12.3 ± 2.0
5,6-diOH-DPAT	-14.8 ± 0.6	-0.7 ± 0.2	4.5 ± 1.1
N 0437	-14.9 ± 1.2	-0.5 ± 0.4	4.1 ± 2.2
7-OH-DPAT	-15.8 ± 0.8	-0.5 ± 0.3	3.7 ± 2.3
5-OH-DPAT	-14.6 ± 0.4	-0.5 ± 0.3	3.6 ± 1.9
LY 141865	-14.3 ± 1.0	-0.5 ± 0.3	3.6 ± 2.0
LY 171555	-14.5 ± 0.6	-0.3 ± 0.2	1.9 ± 1.8
6-OH-DPAT	-12.4 ± 0.8	-0.2 ± 0.2	1.1 ± 1.1
SKF 38393	-14.5 ± 0.6	+1.2 ± 0.3	

Furthermore, several D2 receptor antagonists were tested. (-)-Sulpiride and YM 09151-2, in a concentration of 10 μM, were able to inhibit the DA-induced hyperpolarization of the LGCs. (-)-Sulpiride displayed its full antagonistic activity instantaneously, whereas YM 09151-2 was more active 20 min. after application. (-)-Sulpiride was also able to antagonize the hyperpolarizing effects of 5,6-diOH-DPAT and 6,7-diOH-DPAT. The antagonising effect of (-)-sulpiride is fully abolished after washing for five minutes, whereas the effect of YM09151-2 appears to be irreversible. In addition to (-)-sulpiride and YM 09151-2, four other DA receptor antagonists (spiperone, flupenazine, cis-flupenthixol and domperidone) were tested on the LGCs, but these chemicals did not affect the LGC membrane potential. The effects of all the antagonists on the DA-evoked hyperpolarization, instantaneously and after 20 minutes, are given in Table 2.

Table 2. Effects of DA receptor antagonists on the DA-induced hyperpolarization in LGCs immediately and 20 min after bath application of the antagonists. After dissection of the CNS an LGC was penetrated with a microelectrode to record the membrane potential. DA was applied and the DA-induced hyperpolarization was measured (first column). Then DA was washed out, the antagonist was applied and the DA-induced hyperpolarization was measured immediately (second column). The effect of the antagonist was assessed again after 20 min of preincubation of the antagonist (third column). Finally, the drugs were washed out, and the DA-induced hyperpolarization was measured after 5 min (fourth column). This procedure was performed in 5 different CNS for each antagonist. Data represent mean values ± S.E.M. [DA] = 2 μM, [antagonist]= 10 μM.

Antagonist	Hyperpolarisation (mV)			
	DA	DA + antagonist		DA
		t=0 min.	t=20 min.	
(-)-sulpiride	-9.6 ± 1.0	-3.0 ± 0.3	-4.4 ± 0.7	-11.1 ± 1.1
YM 09151-2	-9.5 ± 0.4	-3.9 ± 0.4	-1.1 ± 0.5	-1.9 ± 0.3
Spiperon	-12.1 ± 0.4	-11.3 ± 0.4	-11.9 ± 0.5	-12.3 ± 0.7
Fluphenazine	-11.7 ± 1.2	-12.7 ± 1.4	-12.5 ± 1.0	-13.6 ± 1.3
is-flupenthixol	-11.1 ± 1.1	-12.3 ± 1.2	-12.8 ± 1.4	-13.0 ± 1.1
Domperidone	-11.4 ± 0.9	-11.6 ± 1.0	-12.9 ± 0.9	-13.2 ± 0.8

From this study, it can be concluded that the DA receptors in the LGCs clearly differ from the mammalian D1 and D2-like receptors. Several aminotetralin derivates, which are mixed D1/D2-R agonists, mimicked the effect of DA on the LGCs. However, in contrast to the mammalian situation, the dipropylaminotetralins were less potent, and the D2-R selective agonists LY 141865 and LY 171555 were without any effect, even in a concentration of 10 μM.

The antagonists tested result in a similar picture. In this case only the benzamides (-)- sulpiride and YM 09151-2 were able to antagonize the DA-induced hyperpolarization, whereas the D2-R selective antagonists domperidone and spiperone, and the mixed D1/D2-R antagonists cis-flupenthixol and fluphenazine were without effect. These results are in agreement with a study on the pharmacological properties of DA receptors present on the neurons B2 and RPeD1 in *Lymnaea* (20), where it was concluded that a D1/D2-R distinction in *Lymnaea* can not be made. Also in the land snail *Helix aspersa*, it has been found that a clear pharmacological distinction between D1 and D2 receptors can not be made. DA induces hyperpolarizing and depolarizing responses, which involve the opening of a potassium and a sodium channel, respectively (25). The D1-R agonists SKF 38391 and dihydroxynomifensine mimick the hyperpolarizing response to DA in a defined celltype. However, these agonists are 100-fold less active than DA. The D2-R agonists quinpirole and RU24213 are without effect. Both the D1-R antagonist SCH23390 and the D2-R antagonist (-)-sulpiride antagonize the DA-induced hyperpolarization (26). Apparently, the recognition sites for several DA (ant)agonists in DA receptors have diverged considerably in the course of evolution.

ELECTROPHYSIOLOGICAL RESPONSES AND SIGNAL TRANSDUCTION OF THE LGC DA-RECEPTOR

Bath application of 1 μM DA induces a prominent hyperpolarization in isolated LGCs. This DA-induced hyperpolarization is accompanied by a decrease in membrane resistance, and is sensitive to variations in extracellular potassium concentration, reversing near the potassium Nernst potential. The potassium channel blocker 4-aminopyridine (4-AP) in a concentration of 0.1 mM blocks the DA-induced hyperpolarization. These results strongly suggest the involvement of potassium channels in the DA-evoked hyperpolarization (22). This conclusion was confirmed in whole cell voltage clamp recordings, where application of DA induced an outward current response carried by K^+-ions (Fig. 3A trace 1). Whole cell voltage clamp experiments were also performed to determine if activation of the DA receptors on the LGCs also induces changes in calcium conductance. Fig. 2A shows the voltage-activated Ca^{++}-current under control conditions. Fig. 2B shows that DA induces inhibition of the Ca^{++}-current. Subtraction of the current remaining during DA application from the current under control conditions yields the current that is suppressed by DA (Fig. 2C). Comparison of Fig. 2B and 2C shows that DA selectively suppresses a slowly inactivating current, while the transient current is not affected. The results are summarized in the current-voltage (IV) relationships of Fig. 2. Comparison of the IV plots of Fig. 2B and 2C also shows a difference between the voltage-dependence of the DA-insensitive Ca^{++}-current (the transient current) and the DA sensitive Ca^{++}-current (the slowly inactivating current). The current suppressed by DA activates at a more negative membrane potential than the DA-insensitive current.

Initial experiments on signal transduction mechanisms underlying the DA-induced hyperpolarization, showed that application of agents that increase cellular cAMP levels in the cell (IBMX, a phosphodiesterase inhibitor, and 8cpt-cAMP, a membrane permeable cAMP derivative) antagonized the DA-induced hyperpolarization. This suggested an

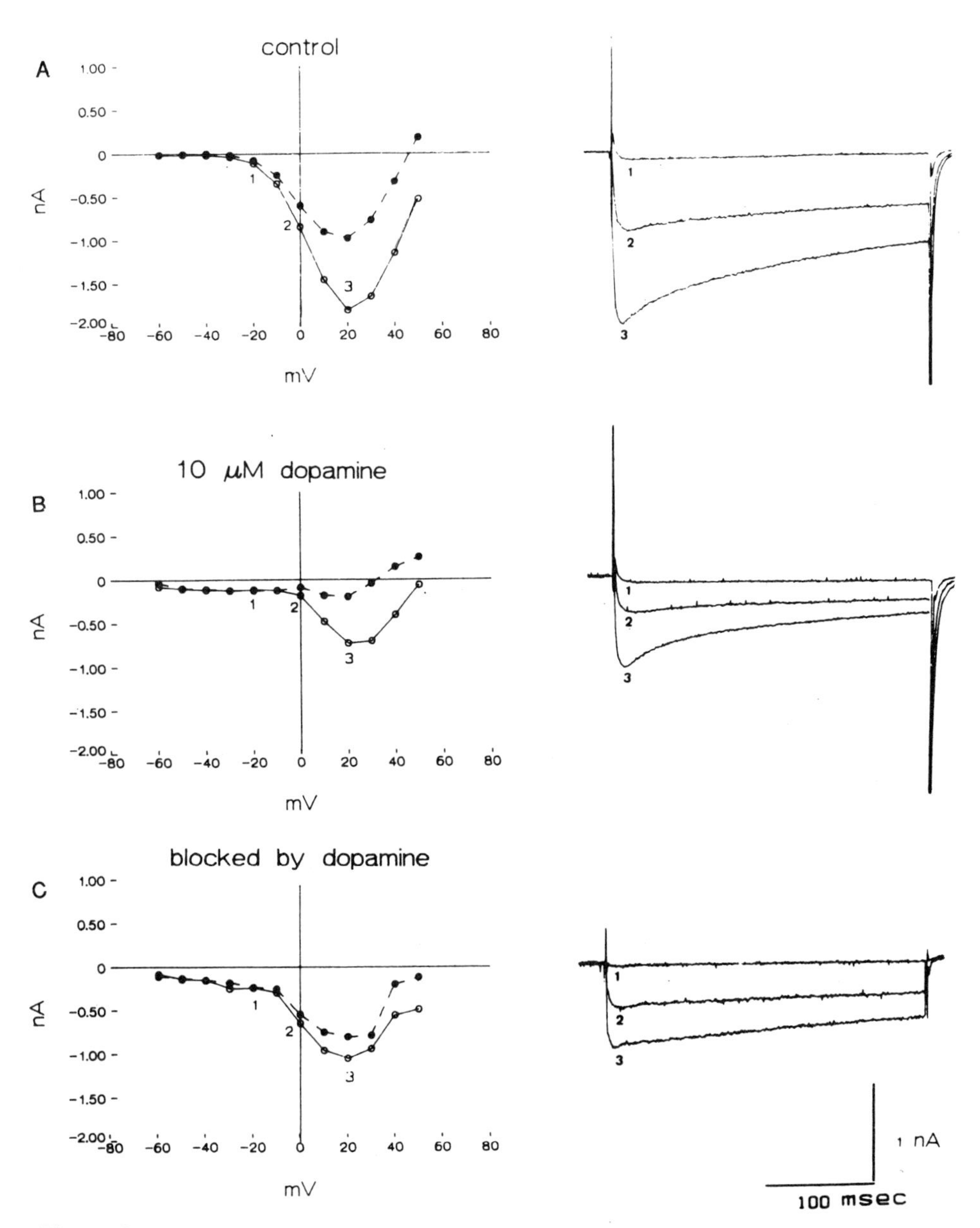

Figure 2. DA-receptor stimulation induces a decrease in the slowly inactivating Ca^{++}- current in the LGCs. On the left, the time-course of the Ca^{++}-current is shown. IV relationships shown on the right are obtained by plotting the current amplitude against the command potential. Open circles indicate the maximal current amplitude measured at each command potential, yielding the IV relationship of predominantly the transient current. Closed circles indicate the current amplitude measured at the end of the current traces (at t=300ms), yielding the IV relationship of the slowly inactivating current. Holding potential is -60 mV. Numbers 1, 2 and 3 refer to a command voltage of respectively -20 mV, 0 mV, and +20 mV. Fig. 2A shows the Ca^{++}-current under control conditions, Fig. 2B shows the decrease in Ca^{++}-current in response to 10 μM dopamine, and Fig. 2C shows the current that is suppressed by DA. In order to isolate the calcium current, TTX, TEA, 4-AP, and Cs^+ were added to the extracellulatr medium. Cs^+ also replaced K^+ in the electrode medium.

inhibitory coupling of the DA-R to adenylyl cyclase (22). This was confirmed in whole cell voltage clamp experiments. By loading the cells with 8cpt-cAMP, the DA-induced outward current was completely suppressed within 10 min (Fig. 3B). Interestingly, the DA-induced inhibiting effect on the calcium current was not affected in the presence of 8cpt-cAMP (Fig. 4), suggesting that the inhibition of Ca^{++}-channels does not proceed via a cAMP-dependent pathway.

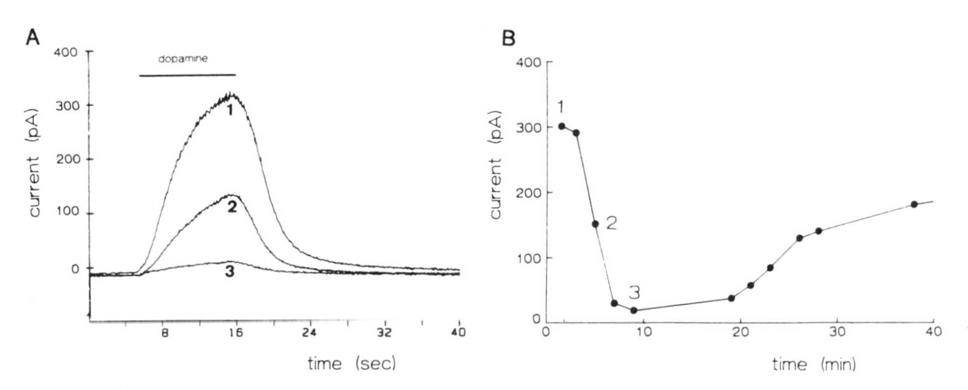

Figure 3. Whole cell voltage clamp experiments performed on isolated LGCs, showing the DA-induced potassium current, and the inhibition of this current by 8cpt-cAMP. Fig. 3A shows the potassium currents activated by DA before (1) and during (2,3) application of 8cpt-cAMP. At time point 2 and 3 the inhibiting effect of 8cpt-cAMP is half-maximal, respectively maximal. Fig. 3B shows the time course of the effect of 8cpt-cAMP on the amplitude of the DA-induced K^{+}-current.

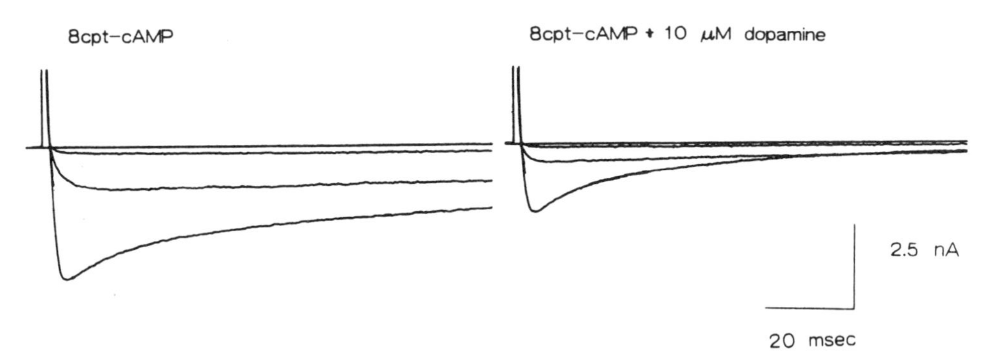

Figure 4. Whole cell voltage clamp experiments performed on isolated LGCs, showing the effect of DA and 8cpt-cAMP on the calcium-current. In order to isolate the Ca^{++}-current, the same conditions were applied as indicated in the legend to Fig. 2. The left figure shows the time course of the Ca^{++}-current under control conditions. The right figure shows that in the presence of 8cpt-cAMP, the DA-induced inhibition of the calcium current is not affected. Holding potential is -60 mV.

DA induces an inhibition of the electrical activity in the LGCs of *Lymnaea stagnalis* (22). Here we show that the alterations in membrane potential induced by DA-R stimulation in the LGCs, are not only mediated by an increase in K^+-conductance, but also by an inhibition of inward Ca^{++}-currents. Especially the slowly-inactivating component of the Ca^{++}-current is reduced.

In the land snail *Helix aspersa* (25, 27) and the see snail *Aplysia california* (28), it has also been reported that DA activates a potassium current. In contradiction to these findings, it has been shown that DA can induce a decrease in potassium conductance (29, 30) in *Helix* neurons. In both *Helix* and *Aplysia*, DA has also been shown to reduce voltage-dependent calcium currents (30, 31, 32). In addition to these DA-induced changes in potassium and calcium conductances, DA receptor stimulation can induce a slow inward sodium current in *Helix* (25) and *Aplysia* (33)

Our data suggest that DA-receptors in the LGCs can both couple to G_i and G_o. The DA-induced increase in K^+-current could be a result of a lowered intracellular cAMP concentration caused by G_i. On the other hand, the inhibiting effect of DA-R stimulation on the Ca^{++}-current seems to be indepent of changes in cAMP concentation, suggesting that this effect is mediated via G_o. Intracellular Ca^{++}-mobilization (via the inositol phosphate metabolism) could be involved in the regulation of the DA-induced changes in Ca^{++}-conductance

Our results indicate an inhibiting effect of cAMP on the potassium current induced by DA, while cAMP itself does not influence the potassium current (data not shown). Therefore we suggest that DA-receptors in the LGCs are coupled to G_i. This is in sharp contrast with earlier studies that have been done to elucidate the role of adenylyl cyclase in molluscan DA receptor mediated signal transduction. Bath application of dibutyryl cAMP to *Helix* neurons has been shown to increase the DA-induced outward K^+-current (27). This current was found to be mimicked by the elevation of intracellular cAMP levels by GTPγS or forskolin (34, 35). Intracellular injection of cAMP could furthermore mimic the inward sodium current evoked by DA (33) and the inward current caused by a DA-induced decrease in K^+-conductance (29). All these results suggest a coupling of DA-receptors to G_s. However, a direct stimulating effect of DA on the cAMP concentration in the cell has not been shown.

In vertebrates, the potentiating effect of DA on potassium currents has been reported to be mediated either by G_i or G_o. In rat pituitary cells, it was shown that injection of specific antibodies raised against G_i, can block the DA induced increase in potassium current (36). On the other hand, it has been reported (37) that stimulation of DA autoreceptors in the mesencephalon raises three different K^+-currents, and that this effect is blocked by antibodies against G_o.

In agreement to what we have shown here for the LGCs, in an other identified *Lymnaea* neuron and in the R15 bursting neuron of *Aplysia*, the inhibiting effect of DA on Ca^{++}-channels has been shown to be independent of cAMP alterations (31, 38). Furthermore, intracellular injection of anti-G_o-antibodies can reduce the DA-induced decrease in Ca^{++}-current in *Helix* neurons (39).

In vertebrates, it has been shown that antibodies raised against G_o can specifically reduce the DA-induced decreases in Ca^{++}-current in rat pituitary cells (36). Similar results were obtained for the DA autoreceptors in the mesencephalon (37). The bovine D2 receptor has been shown to be coupled to G_o as well as to G_i (40).

MOLECULAR BIOLOGY OF THE LGC DA-RECEPTOR

As yet, no data on invertebrate dopamine receptor cDNAs or genes have been published. Such data would nonetheless be highly interesting, not only from an evolutionary point of view but also from a pharmacological point. Evidently, insight in the genomic organization and the number of genes encoding DA-receptors in a number of widely diversing species may lead to a better understanding of the evolution of this important class of neurotransmitter receptors. Furthermore, comparison of the amino acid sequences of invertebrate DA-receptors, with the sequences of vertebrate receptors may be useful to pinpoint domains which play a crucial role in (ant)agonist recognition. As is shown above, *Lymnaea* D2-like receptors respond differently to some DA (ant)agonists as their vertebrate counterparts. Pharmacological differences as seen in the *Lymnaea* receptors, may be correlated to very limited and specific sequence variations. Recently, it was shown that a single amino acid conversion in the human 5HT-1B receptor can cause some major pharmacological differences and in fact can change the human 5HT-1B receptor into a receptor that is pharmacologically indistinguishable from its counterpart in rodents (41). Eventually, such an approach may be of interest for the development of new drugs to be used in the treatment of schizophrenia and Parkinson's disease which show less side-effects than the drugs currently used.

Although DA can induce a biphasic response in the LGCs, it is still not clear wether there are two different DA-receptors present in the LGCs, or wether one receptor is differentially responding to vertebrate D1-R and D2-R antagonists. Knowledge of the cDNA sequence(s) of the *Lymnaea* dopamine receptor(s) can be used to substantiate the presence of one or two receptors responding to DA. This could for instance be done using antisence technology, i.e. the introduction of small oligonucleotides or plasmids into viable LGCs in order to inhibit the transcription of a particular DA receptor gene. The response of such manipulated cells to DA will indicate whether another DA receptor is still present.

To clone the cDNA(s) encoding the DA-R(s) in *Lymnaea*, we have employed a cloning strategy based on the polymerase chain reaction (PCR) and the use of highly degenerated oligonucleotides that recognize conserved regions in transmembrane domain VI and VII of G-protein coupled receptors. These oligonucleotides were used in a PCR to specifically amplify parts of G-protein coupled receptors expressed in the LGCs. Amplified PCR products of the expected size were cloned and sequenced. After data base homology search, one clone, GRL002, exhibited significant homology with the human dopamine D4 receptor, whereas another clone, GRL003, resembled the serotonin 1A receptor. Specific oligonucleotides based on these *Lymnaea* receptor clones were designed to isolate the full length receptor cDNAs from a *Lymnaea* CNS cDNA library. GRL003 is expressed abundantly in the *Lymnaea* CNS (data not shown) and several clones could be isolated from the cDNA library. However, the same receptor has been independently cloned by van Tol et al. (Department of Pharmacology, University of Toronto, Canada; personal communication). Binding studies performed by van Tol have proven that GRL003 indeed is a serotonin receptor. PCR-based screening of the cDNA library with GRL002-specific oligonucleotides resulted in the isolation of several clones which all ended in the same GC-rich sequence encoding a repetitive domain in the putative third cytoplasmatic loop. In order to get a full length receptor sequence, we screened a genomic library, and isolated several positive clones which are currently being analyzed. In contrast to the vertebrate D2-like receptors genes, the GRL002 gene contains no introns. The highests homology between Lymnaea receptors and related vertebrate receptors can be found in the transmembrane

Table 3. Percentage of identity in transmembrane regions III, IV, V, VI and VII of GRL002 and GRL003 as compared with the rat 5HT-1A receptor and the rat D2 receptor. Numbers marked with * indicate incomplete data, as sequence information on TM5 of GRL002 is only available for 50%.

TM	GRL002		GRL003	
	rat 5HT-1A	rat D2	rat 5HT-1A	rat D2
III	61.1	72.2	68.2	50.0
IV	38.1	33.3	38.1	42.8
V	22.0*	44.0*	61.9	23.8
VI	39.1	60.8	65.2	39.1
VII	47.8	65.2	56.2	60.8

regions, which are supposed to be involved in ligand binding. In Table 3 the percentages of identity between transmembrane regions III, IV, V, VI and VII of GRL002 and GRL003 and the rat 5HT 1A and the rat D2 receptor are depicted.

Using a PCR-based, semi-quantitative method to determine the level of expression of GRL002 in the separate ganglia of the CNS of *Lymnaea*, we have found that GRL002 is expressed in all ganglia, but especially in the parietal ganglia (Fig.5). Furthermore it was found that GRL002 is not only expressed in the CNS, but also in the heart (data not shown). DA (as well as 5HT) are known to act as a neurotransmitter on the molluscan heart. The vertebrate D4 receptor has recently been found also to be expressed in the heart (42).

Immunocytochemistry revealed the presence of large amounts of DA in all ganglia, especially the cerebral and pedal ganglia (18). Here we show that GRL002 is also highly expressed in these ganglia. Moreover, expression levels in the parietal ganglia are extremely high. The spatial abundance of GRL002 and DA in the CNS suggest a major function of DA in *Lymnaea stagnalis*. Expression of GRL002 in the different organs of *Lymnaea*, and regulation of GRL002 expression in the different developmental stages will be studied in the near future.

SUMMARY

A detailed description of DA receptors present in the neuroendocrine LGCs in the CNS of the pond snail *Lymnaea stagnalis* is given. These DA receptors show some similarities to D2-like receptors. They exert a pharmacological profile similar, although not identical, to the mammalian D2-R, and after activation they induce a membrane hyperpolarization. This hyperpolarization is mainly caused by increasing outward potassium currents, but also by

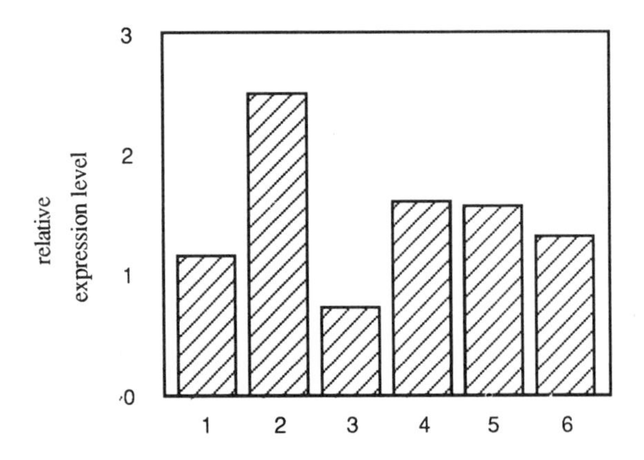

Figure 5. Expression of GRL002 in the ganglia of the *Lymnaea stagnalis CNS*.
1 µg of total RNA was converted into cDNA, and used in a PCR with GRL002-specific oligonucleotides. As an internal control, aldolase (a constitutively expressed gene) was amplified. Products were blotted on to a nylon membrane and hybridized with a labeled oligonucleotide which recognizes a sequence in the amplified part of GRL002. Hybridization signals were scanned, and the amounts of GRL002 transcript are plotted relative to the aldolase signal. 1=visceral ganglion, 2=parietal ganglia, 3=pleural ganglia, 4=cerebral ganglia, 5=pedal ganglia, 6=buccal ganglia.

decreasing inward calcium currents. The former effect is mediated through an adenylyl cyclase dependent mechanism, whereas the latter is uneffected by changes in cAMP. Incidentally a depolarising response was observed suggesting the additional presence of a D1-like receptor. However, there is no clear prove that indeed two different DA receptors are present on the LGCs.

Furthermore, we have developed a strategy to use PCR in neuron-specific cloning of G-protein coupled receptors, and we were able to clone two closely related receptors. One of them appeared to be a 5HT-1A receptor, whereas the other one has a high homology to the vertebrate D2-like receptors. At the moment we are analyzing a genomic clone on which the gene encoding this receptor is present. Unlike the vertebrate D2-like receptors, this *Lymnaea* receptor gene has no introns. In addition, transcript levels of the putative DA receptor in the separate ganglia of the CNS of *Lymnaea* are presented.

REFERENCES

1. Lee, T., Seeman, P., Rajput, A., Farley, I.J. and Hornykiewicz, O., *Nature* **273**; 59-61 (1978).
2. Creese, I., Burt, D.R. and Snyder, S.H., *Science* **192**; 481-483 (1976).
3. Barnes, D.M., *Science* **241**; 415-417 (1988).
4. Stoof, J.C., and Kebabian, J.W., *Life Sci.* **35**; 2281-2296 (1984).
5. Hemmings, H.C., Walaas, S.I., Oiumet, and Greengard, P., *Trends Neurosci.* **10**; 77-82 (1987).

6. Vallar, L. and Meldolesi, J. *Trends Pharmacol. Sci.* **10**; 74-77 (1989).
7. Andersen, P.H. Gingrich, J.A., Bates, M.D., Dearry, A., Falardeau, P., Senogles, S.E. and Caron, M.G., *Trends Pharmacol. Sci.* **11**; 231-236 (1990).
8. Sibley, D.R. and Monsma, F.J. Jr., *Trends Pharmacol. Sci.* **13**; 61-69 (1992).
9. Walker, R.J. and Holde-Dey, L., *Comp. Biochem. Physiol.* **93A**; 25-39 (1989).
10. Gospe, S.M., *Life Sci.* **33**; 1945-1957 (1983).
11. Cawthorpe, D. and Lukowiak, K,. *Neurosc. Lett.* **113**; 345-348 (1990).
12. Tritt, S.H. and Byrne, J.H., *J. Neurophysiol.***48**; 1347-61 (1982).
13. Buckett, K.J., Dockray, G.J., Osborne, N.N. and Benjamin, P.R., *J. Neurophysiol.* **63**; 1413-1425 (1990).
14. Christyskova, M.W., *Neurosci. Behav. Physiol.* **20**; 446-452 (1990).
15. Syed, N.I., Bulloch, A.G. and Lukowiak, K., *Science* **250**; 282-285 (1990).
16. Kyriakides, M.A. and McCrohan, C.R., *J.Neurobiol.* **20**; 635-650 (1989).
17. S.-Rósza, K., *Prog. Neurobiol.* **23**; 7-30 (1984)
18. Werkman, T.R., van Minnen, J., Voorn, P., Steinbusch, H.W., Westerink B.H., de Vlieger, T.A. and Stoof, J.C., *Exp. Brain Res.* **85**; 1-9 (1991).
19. Elekes, K., Kemenes, G., Hiripi, L., Geffard, M. and Benjamin, P.R., J. Comp. *Neurol.* **307**; 214-224 (1991).
20. Audesirk, T.E., *Comp. Biochem. Physiol.* **93**; 115-119 (1989).
21. Smit, A.B., Vreugdenhil, E., Ebberink, R.H.M., Geraerts, W.P.M., Klootwijk, J. and Joosse, J., *Nature* **331**; 535-538 (1988).
22. de Vlieger, T.A., Lodder, J.C., Stoof, J.C. and Werkman, T.R., *Comp. Biochem. Physiol.* **83C**; 429-433 (1986).
23. Stoof, J.C., De Vlieger, T.A., Lodder, J.C. and Werkman, T.R., *Comp. Biochem. Physiol.* **83C**; 429-433 (1986).
24. Werkman, T.R., Lodder, J.C., de Vlieger, T.A. and Stoof, J.C., *Eur. J. Pharmacol.* **139**; 155-161 (1987).
25. Bokisch, A.J. and Walker, R.J., *Comp. Biochem. Physiol.* **84**; 231-241 (1986).
26. Holden-Dye, L. and Walker, R.J., *Comp. Biochem. Physiol.* **93**; 413-419 (1989).
27. Cox, R.T. and Walker, R.J., *Comp. Biochem. Physiol.* **91**; 541-547 (1988).
28. Chesnoy-Marchais, D., *Brain Res.* **304**; 83-91 (1984).
29. Deterre, P., Paupardin-Tritsch, D., Bockaert, J., and Gerschenfeld, H.M., *Proc. Natl. Acad. Sci. USA* **79**; 7934-7938 (1982).
30. Paupardin-Tritsch, D., Colombaioni, L., Deterre, P. and Gerschenfeld, H.M., *J. Neurosci.* **5**; 2522-2532 (1985).
31. Akopyan, A.R., Chemeris, N.K. and Iljin, V.I. *Brain Res.* **326**; 145-148 (1985).
32. Lewis, D.V., Evans, G.R. and Wilson, W.A., *J. Neurosc.***4**; 3014-3020 (1984).
33. Matsumoto, M., Sasaki, K., Sato, M. and Shozushima, K., *J. Physiol.* **407**; 199-213 (1988).
34. Deterre, P., Paupardain-Tritsch, D. and Bockaert, J., *Brain. Res.* **387**; 101-109 (1986).
35. Sawada, M., and Maeno, T., *Jap. J. Physiol.* **37**, 459-478 (1987).
36. Lledo, P.M., Homburger, V. and Bockaert, J., *Neuron* **8**; 455-463 (1992).

37. Liu, L.-X., Chiodo, L.A. and Kapatos, G., *Abstracts of the 22nd Annual Meeting of the Society for Neuroscience* **18**; 1516 (1992).
38. Lemos, J.R. and Levitan, I.R., *J. Gen. Physiol.* **83**; 269-285 (1984).
39. Harris-Warrick, R.M., Hammond, C., Paupardin-Tritsch, D., Homburger, V., Rouot, B., Bockaert, J. and Gerschenfeld, H.M., *Neuron* **1**; 27-32 (1988).
40. Elazar, Z., Siegel, G.and Fuchs, S. *EMBO J.* **8**; 2353-2357 (1989).
41. Oksenberg, D., Marsters, S.A., O'Dowd, B.F., Jin, H., Havlik, S., Peroutka, S.J. and Askenazi, A., *Nature* **360**; 161-163 (1992).
42. O'Malley, K.L., Harmon, S., Lei, T.and Todd, R.D., *New Biol* .**4**; 137-146 (1992).

REGULATION OF GENE EXPRESSION IN RODENT HEPATOCYTES

Kleanthis G. Xanthopoulos

Karolinska Institute
Center for Biotechnology
NOVUM
S- 141 57 Huddinge, Sweden

INTRODUCTION

The processes of cellular differentiation and proliferation, which occur during development, involve many molecular regulatory events. These include the selective induction, expression and regulation of the activity of genes which are likely to encode for DNA-binding proteins or proteins that interact with them and control the transcription of several other genes (Blau, 1988). Thus a cascade of molecular interactions is initiated that will ultimately lead to the adult phenotype. Additional regulatory mechanisms are then required for the maintenance of the differentiated state.

The study of liver development and differentiation offers many advantages over other systems to address these questions at the molecular level. Rodent liver organogenesis has been thoroughly studied at the cellular level. It is now known that at day eight of embryonal development the endodermal cells of the developing foregut are induced by the precardiac mesenchyme and approximately one day later the hepatic endoderm will emerge from the gut endoderm. Subsequently the primary liver diverticulum invades the mesenchyme of the spectrum transversum (For a review see Herbst and Babiss, 1990). During this stage proliferation and differentiation of the hepatic endoderm into hepatoblasts occurs. At day 14-16 more hepatoblasts become hepatocytes, a process that will continue until the liver is fully differentiated. During the last phase of gestation hematopoietic cells migrate from the liver to the bone marrow and the liver now consists of hepatocytes and hematopoietic cells in equal numbers. Glycogen begins to accumulate and full enzymatic activity is gradually detected. Three weeks after birth the liver is fully differentiated. Liver is the second largest organ after skin, and offers an excellent source for large scale biochemical analysis (Herbst and Babiss, 1990).

Finally, it is very important to acquire a fundamental understanding of the mechanisms of normal gene regulation in proliferating, developing and differentiated cells such as hepatocytes in order to comprehend many medical problems associated with liver including neoplasia and tumor formation.

LIVER FUNCTION AND ANATOMY

Liver, the largest gland in the body, is responsible for many vital and complex functions. These include detoxification, bile formation, carbohydrate and fat metabolism, urea formation and inactivation of polypeptide hormones. Liver also provides a variety of essential proteins that are secreted in the serum. The adult organ consists almost exclusively (80-90%) of one parenchymal cell, the hepatocyte, and because most of these cells are tetraploid they collectively contribute up to 90% of the total transcriptional activity that is

New Developments in Lipid-Protein Interactions and Receptor Function
Edited by K.W.A. Wirtz *et al.*, Plenum Press, 1993

monitored in the liver. Furthermore, the ability of the liver to regenerate provides us with a powerful experimental model system to study differentiation that can be induced chemically or surgically. Regulation of the expression of genes that are active in the liver is predominantly controlled at the level of mRNA transcription (Derman et al., 1981). This is mediated partly through the action of a set of hepatocyte-specific trans-acting factors.

In this article we will review the recent developments in this field and present a detailed overview of hepatocyte transcription factors and their potential role in liver development and differentiation.

REGULATORY MOLECULES INVOLVED IN HEPATOCYTE-SPECIFIC GENE EXPRESSION

Gene transcription depends on regulatory cis-acting sequences usually located close to the mRNA start site. These elements, promoters, often in combination with distant regulatory sequences, enhancers, control the expression of tissue-specific genes. Numerous studies over the last several years, using both *in vivo* transient transfection and *in vitro* DNA:protein interaction experiments, have identified a number of hepatocyte-enriched regulatory nuclear DNA-binding proteins. Several such proteins have been isolated, characterized, and their corresponding genes have been cloned (for a recent review see Sladek and Darnell, 1992). These include the Hepatocyte Nuclear Factors, HNF-1, HNF-3 and HNF-4 as well as certain members of the C/EBP family (i.e. C/EBPα and C/EBPβ) and DBP. Two important observations were made: (i) none of these DNA-binding proteins are exclusively expressed in hepatocytes, rather they are present in a limited number of tissues and (ii) they appear to belong to larger families and are able to form heterodimers with other members of the same family. We will discuss these families of nuclear factors in more detail, presenting the current known members of each family, their limited tissue distribution, distinct structural motifs and the DNA sequences they recognize and bind.

The HNF-1 Family

HNF-1 was originally characterized by its ability to bind to the β-fibrinogen gene promoter (Courtois et al., 1987), and has since been described independently by several groups (Monaci et al., 1988; Cereghini et al., 1988; Schorpp et al., 1988). HNF-1 DNA-binding sites are frequently found in the regulatory regions of many genes expressed in hepatocytes, including albumin, transthyretin and α1-antitrypsin (Xanthopoulos et al., 1991). The consensus binding site, 5' ATTAAC 3', appears to be palindromic and the protein is postulated to bind to DNA as a dimer (Johnson, 1990). HNF-1α was isolated from rat liver nuclei as a polypeptide of apparent M_r 88,000 (Curtois et al., 1988) and a cDNA clone was isolated from a rat liver cDNA library (Frain et al., 1989). Primary sequence analysis indicated that the protein contains a highly diverged homeobox DNA-binding domain and a POU domain (Frain et al., 1989). A different form of this DNA-binding activity, termed "variant HNF-1" was detected in gel-shift assays using nuclear extracts from dedifferentiated rat hepatoma cell lines (Baumhueter et al., 1988; Cereghini et al., 1988). Sequence analysis of isolated cDNA clones revealed a 58% identity to the original HNF-1 protein. In addition, extensive homology in the homeodomain (92%), in the pou-specific domain (66 %) and in the dimerization domain (72%) of the molecule was found (Mendel et al., 1991; Cereghini et al., 1991). This protein is currently referred to as HNF-1β and is able to form heterodimers with HNF-1α and activate transcription (Ray-Campos et al., 1991). Both proteins are present in approximately equal amounts in liver and kidney. Two different HNF-1α mRNAs are detected in both tissues presumably originating by differential splicing of the same precursor. Finally, a non-DNA binding protein, called DCoH, was characterized as a cofactor that regulates dimerization of HNF-1α and HNF-1β homodimers.

The CCAAT-Enhancer Binding Protein (C/EBP) Family

Several members of the CCAAT-Enhancer Binding Protein (C/EBP) family of transcription factors have been characterized and their corresponding genes have been cloned. These include C/EBPα cloned from rat (Landschultz et al., 1988) and mouse liver (Xanthopoulos et al., 1989), C/EBPβ originally cloned from human glioblastoma cells (termed NF-IL6, Akira et al., 1990) and subsequently from rat, mouse liver and mouse adipocyte cell lines (termed IL6-DBP, LAP, AGP/EBP, C/EBPβ, CRP II, and rNFIL-6: Poli et al., 1990; Descombes et al., 1990; Chang et al., 1990; Cao et al., 1991; Williams et al., 1991; Metz and Ziff, 1991). Two additional members of the family include C/EBPγ (Roman et al., 1990) and C/EBPδ (Williams et al., 1991 and Cao et al., 1991). However, C/EBPγ and C/EBPδ are present at low concentrations in the liver and so far have not been shown to contribute significantly to liver-specific gene expression.

C/EBPα is a positive transcriptional factor that regulates the expression of several genes expressed in the liver and fat tissue (Costa et al., 1988; Xanthopoulos et al., 1989; Herrera et al., 1989). The C/EBPα gene is regulated at the level of transcription and is expressed in a limited number of tissues. Sequence analysis of C/EBPα cDNA clones revealed a distinct structural motif termed the basic leucine zipper motif. This "bleuzip" motif consists of a basic DNA binding region and a dimerization domain with a few, at least three, leucine residues found every seventh amino acid (Landschulz et al., 1988). Several other nuclear proteins have since shown to include the "bleuzip" motif usually located at the carboxy-termini of the molecules.

The C/EBP family members are structurally related and share extensive homology in their DNA binding and dimerization domains and are classified to belong to the bleuzip class of transcription factors. Recently, two new proteins have been identified that exhibit a negative effect on transcription mediated through specific protein:protein interactions. These include LIP which specifically interacts with C/EBPβ (Descombes and Schibler, 1991) and CHOP which apparently heterodimerizes with C/EBPα (Ron and Habener, 1992). Based on genomic Southern and library screenings under low stringency conditions, it has been postulated that perhaps as many as five to ten additional members of the family are present in the mammalian genome (Williams et al., 1991; Flodby and Xanthopoulos, unpublished data).

Perhaps C/EBPα is the molecule whose potential function appears to be better understood than any of the other hepatocyte-enriched transcriptional activators. The protein is present in high concentrations in terminally differentiated cells such as hepatocytes and adipocytes and it was postulated to have a antiproliferative effect (Landschulz et al., 1988). In accordance with this suggestion C/EBPα protein levels are drastically reduced in regenerating hepatocytes (see Regulation in Disease). Umek et al. 1991, demonstrated in a series of elegant experiments that C/EBPα is a component of the preadipocyte to adipocyte differentiation process. Recently, three independent studies demonstrated that antisense C/EBPα RNA blocks differentiation of cultured 3T3-L1 preadipocyte cells to adipocytes (Samuelsson et al., 1991; Freytag and Geddes, 1992; Lin and Lane, 1992).

Finally, another DNA-binding protein termed DBP, is enriched in rat liver nuclei but is not a formal member of the C/EBP family because it does not heterodimerize with any of the known members. DBP seems to bind in a mutual exclusive manner to the promoter of albumin competing for binding with C/EBPα and C/EBPβ (Mueller et al., 1990).

The HNF-3 Family

HNF-3 DNA-binding activity was initially identified as a nuclear protein interacting with the promoters of the mouse transthyretin and α1-antitrypsin genes (Costa et al., 1989). The consensus HNF-3 binding site, 5' TATTGANTTANC 3', was used as a probe to isolate and clone the first member of the family, termed HNF-3α. Low stringency hybridization screening of a rat liver cDNA library using an HNF-3α probe allowed for the cloning of two additional members, HNF-3β and HNF-3γ (Lai et al., 1991). The HNF-3 family may include as many as 10-15 additional members (J. E. Darnell Jr. personal communication). Members of the HNF-3 family display a great degree of similarity (over 85% identity) around the DNA-binding domain. In addition, blocks of 10 amino acids are shared between the family members clustered around the carboxy-termini of the proteins. A distinct structural feature of the HNF-3 proteins is that they are highly proline-rich with no obvious (known) DNA-binding motifs, and that there is extensive homology with the product of the *forkhead* gene of *Drosophila*. In *Drosophila* this protein occurs in the hindgut, salivary placode and foregut (Weigel et al., 1989). In the adult mouse HNF-3 family members are present in a variety of tissues including liver, large intestine, salivary gland and the esophagus (Lai and Darnell, 1991; Sladek and Darnell, 1992). Other members of the family seem to be expressed in the lung (Xanthopoulos et al., 1991) and brain (J. E. Darnell Jr. personal communication). Based on the stunning homology between the mammalian

HNF-3 and the *Drosophila forkhead* gene, it appears that HNF-3 family members may have a broader developmental role. HNF-3 family members have not been shown to form either homodimers or heterodimers. Studies are under way to determine the developmental expression of the different family members and the structure of the DNA-binding domain of the molecule.

HNF-4

A novel nuclear factor with high affinity for sites in the regulatory elements that control expression of the mouse transthyretin gene was identified by a variety of *in vitro* (Costa et al., 1989) and *in vivo* (Mirkovitch and Darnell, 1991) methods. This DNA-binding activity, named HNF-4, originally recognized to bind to a 5' GGCAAAGGTCCAT 3' sequence. This element was able to mediate hepatocyte-specific transcriptional activation in transient transfection experiments (Costa et al., 1989). The protein was consequently purified from rat liver nuclear extracts and the corresponding cDNA clone was isolated (Sladek et al., 1991). Sequence analysis revealed extensive structural and sequence homology with other members of the steroid-thyroid receptor superfamily (Sladek et al., 1991). The zinc finger DNA-binding domain is up to 67% similar to that of other members of the superfamily (Lai and Darnell, 1991) and the protein appears to bind DNA as a dimer. HNF-4 mRNA is found in a limited amount of tissues, including liver, kidney and intestine (Sladek et al., 1990). The basis for this limited distribution is transcriptional regulation of the gene (Xanthopoulos et al., 1991). Another novel member of the steroid hormone receptor superfamily, ARP-1, has been isolated (Ladias and Karathanasis, 1991), that appears to be ubiquitous, binds to similar sites as HNF-4 and may be a distant member of the HNF-4 family (Ladias and Karathanasis, 1991). So far no ligand has been found for neither molecule and the proteins are therefore classified as orphans nuclear receptors.

An HNF-4 homolog has been identified in the *Drosophila* genome by low stringency hybridization (Zhong and Darnell, unpublished data). The DNA-binding and dimerization/ligand-binding domains are very well conserved (up to 90% amino acid identity). Drosophila eggs contain maternal HNF-4 mRNA for a short period, while the endogenous gene is activated during late organogenesis in the midgut (Zhong and Darnell, unpublished data).

REGULATION IN DISEASE

Several diseases are associated with liver malfunctions and are in most cases well

Table 1. Hepatocyte-enriched nuclear factors.

Family	Member	Mr (Da)	Structural Motif	Tissue Distribution
HNF-1	α	88,000	POU, Homeo	I, K, L, S(1)
	β	72,000		I, K, L(+1),
C/EBP	α	42,000	bleuzip	A, B(*), H(1), I, L, S(1)
	β	32,000		A, I, L, Lg, S(1)
	γ	23,000		Ubiquitous
	δ	28,600		A(1), I, L,
HNF-3	α	50,000	Not Determined	I, L, Lg, St.
	β	47,000	(*forkhead*, Pro-rich)	I, L, Lg, O, St.
	γ	42,000		I, L, O, St., T
HNF-4		53,000	Zinc Finger,	I, K, L
	ARP-1	47,000	Steroid/Thyroid nuclear receptor	Ubiquitous

ARP-1 is another steroid/thyroid orphan receptor with similar DNA-binding site specificity as HNF-4. Mr: Molecular Mass

A: adipose tissue; B:Brain; H: heart; I: intestine; K:kidney; L: liver; Lg: lung; O: ovary; S:spleen; T:testes. (1) Indicates very low or (+1) low amounts or (*) expression in selected areas of the given tissue.

characterized at the cellular level. Using molecular probes for the genes encoding liver-enriched transcriptional factors we and others have tried to comprehend the alterations at the molecular level that affect liver gene expression under non-physiological conditions. We will discuss three cases where such experiments have been performed for (i) regenerating liver following partial hepatectomy (ii) experimentally induced rat liver carcinogenesis and (iii) analysis of transcriptional rates in c^{14Cos} albino deletion mutant mice.

Partial hepatectomy of the adult fully differentiated liver in rodents induces the remnant liver cells to divide in a synchronized manner (Johansson and Andersson, 1990). Normal quiescent hepatocytes are thus forced to proliferate rapidly while retaining almost all metabolic activities and hepatic functions (Friedman et al., 1984). Gene expression during liver regeneration has been studied in the rat liver using a variety of genes including the genes for HNF 1-4, and the different C/EBP isoforms. The results of these experiments indicate that drastic and rapid changes are introduced at the level of transcription, most notably with C/EBPα whose transcriptional rate is reduced by several fold during liver regeneration (Michoulon et al., 1992; Flodby and Xanthopoulos, manuscript in preparation).

Persistent nodules occur in rat liver after treatment with carcinogens and are regarded as the direct precursors for hepatocellular carcinomas. Using the Resistant hepatocyte model (RH-model) which consist of initiation with a single dose of diethylnitrosamine (DEN) and promotion with 2-Acetylaminofluorence (2-AAF) followed by partial hepatectomy, it was found that dramatic changes occurred at the level of protein translation or stability affecting mainly C/EBPα, C/EBPβ, HNF-1α and HNF-4. The nuclear levels of these proteins, as detected by specific antisera in Western blots, were significantly decreased in persistent nodules and even more drastically decreased in hepatocellular carcinomas (Flodby and Xanthopoulos, manuscript in preparation).

Finally, livers of newborn or fetal c^{14Cos} albino deletion mutant mice were used to analyze the transcription rates of the same genes. This deletion which is believed to include the *hepatocyte specific developmental regulation locus* (hsdr-1) has a lethal effect within a few hours after birth. The results of these experiments showed that C/EBPα, HNF-1α, HNF-4 as well as albumin and α-fetoprotein were affected (Tönjes et al., 1992), and that the hsdr-1 deletion had a pleiotropic effect operating both at the levels of transcription and mRNA stability.

In summary, changes in liver gene expression under the above described conditions appear to involve complex mechanisms that operate at different levels of control including transcriptional and posttranscriptional processes.

HOW ARE THE FACTORS INVOLVED IN HEPATOCYTE-SPECIFIC GENE EXPRESSION REGULATED?

Mammalian development, as is the case with *Drosophila* development, appears to involve a cross-regulatory network of interactions between different regulatory molecules including transcription factors. The experimental path for the identification and characterization of "early" developmental signals requires a thorough understanding of the mechanisms that regulate the regulators (for a comprehensive review see Falvey and Schibler, 1991). In a series of recent experiments designed to determine the levels at which these factors are regulated, it was shown that modulation of production of a regulatory protein can be brought about through transcriptional, post-transcriptional and post-translational modification events (Mueller et al., 1990; Xanthopoulos et al., 1991).

The majority of these genes are themselves regulated at the level of transcription (Xanthopoulos et al., 1991) suggesting that a potential cascade may play an important role in mammalian development and differentiation. The cloning of genomic sequences encoding hepatocyte-enriched transcriptional factors has already produced several interesting results. For example the mouse C/EBPα promoter in addition to some novel regulatory sites contains a C/EBP binding site which is able to mediate a potent transactivation effect when a C/EBPα expression vector is co-transfected in cultured cells (Antonson and Xanthopoulos, unpublished data). Evidence for autoregulation of the HNF-3β mouse gene was also shown in similar types of experiments (Pani et al., 1992). Finally, two independent studies suggested that HNF-4 is an essential positive regulator of HNF-1α capable of transactivating the HNF-1α promoter in cotransfection assays and *in vitro* transcription experiments (Tian and Schibler, 1991; Kuo et al., 1992).

Although control of gene expression at the level of transcription is the predominant mechanism, other levels of regulation are active in determining the regulation of gene expression in hepatocytes. For example, differential splicing is likely to be involved in determining the abundance of the two variant forms of HNF-1 α mRNA in liver and kidney cells. In addition, the tissue distribution of two liver-enriched transcriptional factors, DBP and C/EBPβ (LAP) is regulated at the level of translation and/or protein stability (Mueller et al., 1990; Descombes et al., 1990). Phosphorylation of C/EBPβ (LAP) as a response to an extracellular stimulus results in the translocation of the molecule from the cytoplasm into the nucleus and consequent activation of its target genes (Metz and Ziff, 1991; Wegner, et al., 1992).

Finally, nuclear inhibitors such as LIP, which is encoded by a transcript derived from C/EBPβ (LAP) larger transcript, and CHOP add another level of control of hepatocyte-specific gene expression through specific protein:protein interactions that mediate a negative transcriptional effect.

CONCLUSIONS

A wealth of information concerning the interplay of cis-acting elements present in the regulatory regions of genes expressed mainly in the liver and their corresponding trans-acting factors has been accumulated. It is becoming evident that hepatocyte-specific transcription of genes is regulated, at least in part, by DNA sequences composed of multiple domains which have an additive effect on the rate of transcription of a given gene. Recent data indicate that the regulation of gene-expression in hepatocytes is controlled by a limited amount of hepatocyte-enriched transcription factors which operate co-ordinately and in combination with the basal transcription factors to determine the rate of transcription of a target gene in the liver.

Different combinations of these factors operate on every gene or may act differentially on a variety of genes, thus limiting the need for the existence of many different factors. This combinatorial effect greatly increase the flexibility of the transcriptional machinery and allows it to operate safely with a limited amount of different molecules. In addition, functionally important sequences that represent binding sites for basal transcription factors such as AP-1 and NF-Y are very often found, and are equally important in maintaining high levels of transcriptional activity. Members of the HNF-1 and C/EBP families are able to heterodimerize with each other. Heterodimer formation *in vivo* may prove to be an important mechanism by which the cell can increase significantly the arrangements of the different subunits and thus create a variety of complexes with unique specificities and affinities.

Why is the hepatocyte expressing a set of genes distinctly different from those in kidney cells? Several of the hepatocyte-enriched factors are present in equal or greater amounts in kidney (i.e. HNF-1 and HNF-4), however, the different cell types maintain different levels or gradients of a *set* of factors which in combination determine the genes that are transcriptionally active in a specific tissue. When some of the regulatory molecules are not present in sufficient amounts the gene that requires that set of factors is not activated or maintains low rates of expression. It should be noted that it is perhaps true that a certain cell type should not be viewed as a box with "on/off factor switches" but rather as a system that maintains and controls a delicate balance of levels of transcription factors, co-factors and modulators. Therefore, cell-type specific transcriptional control appears to be controlled by a cascade of combinatorial events through the interaction of the current known factors and their respective isoforms. Another way of modulating the activity of a positive acting regulatory molecule is through a specific repressor or cofactor. This additional level of regulation is introduced by the action of specific nuclear protein inhibitors that operate in hepatocytes such as CHOP (Ron and Habener, 1992) and LIP (Descombes and Schibler, 1991) and co-factors such as DCoH (Mendel, et al., 1991).

It appears that the mechanisms regulating hepatocyte-specific gene expression involve a great level of complexity that not only includes these specific group of factors but is controlled at the level of higher chromatin order, the state of methylation of the locus and the influence of negative developmentally acting regulatory molecules that may prove to be as important.

Determining the physiological function of most of these proteins will be crucial to understanding their specificity *in vivo*. Experiments designed to introduce mutations at specific sites in the mouse genome by homologous recombination, if successful, will probably demonstrate the functional importance of these genes and their role in liver development and differentiation.

REFERENCES

Akira, S., Isshiki, H., Sugita, T., Tanabe, O., Kinoshita, S., Nishio, Y., Nakajima, T., Hirano, T., and Kishimoto, T., 1990, A nuclear factor for IL-6 expression (NF-IL6) is a member of a C/EBP family, *EMBO J.* 9: 1897.

Baumhueter, S., Courtois, G., and Crabtree, G.R., 1988, A variant nuclear protein in dedifferentiated hepatoma cells binds to the same functional sequences in the β-fibrinogen gene promoter as HNF-1, *EMBO J.* 7: 2485.

Birkenmeier, E. H., Gwynn, B., Howard, S., Jerry, J., Gordon, J. I., Landschulz, W. H., and McKnight, S. L., 1989, Tissue-specific expression, developmental regulation, and genetic mapping of the gene encoding CCAAT/enhancer binding protein, *Genes Dev* . 3: 1146.

Blau, H., 1988, Hierarchies of regulatory genes may specify mammalian development, *Cell 53:673.*

Cao, Z., Umek, R. M., and McKnight, S. L. , 1991, Regulated expression of three C/EBP isoforms during adipose conversion of 3T3-L1 cells, *Genes Dev.* 5: 1538.

Cereghini, S., Blumenfeld, M., and Yaniv, M.A., 1988, A liver-specific factor essential for albumin transcription differs between differentiated and dedifferentiated rat hepatoma cells, *Genes Dev.* 2: 957.

Chang, C.-J., Chen, T.-T., Lei, H.-Y., Chen, D.-S., and Lee, S.-C., 1990, Molecular cloning of a transcription factor, AGP/EBP, that belongs to members of the C/EBP family, *Mol. Cel Biol.* 10: 6642.

Christy, R. J., Yang, V. W., Ntambi, J. M., Geiman, D. E., Landschultz, W. H., Friedman, A. D., Nakabeppu, Y., Kelly, T. J., and Lane, M. D., 1989, Differentiation-induced gene expression in 3T3-L1 preadipocytes: CCAAT/enhancer binding protein interacts with and activates the promoters of two adipocyte-specific genes, *Genes Dev.* 3:1323.

Christy, R. J., Kaestner, K. H., Geiman, D. E., and Lane, M. D. , 1991, CCAAT/enhancer binding protein gene promoter: Binding of nuclear factors during differentiation of 3T3-L1 preadipocytes, *Proc. Natl. Acad. Sci. USA* 88: 2593.

Costa, R. H., Grayson, D.R., Xanthopoulos, K. G., and Darnell, J.E. Jr., 1988, A liver-specific DNA-binding protein recognizes multiple nucelotide sites in regulatory regions of transthyretin, α1-antitrypsin, albumin and SV40 genes, *Proc. Natl. Acad. Sci. USA* 85: 3840.

Costa, R. H., Grayson, D.R., and Darnell, J.E. Jr., 1989, Multiple hepatocyte-enriched nuclear factors function in the regulation of transthyretin and α1-antitrypsin genes, *Mol. Cell Biol. 9: 1415.*

Courtois, G., Morgan, J.G., Campbell, L.A., Fourel, G., and Crabtree, G.R., 1987, Interaction of a liver-specific nuclear factor with the firbrinogen and α1-antitrypsin promoters, *Science 238: 688.*

Courtois, G., Baumhueter, S., and Crabtree, G.R., 1988, Purified hepatocyte nuclear factor 1 interacts with a family of hepatocyte-specific promoters, *Proc. Natl. Acad. Sci. USA* 85: 7937.

Derman, E., Krauter, K., Walling, L., Weinberger, C., Ray, M., and Darnell, J.E. Jr., 1981, Transcriptional control in the production of liver-specific mRNAs, *Cell* 23: 731.

Descombes, P., Chojkier, M., Lichtsteiner, S., Falvey, E., and Schibler, U., 1990, LAP, a novel member of the C/EBP gene family, encodes a liver-enriched transcriptional activator protein, *Genes Dev.* 4: 1541.

Descombes, P., and Schibler, U., 1991, A liver-enriched transcriptional activator protein, LAP, and a transcriptional inhibitory protein, LIP, are transcribed from the same mRNA, *Cell* 67: 569.

Falvey, E., and Schibler, U., 1991, How are the regulators regulated?, *FASEB J.* 5:309.

Frain, M., Swart, G., Monaci, P., Nicosia, A., Stämpfli, S., Frank, R., and Cortese, R., 1989, The liver-specific transcription factor LF-B1 contains a highly diverged homeobox DNA binding domain, *Cell* 59: 145.

Freytag, S. O., and Geddes, T. J., 1992, Reciprocal regulation of adipogenesis by Myc and C/EBPα, *Science 256: 379.*

Friedman, J.M., Chung, E.Y., and Darnell, J.E. Jr., 1984, Gene expression during liver regeneration, *J. Mol. Biol.* 179:37.

Herbst, R. S., and Babiss, L. E., 1990, Regulation of gene expression during development and regeneration, in: "Mechanisms of Differentation" P.B., Fisher, and M., Phil, eds., CRC Press, Boca Raton

Herrera, R., Ro, H. S., Robinson, G. S., Xanthopoulos, K. G., and Spiegelman, B. M., 1989, A Direct Role for C/EBP and AP-1-Binding Site in Gene Expression Linked to Adipocyte Differentiation, *Mol. Cell Biol. 9: 533.*

Johansson, S., and Andersson, G., 1990, Similar induction of the hepatic EGF receptor in vivo by EGF and partial hepatectomy, *Biochem. Biophys. Res. Comm.* 166:661.

Johnson, P.F., 1990, Transcriptional activators in hepatocytes, *Cell Growth Differ.* 1:47.

Kuo, C.J., Conley, P. B., Chen, L., Sladek, F.M., Darnell, J.E. Jr., and Crabtree, G.R., 1992, A transcriptional hierarchy involved in mammalian cell-type specification, *Nature* 355:457.

Ladias, J. A. A., and Karathanasis, S.K., 1991, Regulation of the apolipoprotein A1 gene by ARP-1, a novel member of the steroid receptor superfamily, *Science* 251: 561.

Lai, E., Prezioso, V.R., Tao, W., Chen, W.S., and Darnell, J.E. Jr., 1991, Hepatocyte Nuclear Factor 3a belongs to a gene family in mammals that is homologous to the Drosophila homotic gene forkhead, *Genes Dev* 5: 416.

Landschulz, W. H., Johnson, P. F., Adashi, E. Y., Graves, B. J., and McKnight, S. L., 1988, Isolation of a recombinant copy of the gene encoding C/EBP, *Genes Dev.* 2: 786.

Lin, F. T., and Lane, D.M., 1992, Antisense CCAAT/enhancer-binding protein RNA suppresses coordinate gene expression and triglyceride accumulation during differentiation of 3T3-L1 preadipocytes, *Genes Dev.* 6: 533.

Mendel, D. B., Hansen, L.P., Graves, M.K., Conley, P.B., and Crabtree, G.R., 1991, HNF-1α and HNF-1β (vHNF-1) share dimerization and homeo domains, but not activation domains, and form heterodimers in vitro., *Genes Dev* 5: 1042.

Mendel D.B., K., P.A., Conley, P.B., Graves, M.K., Hansen, L.P., Admon, A., and Grabtree, G.R., 1991, Characterization of a cofactor that regulates dimerization of a mammalian homeodomain protein, *Science* 254: 1762.

Metz, R., and Ziff, E., 1991, cAMP stimulates the C/EBP-related transcription factor rNFIL-6 to trans-locate to the nucleus and induce c-fos transcription, *Genes Dev.* 5: 1754.

Michoulon, D., Rana, B., Bucher, N.L.R., and farmer, S.R., 1992, Growht-dependent inhibition of CCAAT Enhancer-Binding Protein (C/EBPα) gene expression during hepatocyte proliferation in the regenerating liver and in culture, *Mol. Cell Biol.* 12:2553.

Monaci, P., Nicosia, A., and Cortese, R., 1988, Two different liver-specific factors stimulate in vitro transcription from the human α1-antitrypsin promoter, *EMBO J.* 7: 2075.

Mueller, C. R., Maire, P., and Schibler, U., 1990, DBP, a liver-enriched transcriptional activator, is expressed late in ontogeny and its tissue specificity is determined posttranscriptionally, *Cell* 61: 279.

Pani, L., Qian, X., Clevidence, D., and Costa, R.H., 1992, The restricted promoter activity of the liver transcription factor HNF-3β invloves a cell-specific factor and positive autoactivation, *Mol. Cell Biol.* 12:552.

Poli, V., Mancini, F. P., and Cortese, R., 1990, IL-6DBP, a Nuclear Protein Involved in Interleukin-6 Signal Transduction, Defines a New Family of Leucine Zipper Proteins Related to C/EBP, *Cell* 63: 643.

Rey-Campos, J., Chouard, T., Yaniv, M., and Cereghini, S., 1991, vHNF-1 is a homeoprotein that activates transcription and forms heterodimers with HNF-1, *EMBO J.* 10: 1445.

Roman, C., Platero, J.S., Shuman, J., and Calame, K., 1990, Ig/EBP-1: a ubiquitously expressed immunoglobulin enhancer binding protein that is similar to C/EBP and heterodimerizes with C/EBP, *Genes Dev.* 4: 1404.

Ron, D., and Habener, J. F., 1992, CHOP, a novel developmentally regulated nuclar protein that dimerizes with transcription factor C/EBP and LAP and functions as a dominant-negative inhibitor of gene transcription, *Genes Dev.* 6: 439.

Samuelsson, L., Strömberg, K., Vikman, K., Bjursell, G., and Enerbäck, S., 1991, The CCAAT/enhancer binding protein and its role in adipocyte differentiation: evidence for direct involvement in terminal adipocyte development, *EMBO J.* 10: 3787.

Schorpp, M., Dobbeling, U., Wagner, U., and Ryffel, G.U., 1988, 5'-Flanking and 5'-proximal exon regions of the two Xenopous albumin genes, *J. Mol. Biol.* 199: 83.

Sladek, F. M., Zhong, W., Lai, E., and Darnell, J.E. Jr., 1990, Liver-enriched transcription factor HNF-4 is a novel member of the steroid hormone receptor superfamily, *Genes Dev 4: 2353.*

Sladek, F. M., and Darnell, J.E. Jr., 1992, Mechanisms of liver-specific gene expression, *Current Opin. Gen. Dev.* 2: 256.

Tian, J. M., and Schibler, U., 1991, Tissue-specific expression of the gene encoding hepatocyte nuclear factor 1 may involve hepatocyte nuclear factor 4, *Genes Dev.* 5: 2225

Tönjes, R. R., Xanthopoulos, K.G., Darnell, J.E.., Jr., and Paul, D., 1992, Transcriptional control in hepatocytes of normal and c14CoS albino deletion mice, *EMBO J.* 11: 127.

Umek, R. M., Friedman, A. D., and McKnight, S. L., 1991, CCAAT/enhancer binding protein: A component of a differentiation switch, *Science* 251: 288.

Weigel, D., Jurgens, G., Kuttner, F., Seifert, E., and Jäckle, H., 1989, The homeotic gene forkhead encodes a nuclear protein and is expressed in the terminal regions of the Drosophila embryo, *Cell* 57:645.

Wegner, M., Cao, Z., and Rosenfeld, M.G., 1992, Calcium-regulated phosphorylation within the leucine zipper of C/EBPβ, *Science* 256: 370.

Williams, S. C., Cantwell, C. A., and Johnson, P. F., 1991, A family of C/EBP-related proteins capable of forming covalently linked leucine zipper dimers in vitro, *Genes Dev.* 5: 1553.

Xanthopoulos, K. G., Mirkovitch, J., Decker, T., Kuo, C. F., and Darnell, J.E., Jr., 1989, Cell-specific transcriptional control of the mouse DNA-binding protein mC/EBP, *Proc. Natl. Acad. Sci. USA 86: 4117.*

Xanthopoulos, K. G., Prezioso, V. R., Chen, W., Sladek, F., Cortese, R., and Darnell, Jr., J. E., 1991, The different tissue transcription patterns of genes for HNF-1, C/EBP, HNF-3, and HNF-4, protein factors that govern liver-specific transcription, *Proc. Natl. Acad. Sci. USA 88: 3807.*

TRANSCRIPTION FACTOR CREM: A KEY ELEMENT OF THE NUCLEAR RESPONSE TO cAMP

Denis Masquilier, Brid M. Laoide, Véronique Delmas,
Rolf P. de Groot, Nicholas S. Foulkes, Enrico Benusiglio,
Carlos A. Molina, Florence Schlotter
and Paolo Sassone-Corsi

Laboratoire De Génétique Moléculaire des Eucaryotes, CNRS, U184
INSERM, Institut de Chimie Biologique, Faculté de Médecine, 11, rue
Humann, 67085 Strasbourg, France

INTRODUCTION

A prerequisite for normal cell growth and differentiation is that each cell must be able to receive, interpret and respond appropriately to signals from other cells and the environment. The plasma membrane is the external interface of the cell and bears many elements which are required for the primary analysis of such signals. Among these elements are membrane receptors which are able to interact specifically with a variety of compounds. Binding of a ligand to its receptor initiates a cascade of events which modulate a variety of cellular functions including the control of gene expression. By altering the spectrum of genes expressed, the cell appropriately modifies its physiology for a given stimulus. Aberrations in this process can lead to deregulated cell proliferation and ultimately tumorigenesis.

Two major signal transduction pathways exist, which use cyclic adenosine monophosphate (cAMP) and diacylglycerol (DAG) as secondary messengers (Nishizuka, 1986; Borrelli et al., 1992). Each pathway is characterised by its specific protein kinase (protein kinase A (PKA) and protein kinase C (PKC), respectively) and its ultimate target for transcriptional control (cAMP-response element (CRE) and TPA-response element (TRE), respectively). This article will focus on the cAMP/PKA pathway and on one

New Developments in Lipid-Protein Interactions and Receptor Function
Edited by K.W.A. Wirtz *et al.*, Plenum Press, 1993

nuclear effector which appears to dynamically modulate gene expression, the transcription factor CREM.

THE NUCLEAR RESPONSE TO cAMP

The analysis of promoter sequences of several genes allowed the identification of promoter elements which could mediate the transcriptional response to increased levels of intracellular cAMP. A number of sequences have been identified of which the best characterised is the CRE (cAMP-response element) (Comb et al., 1986; Delegeane et al., 1987; Roesler et al., 1988; Sassone-Corsi, 1988). A consensus CRE site constitutes an 8 bp palindromic sequence (TGACGTCA). Several genes which are regulated by a variety of endocrinological stimuli contain similar sequences in their promoter regions, although at different positions (Borrelli et al., 1992; Roesler et al., 1988). A comparison of the CRE sequences identified to date, shows that the 5'-half of the palindrome, TGACG is the best conserved, whereas the 3' TCA motif is less constant. The binding site specificity appears to require 18-20 bp, since the five or so bases flanking the core consensus have been shown to dictate, in some cases, the permissivity of transcriptional activation. In many genes the CRE sequence is located in the first 200 bp upstream from the cap site. In most cases there is only one CRE element per promoter, although there are notable exceptions. The promoter of the α-chorionic gonadotropin gene, for instance, contains two identical, canonical CREs in tandem, between positions -117/-142 (Delegeane et al., 1987) and the promoter of the pituitary-specific transcription factor GHF-1/Pit-1 contains two different CREs between positions -200/-150, which are separated by a 40 bp spacer. The proto-oncogene c-*fos* contains a powerful CRE at position -60 (Sassone-Corsi et al., 1988), but other CRE-like sequences are also present within the gene regulatory region (Berkowitz et al., 1989), although the individual contribution of these elements to c-*fos* cAMP-inducibility has yet to be determined.

The CRE consensus sequence also appears in the context of other promoter elements. These include the ATF sites in the early promoters of adenovirus (Sassone-Corsi, 1988; Lee et al., 1989; Lillie and Green, 1989; Liu and Green, 1990), the 21bp *tax*-dependent enhancer of the HTLVI virus LTR (Maekawa et al., 1991) and the X-box motif associated with MHC class II genes (Liou et al., 1990).

It is tempting to speculate that more than one class of sequence element can mediate the response to cAMP, possibly in a developmental- or cell-specific manner. In further support of this notion, another non-CRE element, which also mediates cAMP induction is the AP-2 recognition site.

TRANSCRIPTION FACTORS AS TARGETS OF THE PKA-DEPENDENT SIGNALLING PATHWAY

An important step toward the understanding of cAMP-regulated gene transcription has been made with the cloning of cDNAs encoding CRE-binding proteins. The first cDNAs to be characterised encoded CREB (CRE-binding protein). They were isolated from human placental and rat PC12 cell libraries (Hoeffler et al., 1988; Gonzalez et al.,

1989) however it now seems that CREB is ubiquitously expressed suggesting a housekeeping role for this factor (Foulkes et al., 1991a). Subsequent extensive studies have elucidated many aspects of the structure-function relationship of this factor.

CREB belongs to the leucine-zipper family of proteins (Landschulz et al., 1988); at its C-terminus it contains four leucines arranged in a heptad repeat which constitute an α-helical coiled structure (Busch and Sassone-Corsi, 1990). It has been demonstrated that, as in the case of Fos, Jun and C/EBP, the leucine zipper is responsible for the dimerization of the protein and that dimerization is a prerequisite for DNA-binding. A model proposes a *bipartite* DNA-binding domain, where dimerization ensures the correct orientation of the adjacent basic regions in order to allow their optimal contact with the recognition sequence. It has been suggested that the basic region, composed of 50% lysine and arginine residues,

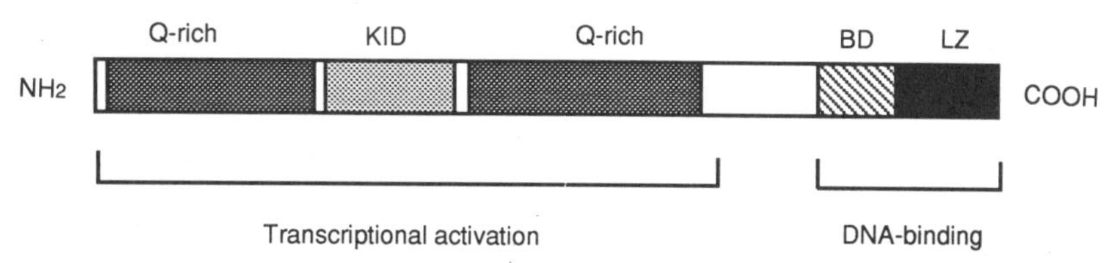

Figure 1. General structure of the transcriptional activator CREB. The protein is 341aa long and can be divided in two parts: 1) A domain required for DNA-binding (constituted by the leucine-zipper dimerization motif (LZ) and a basic region (BD)); 2) The transcriptional activation region, divided in the KID (Kinase-inducble domain) and the glutamine-rich domains (Q-rich). See text for details (see also Habener, 1990; Borrelli et al., 1992).

is infact divided into two sub-domains containing clusters of basic residues separated by a "spacer" of alanines, conserved among all leucine-zipper transcription factors (Busch and Sassone-Corsi, 1990). In this model, these two regions recognize the two halves of the palindromic recognition sequence. The positively charged aminoacids in the basic region are clustered on one face of the two helices of a helix-bend-helix structure. The two positively charged α-helices lie in the major groove of the DNA helix positioned so that the positive charges are in contact with the negative charges of the phosphate backbone (Landschulz et al., 1988; Busch and Sassone-Corsi, 1990). The structure of the DNA binding domain of CRE-binding proteins appears to be conserved (Borrelli et al., 1992). It is now apparent that the leucine zipper dimerization domain, as well as mediating homodimer formation, also permits specific heterodimerization between other CRE/ATF factors. This implies that the leucine zipper plays a central role in defining the interactions between different nuclear targets of the signal transduction pathway.

THE ROLE OF PHOSPHORYLATION

In Fig. 1 we present the general structure of the predicted CREB protein. The transcriptional activation domain consists of several independent regions. The first, indicated as KID domain, contains several consensus phosphorylation sites for kinases such as PKA, PKC, glycogen synthase kinase-3 and casein kinases (CK) I and II. The α-region is a 14 aminoacid peptide which lies N-terminal to the KID domain and is encoded by an alternatively spliced exon. Flanking the KID domain and the α-region are two protein domains rich in glutamine residues. Phosphorylation by PKA is necessary for transcriptional activation by CREB and thus constitutes the link with the cAMP signal transduction pathway (Gonzalez et al., 1991; Habener, 1990).

The KID domain of CREB has been shown to be the site of phosphorylation by both PKA and PKC *in vitro* (Yamamoto et al., 1988). The PKC phosphorylation is somewhat surprising and unexplained. The PKA effect has also been confirmed by using extracts from PC12 cells and scoring for CREB-dependent activation of the somatostatin promoter. Although interesting, this result has not been reproduced by others using a HeLa nuclear extract (Merino et al., 1989). This could be due to a lack of some cell-specific functions, but HeLa cells have been shown to naturally contain CREB and PKA (Foulkes et al., 1991a). Also PKA appears to modulate CRE-binding of proteins in nuclear extracts with variable efficiency, depending upon the cell type (Auwerx and Sassone-Corsi, 1991) (R. deGroot, in preparation). Mutagenesis data indicate that a serine residue at position 133 in CREB, is a phosphoacceptor for PKA (Yamamoto et al., 1988; Gonzalez and Montminy, 1989). This serine is part of a PKA consensus recognition site RRPSY, indeed, a serine to alanine mutation impairs transcriptional activation potential of CREB showing the importance of the site. However the presence of other phosphorylation sites strongly suggests the involvement of additional phoshorylation as yet poorly understood. It is conceivable that the various phosphorylation sites could cooperate to elicit the final regulatory function. Additional consensus sites for kinases, such as CKI and II and glycogen synthase kinase-3 (GSK-3), are present in the same area, but no information is yet available about their function. Further complexity in the function of the KID domain is implied by the identification of the motif DLSSD, C-terminal to the serine 133 residue, which cooperates with the PKA site in transcriptional activation (Gonzalez et al., 1991). The presence and the relative spacing of the two elements is essential for activation. The inclusion of the α-region, N-terminal to the KID domain strongly enhances PKA-induced transcriptional activation. Several lines of evidence predict that it constitutes an amphipathic α-helix however the precise role played by this motif remains unclear. Preliminary cross linking data implicates this region in subunit interactions within the CREB homodimer. It may also be involved in orientating the KID domain relative to the N-terminal glutamine rich domain (Gonzalez et al., 1991).

In two regions flanking the KID domain there are about three-times more glutamine residues than in the remainder of the protein. Glutamine-rich regions have been found in the activation domain of other factors, such as Oct-1 and -2 and Sp1. The current notion is that they mediate direct interactions with other components of the transcriptional machinery, although the precise nature of these interactions is still unclear. Recent data have demonstrated that at least in CREB the N-terminal glutamine domain is an absolute requirement for transactivation.

The current model to explain the CREB activation domain function is based upon allosteric conformational changes mediated by phosphorylation of the KID domain. In this model, phosphorylation triggers the activation of the glutamine-rich domains by a distant conformational change (Gonzalez et al., 1991).

A FAMILY OF TRANSCRIPTION FACTORS

Since the cloning of CREB, at least nine other CRE-binding factor cDNAs have been isolated. They were obtained by screening a variety of cDNA expression libraries with CRE and ATF sites, the HTLV-I LTR 21bp repeat and the MHC class II X-box sequences. In many cases the same cDNAs were isolated by independent laboratories using different strategies and so there has been some confusion over nomenclature; certain clear features have, however, emerged. (1) All share the same DNA binding motif: the basic domain and leucine zipper. (2) Based upon common regions of homology, these factors can be divided into subfamilies. For example CREB, ATF-1 and CREM share extensive regions of homology (Foulkes et al., 1991a; Hai and Curran, 1991), but are clearly distinct from CRE-BP1 which is part of another subfamily. (3) Alternative splicing of the transcripts seems common. For example, mXBP/CREBP2 is likely to be a splicing variant of CREBP1: this mouse clone carries a deletion relative to the human CRE-BP1, which lies within a proline-serine-threonine rich domain; outside of this region, the clones are >99% identical at the aminoacid level (Maekawa et al., 1989; Ivashkiv et al., 1990). Also there are reports of additional splicing variants of CRE-BP1 (Ivashkiv et al., 1990) and CREB (the α-region (Habener, 1990)). Alternative splicing is also dramatically evident in the CREM gene (see below) (4) The different factors are able to heterodimerize with each other but only in certain combinations. A "dimerization code" exists which seems to be a property of the leucine-zipper structure of each factor. For example, ATF1 can dimerize with CREB but not with CREBP1 or ATF3 while CREBP1 can heterodimerize with ATF3 but not with CREB (Hai et al., 1989; Hurst et al., 1991; Hoeffler et al., 1991).

It has also been demonstrated that some ATF/CREB factors are able to heterodimerize with Fos and Jun, again with a dimerization code. Heterodimerization seems to affect the stability of specific heterodimer/DNA complexes. Fos/Jun heterodimers bind with a greater affinity to a TRE (the element which is the nuclear target of the DAG/PKC pathway) than a CRE; however, heterodimers between Jun and the ATF3, ATF4 and CREBP1 bind more strongly to a CRE than a TRE (Hai and Curran, 1991). This property is likely to reside in the similarity between the CRE (TGACGTCA) and TRE (TGACTCA) sequences (Sassone-Corsi et al., 1990). This property of heterodimerization allows for more complexity in gene regulation by these factors. In addition, the possibility for targets of the cAMP/PKA pathway to interact with Fos and Jun provide an important route for different intracellular signals to be integrated at the level of transcriptional control (Deutsch et al., 1988; Masquilier and Sassone-Corsi, 1992; deGroot and Sassone-Corsi, 1992).

THE CREM GENE

The discovery of the CREM gene (cAMP- response element modulator) opened a new dimension in the study of transcriptional response to cAMP (Foulkes et al., 1991a). This is due to the remarkable organization of the CREM gene, which offers clues to the understanding of the generation of functional diversity in transcription factors.

The CREM gene was isolated from a mouse pituitary cDNA library screened at low stringency with oligonucleotides corresponding to the leucine zipper and basic region of CREB. The most striking feature of the CREM cDNA is the presence of two DNA-binding domains (Fig. 2). The first is complete and contains a leucine zipper and basic region very similar to CREB; the second is located in the 3' untranslated region of the gene, out of phase with the main coding region, and contains a half basic region and a leucine zipper more divergent from CREB. Four mRNA isoforms were identified that appear to be obtained by differential cell-specific splicing. Alternative usage of the two DNA binding domains was demonstrated in various tissues and cell types, where quite different patterns of expression were found (Foulkes et al., 1991a; 1992).

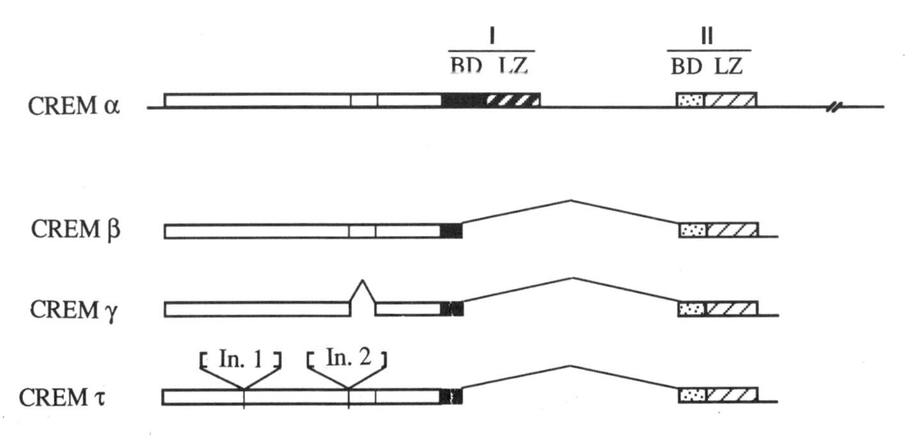

Figure 2. Structure of the CREM gene and of the various isoforms obtained by alternative splicing. Two alternative DNA-binding domains (I and II) can be used. While CREMα, β and γ act as antagonists of cAMP-induced transcription, CREMτ, in virtue of the presence of the two glutamine-rich domains, act as an activator. Detailed description is reported in the text and in Foulkes et al., 1991a; 1992.

CREM expression appears to be finely regulated, both transcriptionally and posttranscriptionally. In fact, not only cell- and tissue-specific expression is observed, but also the production of various isoforms. Three major isoforms were initially characterized, which revealed alternative usage of the two DNA binding domains (α and β isoforms, see Fig. 2), as well as a small deletion of 12 aminoacids (γ isoform). The potential for even more complexity of CREM regulation is hinted at by the possible usage of alternative poly

(A) addition sites and by the presence of ten AUUUA sequences in the 3' untranslated region, elements thought to be involved in mRNA instability (Shaw and Kamen, 1986). The strict cell- and tissue-specific expression of CREM is indicative of a pivotal function in the regulation of cell-specific cAMP response. This suggests that CREM occupies a central control point in the pituitary, since it is known that the physiology of this gland is finely regulated by a multiplicity of hormones whose coupled signal transduction pathways involve adenylyl cyclase. Interestingly, other well described examples of cell-specific splicing include the genes encoding neuronal peptides and hormones in brain and pituitary cells. Thus, it appears clear that cell-specific splicing is a crucial mechanism for CREM regulation, which modulates the DNA-binding specificity and activity of the final CREM products.

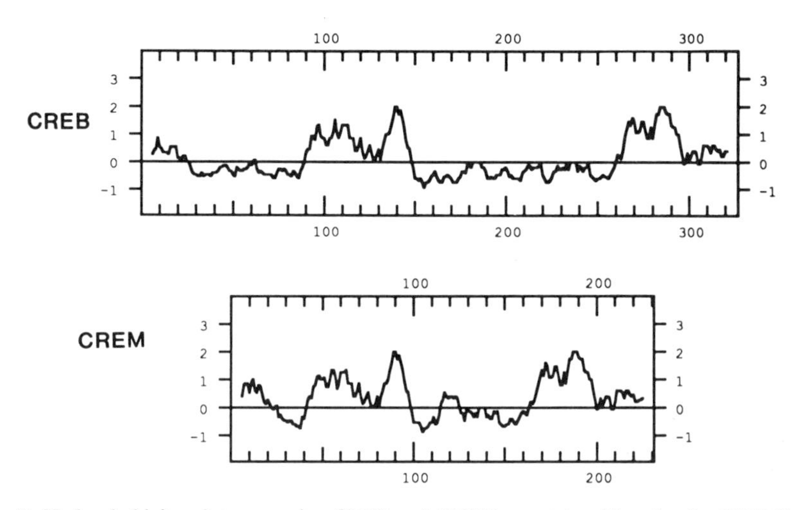

Figure 3. Hydrophobicity plot comparing CREB and CREMα proteins. The plot for CREMβ would be equivalent to CREMα due to the similarity in the DNA binding domains. The CREMα and β products are about 100aa shorter than CREB; the conservation of the DNA-binding domains and of the KID domain region is evident; the glutamine-rich regions flanking the KID domain are not present in CREMα and β, while they are inserted in CREMτ. Along the x-axis is measured the aminoacid position while along the y-axis is plotted the hydrophobicity index.

The CREM products share extensive homology with CREB, especially in the DNA-binding domains and the KID domain region. In Fig. 3 a hydrophobicity plot compares the profiles of CREB and CREMα; CREMβ has a very similar profile, since the difference with CREMa resides only in the DNA-binding domain. As is clear from Fig. 3, the two proteins have the same basic structure although strikingly CREMα and β are much smaller proteins. Indeed, sequence comparison indicates that the two glutamine-rich regions are absent in CREM (see Fig. 1), despite the perfect conservation of the KID domain.

The CREM proteins specifically bind to CREs with the same properties as CREB. This is not surprising, considering the homology between these proteins in the DNA-binding domains. CREM proteins containing either DNA-binding domain I or II, heterodimerize with CREB (Foulkes et al., 1991a), although it appears that CREMα-CREB heterodimer formation is more favoured than CREMβ-CREB (B. Laoide, unpublished observation). These properties suggest that CREM proteins might occupy CRE sites as CREM dimers or as CREM-CREB heterodimers, thus generating complexes with altered transcriptional functions. Infact CREMα and β products act by impairing CRE-mediated transcription, and as such are considered as antagonists of cAMP-induced expression (Foulkes et al., 1991a; 1991b). In transfection experiments, using CRE reporter plasmids, it was demonstrated that CREM proteins block the transcriptional activation obtained by the joint action of CREB and the catalytic subunit of the cAMP-dependent protein kinase A (Yamamoto et al., 1990). These observations strongly support the notion that CREMα and β proteins negatively modulate CRE promoter elements in vivo. An important question is how CREM proteins work. The two most likely models are as follows. According to the first scenario, CREM proteins dimerize and bind to CRE sites. Down-regulation is achieved by the occupation of these sites, which are unavailable for CREB. Similarly, if CREB is already bound, CREM proteins might squelch them because of their possible higher affinity for a specific site. According to the second model, CREM proteins are able to dimerize with CREB to generate non-functional heterodimers. Negative regulation is achieved by titrating active CREB molecules; CREMα and β proteins thereby act as activator traps. Since both CREM dimers and CREB-CREM heterodimers bind to CRE sites, both models are justified and both mechanisms may operate.

CREM: ANTAGONISTS AND ACTIVATOR FROM THE SAME GENE.

As well as antagonists, the CREM gene encodes an activator of cAMP-dependent transcription (Foulkes et al., 1992; Foulkes and Sassone-Corsi, 1992). In adult testis a novel CREM isoform (CREMτ) has been identified. CREMτ differs from the previously characterized antagonist CREMβ by the coordinate insertion of two glutamine-rich domains which confer transcriptional activation function to the protein. This is particularly interesting since mammalian spermatogenesis consist of a series of complex developmental processes. Germ cells arise directly from primitive ectoderm while testis somatic components originate from the mesodermal layers of the peritoneum.

The germ cells proliferate extensively, maturing into spermatogonia, which then proceed down the pathway of spermatogenesis. They enlarge and mature into primary spermatocytes, meiosis ensues and after the second reduction division, the haploid, round spermatids are transformed into spermatozoa by the process of spermiogenesis which involves an extensive biochemical and morphological restructuring. Transcriptional activation of several genes occurs during key stages of this differentiation process. Some genes are likely to play crucial roles during spermatogenesis, such as the oncogenes c-*mos*, c-*kit*, c-*abl*, *Wnt*-1 and N-*ras* ; the transcription factors *Hox*-1.4, Oct-4 and Zfp-35; heat shock protein *hsp*-70, inhibin, the peptide hormone POMC and protamines.

The multistep process which leads to the maturation of germ cells is under the control of the pituitary/hypothalamic axis. This flow of biochemical information is directly regulated by the adenylate cyclase signal transduction pathway. During spermatogenesis there is an abrupt switch in CREM expression (see Fig. 4). In premeiotic germ cells CREM is expressed at low levels in the antagonist α and β forms. Subsequently, from the pachytene spermatocyte stage onwards, a splicing event generates exclusively the CREMτ activator. This new transcript accumulates at extremely high levels. This splicing-dependent reversal in CREM function represents an important example of developmental

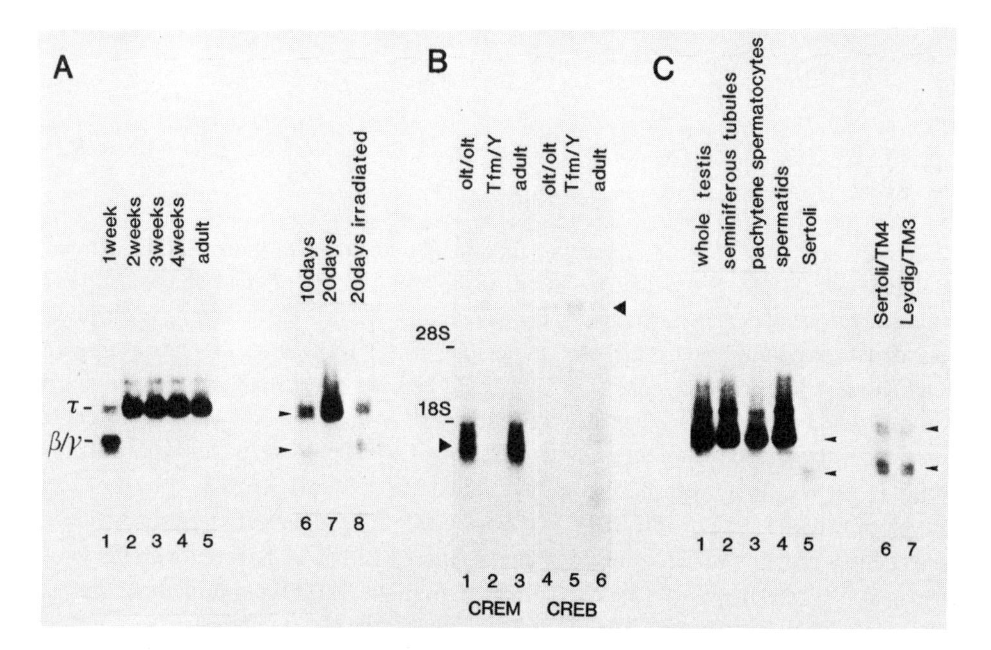

Figure 4 . A. PCR analysis of prepubertal testis CREM mRNA. Lanes 1 to 5: total RNA from the testis of 1, 2, 3 and 4 week old mice. In lane 5 is an adult testis positive control. Lanes 6 and 7 contain samples from normal animals of 10 and 20 days old, respectively. Lane 8 contains a sample from a 20 day old animal irradiated in-utero. **B.** Northern blot analysis of mutant mice testis RNA. Samples from adult mice mutants; oligotriche (olt/olt) (lanes 1 and 4) and testicular feminization (Tfm/Y) (lanes 2 and 5) and from a normal adult control (adult) (lanes 3 and 6) were included. **C.** CREM expression in RNAs from isolated testis cell types. Cells from disrupted testis were fractionated on BSA density gradient. Total RNA was prepared from pachytene spermatocytes, spermatids and Sertoli cells. In addition RNA was prepared from isolated seminiferous tubules and the mouse Sertoli and Leydig cell lines, TM3 and TM4.

modulation in gene expression. The fact that CREM is able to encode both activator and repressors by a dynamic splicing program (Foulkes and Sassone-Corsi, 1992), is further evidence that it occupies a pivotal position within the transcriptional response to cAMP.

ROLE OF CREM DURING SPERMATOGENESIS

The dramatic pattern of CREM expression begs the question as to the role of this abundant transactivator in testis, particularly when CREB is continued to be expressed at normal, low levels. The expression of testis-specific isoforms of the catalytic and regulatory subunits of PKA implies that in spermatogenic cells the cAMP signal transduction pathway needs to fulfill a specialized function and CREMτ may represent a nuclear target of this pathway. An alternative hypothesis could be that CREMτ is actually incorporated into the structure of the sperm head and plays a role after fertilization. Indeed it is established that the presence of a male pronucleus is essential for early zygote development. This would be consistent with the spermatozoon contributing transcription factors synthesised previously during spermiogenesis.

AN INTERNAL AUG IS USED TO GENERATE A NOVEL REPRESSOR, S-CREM

A 21kD CREM-immunoreactive protein is detectable in mouse brain extracts (Delmas et al., 1992). We wished to determine the origin of this protein. Since no CREM transcript has been detected to date which could encode a 21kD CRE-binding protein, we considered the possibility that the AUG at position aa152 in CREMτ, which constitutes a good Kozak consensus sequence (Fig. 5A) could be used alternatively to the canonical initiation AUG at position +1. The use of an alternative AUG in the CREM transcript could constitute a novel mechanism for generating this new CREM product, with possibly altered activity. To test this hypothesis we first translated CREMτ mRNA in an *in vitro* reticulocyte lysate system (Fig. 5B). Synthesis using this single transcript does indeed generate two protein products, one corresponding to CREMτ and the other with the same size as S-CREM (lanes 1, 4, 6). When mRNA from the CREB transcript was used no smaller product is visible (lane 3). It is worth noting that the CREB aminoacid sequence is homologous to CREMτ in the domain containing the putative AUG initiation codon, but the methionine in CREM is substituted by a valine residue in CREB. In addition, mRNA synthesized from S-CREM, generates a protein which comigrates with the smaller proteins produced with CREMτ mRNA (compare lanes 4 and 6). To unambiguously determine whether the use of the internal AUG could be responsible for the generation of S-CREM, we mutated the ATG sequence in the CREMτ cDNA to ATC, encoding an isoleucine instead of a methionine (CREMτ-ATC in Fig. 5A). RNA from CREMτ-ATC failed to generate the S-CREM protein (lane 5), demonstrating the requirement of the internal AUG for initiation.

Next, we determined whether internal initiation could occur *in vivo*, and thus explain the 21kD protein present in adult brain tissues. We transfected cultured cells with expression plasmids containing either wild-type CREMτ, CREMτ-ATC or the S-CREM coding sequences. The products were scored either by Western blot (Fig. 5C, lanes 1-5) or by immunoprecipitation of ^{35}S labeled cells (lanes 6-9). The results confirm that the internal AUG is responsible for the production of S-CREM *in vivo*. It is important to note that S-CREM shows the same size whether it is generated *in vitro*, in bacteria or in transfected cells, and that it comigrates exactly with the 21kD product from mouse brain (Fig. 5B, lanes 1, 4, 6; and compare Fig. 5C, lanes 1, 3-5). Taking into account all these

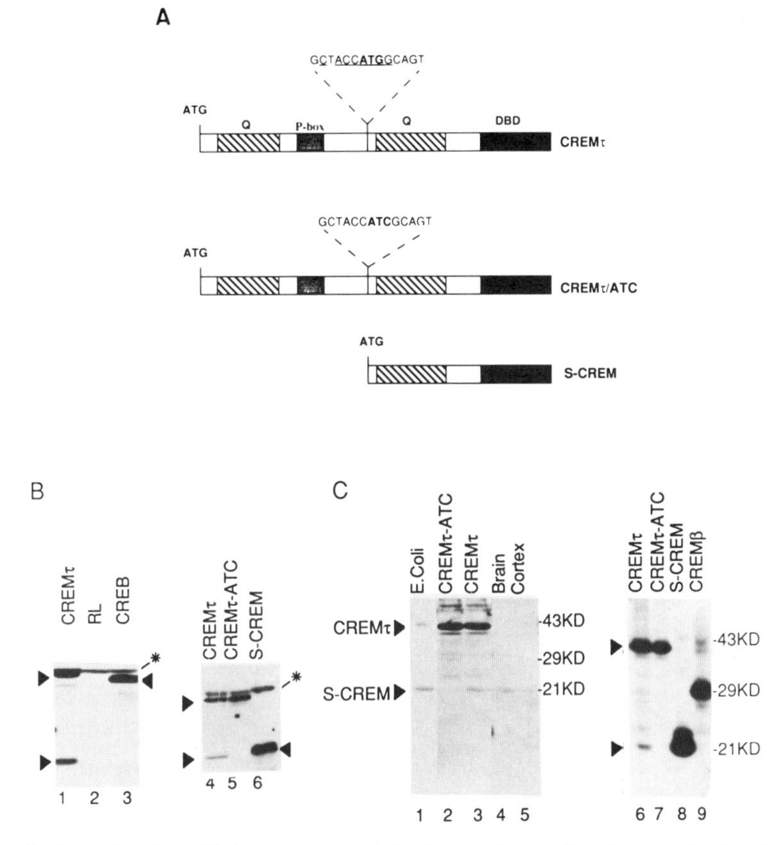

Figure 5. An internal AUG is used as an initiation codon in CREM mRNA. **A)** Schematic representation of the CREMτ-ATC mutant engineered by *in vitro* mutagenesis. The Kozak consensus is underlined. DBD: DNA-binding and dimerization domain; Q: Glutamine-rich domains; P-box: phosphorylation acceptor sites. **B)** *In vitro* generated CREM and CREB proteins. Lanes 1 and 4: Proteins generated using a CREMτ *in vitro* synthesized transcript. Lane 3: CREB; Lane 5: CREMτ-ATC; Lane 6: S-CREM; Lane 2: translation control (RL) with no specific mRNA. **C)** Analysis of *in vivo* synthesized CREM proteins. Lanes 2-5: Western blot analysis. Lanes 6-9: immunoprecipitation after ^{35}S labeling of the cells. Lanes 2, 3 and 6-9 carry proteins from transfected cultured cells; Lanes 4 and 5 show the 21kD protein from brain and cortex, respectively. In lane 1 there are bacterially produced CREMτ and S-CREM proteins as controls.

data we postulate that S-CREM and the 21kD brain protein are the same. S-CREM acts as a repressor of cAMP-induced transcription (Delmas et al., 1992).

CONCLUSIONS

The first cDNA clones which were characterised from the CREM gene (cAMP response element modulator) encode antagonists of cAMP-induced transcription (Foulkes et al., 1991a). The central role of splicing in the regulation of this gene was hinted at by the presence of two alternative DNA-binding domains which are used differentially in a cell-specific fashion. The CREM antagonists are able to bind to CRE (cAMP response element) sites as homodimers and also as heterodimers with the transcriptional activator CREB (cAMP response element binding protein). It is postulated that either by blocking the CRE site for CREB binding or by forming non-functional heterodimers, CREM exerts down-regulation. In cotransfection experiments, substoichiometric amounts of CREM cause significant downregulation of CRE-mediated activation (Foulkes et al., 1991a), thus similarly to mTFE3-S, this would imply that heterodimerization is a more plausible model. The CREM antagonists share extensive homology with CREB but they lack two glutamine-rich domains, which have been shown to be necessary for transcriptional activation. Now it is clear that the CREM gene also encodes an activator of transcription (Foulkes et al., 1992). In the adult testis, an isoform CREMτ has been identified which resembles in structure one of the antagonist forms (CREMβ) but includes two exons that encode two glutamine-rich domains. This form has been demonstrated to transactivate transcription from a CRE site. In adult testis, the CREMτ isoform is expressed alone and it constitutes an abundant species in late spermatocytes and spermatids. The CREM mRNA isoforms are a graphic illustration of how alternative splicing can modulate the function of a transcription factor in a tissue- and developmental-specific manner.

Acknowledgments

We wish to thank E. Borrelli, J.R. Naranjo, B. Mellström, P. Pévet, U. Schibler and S.G. McKnight for help, discussions and gifts of material. B.M.L. is supported by a fellowship from the Human Frontier Science Program; R.P. deG. by the Dutch Cancer Society; N.S.F. by the Association pour la Recherche contre le Cancer; C.A.M. by the INSERM; the work performed in our laboratory is supported by grants from CNRS, INSERM, ARC and Rhône-Poulenc Rorer.

REFERENCES

Auwerx, J., and Sassone-Corsi, P., 1991, IP-1: a dominant inhibitor of Fos/Jun whose activity is regulated by phosphorylation. *Cell* 64: 983.

Berkowitz, L.A., Riabowol, K.T., and Gilman, M.Z., 1989, Multiple sequence elements of a single functional class are required for cyclic AMP responsiveness of the mouse c-*fos* promoter. *Mol. Cell. Biol.* 9:4272.

Borrelli, E., Montmayeur, J.P., Foulkes, N.S., and Sassone-Corsi, P., 1992, Signal transduction and gene control: the cAMP pathway. *Critical Rev Oncogenesis* 3: 321.

Busch, S.J., and Sassone-Corsi, P., 1990, Dimers, leucine zippers and DNA binding domains. *Trends Genet.* 6: 36.

Comb, M., Birnberg, N.C., Seasholtz, A., Herbert, E., and Goodman, H.M., 1986, A cyclic-AMP and phorbol ester-inducible DNA element. *Nature* 323: 353.

deGroot, R.P., and Sassone-Corsi, P., 1992, Activation of Jun/AP-1 by protein kinase A. *Oncogene* 7:2281.

Delegeane, A., Ferland, L., and Mellon, P.L., 1987, Tissue specific enhancer of the human glycoprotein hormone α–subunit gene: dependence on cyclic AMP-inducible elements. *Mol. Cell. Biol.* 7: 3994.

Delmas, V., Laoide, B.M., Masquilier, D., de Groot, R.P., Foulkes, N.S., and Sassone-Corsi, P., 1992, Alternative usage of initiation codons in mRNA encoding the cAMP-responsive-element modulator (CREM) generates regulators with opposite functions. *Proc Natl Acad Sci USA* 89: 4226.

Deutsch, P.J., Hoeffler, J.P., Jameson, J.L. and Habener, J.F., 1988, Cyclic AMP and phorbol ester-stimulated transcription mediated by similar DNA elements that bind distinct proteins. *Proc. Natl. Acad. Sci. USA.* 85:7922.

Foulkes, N.S., Borrelli, E., Sassone-Corsi, P., 1991a, CREM gene: use of alternative DNA binding domains generates multiple antagonists of cAMP-induced transcription. *Cell* 64: 739.

Foulkes, N.S., Laoide, B.M., Schlotter, F., and Sassone-Corsi, P., 1991b, Transcriptional antagonist CREM down-regulates c-*fos* cAMP-induced expression. *Proc. Natl. Acad. Sci. USA* 88: 5448.

Foulkes, N.S., Mellström, B., Benusiglio, E., and Sassone-Corsi, P., 1992, Developmental switch of CREM function during spermatogenesis: from antagonist to transcriptional activator. *Nature* 355: 80.

Foulkes, N.S., and Sassone-Corsi, P., 1992, More is better: activators and repressors from the same gene. *Cell* 68: 411.

Gonzalez, G.A., Menzel, P., Leonard, J., Fischer, W.H., and Montminy, M.R., 1991, Characterization of motifs which are critical for activity of the cyclic AMP-responsive transcription factor CREB. *Mol. Cell. Biol.* 11: 1306.

Gonzalez, G.A., Yamamoto, K.K., Fischer, W.H., Karr, K., Menzel, P., Briggs, III W., Vale, W.W., and Montminy, M.R., 1989, A cluster of phosphorylation sites on the cAMP-regulated nuclear factor CREB predicted by its sequence. *Nature* 337: 749.

Gonzalez, G.A., and Montminy, M.R., 1989, Cyclic AMP stimulates somatostatin gene transcription by phosphorylation of CREB at ser 133. *Cell* 59: 675.

Habener, J., 1990, Cyclic AMP response element binding proteins: a cornucopia of transcription factors. *Mol. Endocrionol.* 4: 1087.

Hai, T.-Y., and Curran, T., 1991, Cross-family dimerization of transcription factors Fos:Jun and ATF/CREB alters DNA binding specificity. *Proc. Natl. Acad. Sci. USA* 88: 3720.

Hai, T.-Y., Liu, F., Coukos, W.J., and Green, M.R., 1989, Transcription factor ATF cDNA clones: an extensive family of leucine zipper proteins able to selectively form DNA binding heterodimers. *Genes & Dev.* 3: 2083.

Hoeffler, J.P., Meyer, T.E., Yun, Y., Jameson, J.L., and Habener, J.F., 1988, Cyclic AMP-responsive DNA-binding protein: structure based on a cloned placental cDNA. *Science* 242: 1430.

Hoeffler, J.P., Lustbader, J.W., and Chen, C.-Y., 1991, Identification of multiple nuclear factors that interact with cyclic adenosine 3' 5'-monophosphate response element-binding protein and activating transcription factor-2 by protein-protein interactions. *Mol. Endocrinol.* 5: 256.

Hurst, H.C., Totty, N.F., and Jones, N.C., 1991, Identification and functional characterization of the cellular activating transcription factor-43 (ATF-43) protein. *Nucl. Acids Res.* 19: 4601.

Ivashkiv, L.B., Liou, H.C., Kara, C.J., Lamph, W.W., Verma, I.M., and Glimcher, L.H., 1990, mXBP/CRE-BP2 and c-Jun form a complex which binds to the cAMP, but not to the 12-0-tetradecanoyl phorbol-13-acetate response element. *Mol. Cell. Biol.* 10: 1609.

Landschulz, W.H., Johnson, P.F., and McKnight, S.L., 1988, The leucine zipper; a hypothetical structure common to a new class of DNA binding proteins. *Science* 240: 1759.

Lee, K.A.W., Fink, S.J., Goodman, R.H., and Green, M.R., 1989, Distinguishable promoter elements are involved in transcriptional activation by E1A and cyclic AMP. *Mol. Cell. Biol.* 9: 4390.

Lillie, J.W., and Green, M.R., 1989, Transcription activation by the adenovirus E1a protein. *Nature* 338: 39.

Liou, H.-C., Boothby, M.R., Finn, P.W., Davidon, R., Nabavi, N., Zeleznik-Le, N.J., Ting, J.P.-Y., and Glimcher, L.H., 1990, A new member of the leucine zipper class of proteins that binds to the HLA DRa promoter. *Science* 247: 1581.

Liu, F., and Green, M.R., 1990, A specific member of the ATF transcription factor family can mediate transcription activation by the adenovirus E1a protein. *Cell* 61: 1217.

Maekawa, T., Matsuda, S., Fujisawa, J.-I., Yoshida, M., and Ishii, S., 1991, Cyclic AMP response element-binding protein, CRE-BP1, mediates the E1A-induced but not the *tax*-induced trans-activation. *Oncogene* 6: 627.

Maekawa, T., Sakura, H., Kanei-Ishii, C., Sudo, T., Yoshimura, T., Fujisawa, J., Yoshida, M., and Ishii, S., 1989, Leucine zipper structure of the protein CRE-BP1 binding to the cyclic AMP response element in brain. *EMBO J.* 8: 2023.

Masquilier, D., and Sassone-Corsi, P., 1992, Transcriptional cross-talk: nuclear factors CREM and CREB bind to AP-1 sites and inhibit activation by Jun. *J. Biol. Chem.* (in press).

Merino, A., Buckbinder, L., Mermelstein, F.H., and Reinberg, D., 1989, Phosphorylation of cellular proteins regulates their binding to the cAMP-response element. *J. Biol. Chem.* 264: 21266.

Nishizuka, Y., 1986, Studies and perspectives of protein kinase C. *Science* 233: 305.

Roesler, W.J., Vanderbark, G.R., and Hanson, R.W., 1988, Cyclic AMP and the induction of eukaryotic gene expression. *J. Biol. Chem.* 263: 9063.

Sassone-Corsi, P., Visvader, J., Ferland, L., Mellon, P.L. and Verma, I.M., 1988, Induction of proto-oncogene *fos* transcription through the adenylate cyclase pathway: characterization of a cAMP-responsive element. *Genes & Dev.* 2: 1529.

Sassone-Corsi, P., 1988, Cyclic AMP induction of early adenovirus promoters involves sequences required for E1A-transactivation. *Proc. Natl. Acad. Sci. USA* 85: 7192.

Sassone-Corsi, P., Ransone, L.J., and Verma, I.M., 1990, Cross-talk in signal transduction: TPA-inducible factor Jun/AP-1 activates cAMP-responsive enhancer elements. *Oncogene* 5:427.

Shaw, G., and Kamen, R., 1986, A conserved AU sequence from the 3' untranslated region of GM-CSF mRNA mediates selective mRNA degradation. *Cell* 46:659.

Yamamoto, K.K., Gonzales, G.A., Briggs, III W.H., and Montminy, M.R., 1988, Phosphorylation-induced binding and transcriptional efficiency of nuclear factor CREB. *Nature* 334: 494.

Yamamoto, K.K., Gonzales, G.A., Menzel, P., Rivier, J., and Montminy, M.R. 1990, Characterisation of a bipartite activator domain in transcription factor CREB. *Cell* 60: 611.

THREE-DIMENSIONAL STRUCTURE OF THE HUMAN RETINOIC ACID RECEPTOR-β DNA BINDING DOMAIN: IMPLICATIONS FOR DNA-BINDING

Ronald M. A. Knegtel[1], Masato Katahira[1], Johannes G. Schilthuis[2,3], Rolf Boelens[1], Douglas Eib[1], Paul T. van der Saag[2] and Robert Kaptein[1]

[1]Department of Chemistry, University of Utrecht, Padualaan 8, 3584 CH Utrecht, The Netherlands
[2]Hubrecht Laboratory, Netherlands Institute for Developmental Biology Uppsalalaan 8, 3584 CT, Utrecht, The Netherlands
[3]Present address: Ludwig Institute for Cancer Research, Courtauld Building, 91 Riding House Street, London W1P8BT, UK

INTRODUCTION

All-trans-retinoic acid (RA), a vitamin A derivative, plays a crucial role in vertebrate development and differentiation (Thaller & Eichele, 1987). Retinoic acid acts through binding to nuclear retinoic acid receptors (RARs). RARs are members of a superfamily of ligand-inducible nuclear transcription factors which comprises receptors for steroid and thyroid hormones and vitamin D3 (Evans, 1988). Until now three different RAR genes have been identified, RARα, β, and γ (Petkovich et al., 1987, Giguere et al., 1987, de Thé et al., 1987, Brand et al., 1988, Krust et al.. 1989). These proteins are structurally organized in separate domains, labelled A through F (cf. Figure 1). The C domain is the highly conserved DNA-binding domain (DBD) of about 70 amino acid residues and the E domain of about 230 amino acid residues is responsible for ligand binding (Beato, 1989). The C domain recognizes response elements, upstream of receptor target genes. It contains nine conserved cysteine residues, eight of which coordinate two zinc atoms (Beato, 1989). The presence of a zinc binding domain is reminiscent of the "zinc finger" motif found in *Xenopus* Transcription Factor IIIA (TFIIIA) (Miller et al., 1985) as well as similar domains found in retroviral DNA binding proteins (Green & Berg, 1989) but the DBD of these proteins is structurally different from that of the nuclear hormone receptors (Berg, 1989,

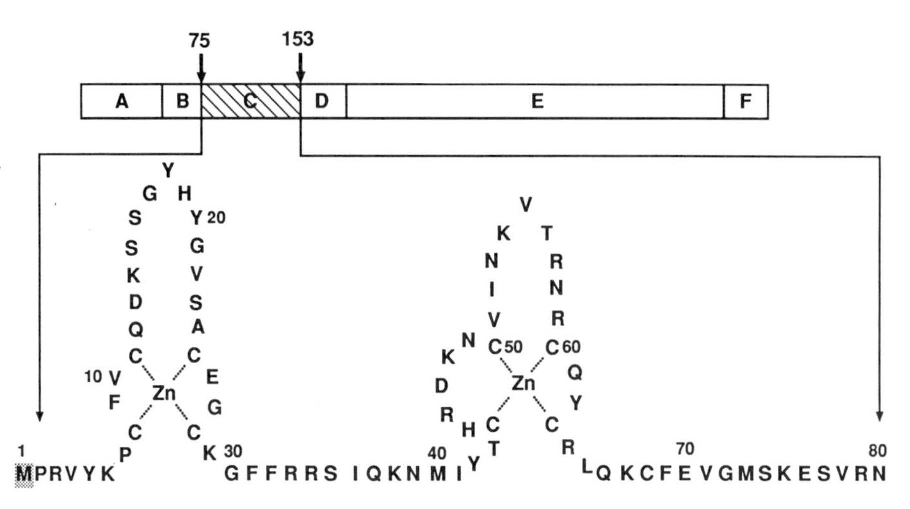

Figure 1. Sequence of the human retinoic acid receptor β DNA binding domain and its position in the complete receptor sequence. Arrows indicate the sites where restriction sites were introduced, as well as translation start and stop codons. The 79 residue C-region or DNA binding domain of the molecule has been indicated by shading. The zinc coordinating cysteins are indicated. The first Met residue is not part of the wild-type DNA binding domain sequence.

Härd et al., 1990, Luisi et al., 1991). The high degree of amino acid sequence homology observed between the RAR DBD and that of other superfamily members (Benbrook et al., 1988) suggests a tertiary fold similar to that found in both the glucocorticoid (Härd et al., 1990) and the estrogen receptors (Schwabe et al., 1990) for which the solution structures have been determined by NMR. Recently, the structure of the GR DBD bound to its response element (RE) as a dimer has been elucidated by X-ray crystallography (Luisi et al., 1991). The mode of binding of the GR DBDs, as seen in the crystal structure of the complex, confirmed models that were proposed for the GR and ER on the basis of their solution structures and biochemical data (Härd et al., 1990, Schwabe et al., 1990). The only difference was an additional short stretch of distorted α-helix observed in the second finger, which appears to be rather flexible in the protein free in solution.

Within the superfamily of nuclear hormone receptors a distinction can be made between two classes which recognize different types of response elements. The GR and ER bind only to inverted repeats of their respective response elements with a 3 bp spacing in between. The TRs and RARs, however, are able to bind both to inverted repeats (with no spacer) and direct repeats (with a 4 or 5 bp spacing, respectively) of their consensus response elements. Although different spacer requirements are found, the consensus sequences for the GRE and ERE are similar to those of the RARE and TRE, which suggests a similar mode of binding to the DNA, involving residues located near the two C-terminal cysteines in the first zinc finger (Mader et al., 1989, Umesono & Evans, 1989, Danielsen et al., 1989, Luisi et al., 1991). Recently it was shown that both the RAR and the TR preferably bind to their response elements as heterodimers with retinoid X receptors (RXR) (Yu et al. 1990, Zhang et al., 1992, Kliewer et al. 1992, Leid et al. 1992), in contrast to the GR and ER which form homodimers upon binding.

Structural information on the human RAR-β DBD might give more insight in the preference of the RARs for inverted repeats of the consensus response element with zero bp spacing and direct repeats with a five bp spacing. Possibly, the binding properties of RARs could be explained by protein-protein interactions between different DBDs as was observed for the GR DBD (Härd et al., 1990, Luisi et al., 1991). In order to elucidate the solution structure of the RAR DBD several NMR experiments were performed and both nitrogen-15 and proton resonance assignments for all but the first three residues were obtained and are reported elsewhere (Katahira et al., 1992). Here we report on the determination of the solution structure by NMR of a 80 residue protein fragment containing the human RAR-β DNA binding domain (Figure 1) and its DNA binding properties. The consequences for the interaction of the RAR with its response elements are discussed.

RESULTS AND DISCUSSION

Molecular Cloning and DNA Binding Properties of the hRAR-β DBD

In order to obtain sufficient amounts of pure hRAR-β DBD to allow structure determination by NMR, a recombinant vector (pET3) coding for residues 75 to 153 of the hRAR-β was constructed. The encoded polypeptide comprised the 79 residue DNA binding domain C of the hRAR-β with an additional N-terminal methionine. This region contains two putative zinc fingers homologous to other nuclear hormone receptors, as is illustrated in Figure 1. The polypeptide was overexpressed in *Escherichia coli* using the T7 expression system (Studier & Moffat, 1986) and subsequently isolated to 95 % purity (results not shown).

To demonstrate that the bacterially expressed RAR-DBD peptide still contained the sequence specific DNA binding properties of its parent molecule, purified RAR DBD was used in a DNase I protection assay. Incubation of a ^{32}P labelled DNA fragment that contains a RA responsive element from the hRARβ gene (RARE, De Thé *et al.*, 1990) with the RAR-DBD peptide results in protection of the sequence spanning the RARE against subsequent digestion by DNase I. The results of such an experiment are shown in Figure 2A for both the upper and the lower strand of the labelled probe. The protected region includes the direct repeat that makes up the RARE. To further show that this protection was the result of sequence specific DNA binding, two other probes were used in similar experiments. The first, designated RARE-m1 contains a RARE that has been mutated in one of the two half sites (see Figure 2B for its exact sequence). This change results in complete disappearance of protection specifically around the mutated half site. Mutation of both half sites completely abolishes protection against DNase I attack by RAR-DBD, as shown by the results of a protection assay using the probe RARE-m2. These results clearly show that protection against DNase I digestion is the result of sequence specific binding of the RAR-DBD peptide to its cognate sequence.

Although the region of the non-mutated probe protected against DNase I attack completely spans the RARE sequence, there are two short stretches on both strands within this region that are not protected or only partly protected. These are indicated in Figure 2B where the protected area is mapped on the probe's sequence. On the map of the m1 probe a corresponding gap can be seen in the protection of the non-mutated half site. This feature of the DNase I protection assays can be understood if it realized that DNase I interacts with the minor groove of DNA (Suck & Oefner, 1986). In the case of a single half site or two separated half sites the DNase I molecule could attack the DNA minor groove opposite to the RAR DBD binding site.

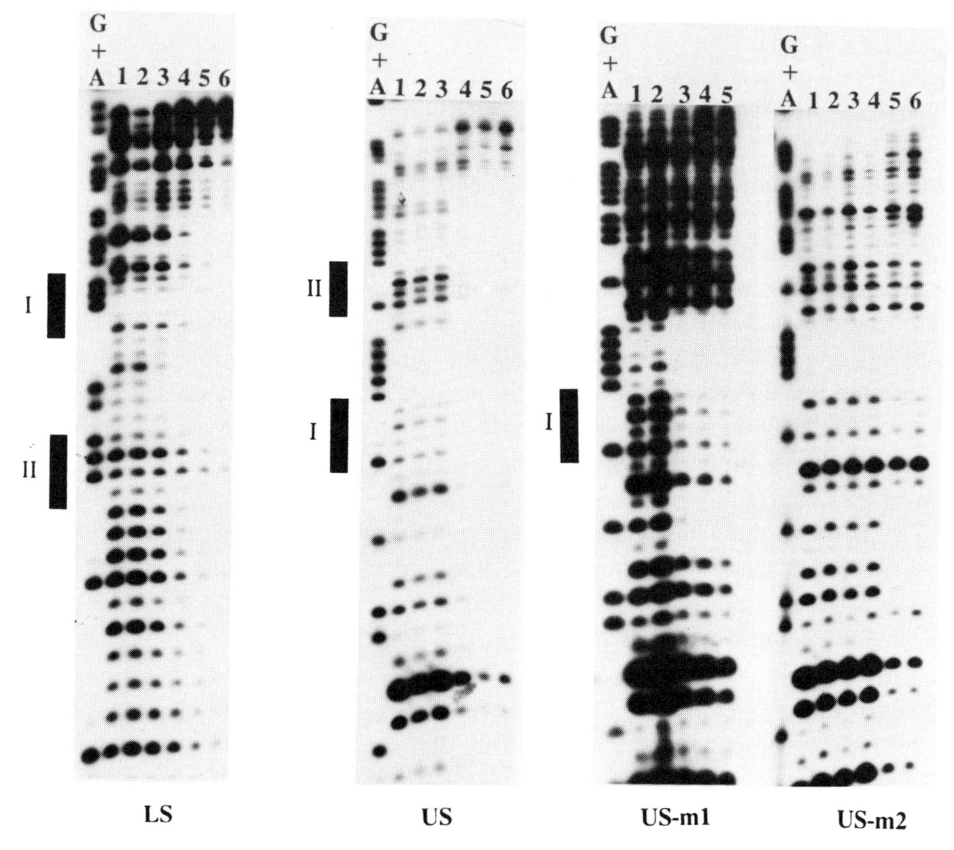

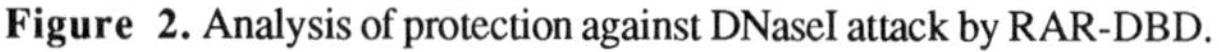

Figure 2. Analysis of protection against DNaseI attack by RAR-DBD.

A. Footprint assays using probes containing RAR binding sites. Synthetic dsDNA fragments were used as probes in a DNaseI protection assay, radioactively labelled on the upper (US) or lower (LS) strand as indicated over the various experiments shown. Names of probes correspond to the sequences given below in panel B. Lanes marked G+A contain a purine specific sequence reaction (Maxam & Gilbert, 1980) performed on the same DNA fragment that was used in the protection assay as sequence specific size markers. Lanes 1 and 2 contain labelled probe in binding mix with no added RAR-DBD protein, digested with 1/2 or 2 units of DNaseI, respectively. Lanes 3-6 contain probes, bound to an increasing amount of RAR-DBD protein, digested with 2 U of enzyme. Lanes 3: 50 ng; lanes 4: 100 ng; lanes 5: 150 ng and lanes 6: 200 ng of protein.

As the purified hRAR-β DBD proved to be functionally intact, we proceeded with recording nuclear magnetic resonance (NMR) spectra. Sequence specific assignments of nitrogen-15 and proton resonances for all but the first three N-terminal residues were obtained (Katahira et al., 1992). The proton resonance assignment allowed us to assign medium and long range NOEs which are essential for tertiary structure determination.

The Three-Dimensional Structure of the h-RARβ DBD

From the 2D NOE spectra recorded in ^1H_2O with mixing times of 50, 100, 150 and 200 ms a total of 2202 NOE crosspeaks was obtained, 1966 of which could be assigned unambiguously. The integrated crosspeaks were translated into distance restraints by determining the initial rates of the buildup of NOE intensities. To all distances an error of 10 % and appropriate pseudo atom corrections (Wüthrich, 1986) were added. Fifteen NOEs for which no good build up curves could be obtained, due to overlap or noise, were assumed to represent distances of ≤ 5 Å. A total of 989 distance restraints was obtained, derived from 335 intra-residue, 265 sequential , 165 medium range and 224 long range NOEs. Lower bounds were set to 2.0 Å. All valines except Val 4 and Val 71, which have degenerate γ-methyl resonances, could be assigned stereospecifically. On basis of the covalent structure, 794 distance restraints were found to be non-redundant. All NOEs are summarized in Figure 3 where the α-helices can be recognized by the presence of i,i+3 and i,i+4 NOEs parallel to the diagonal. Off-diagonal intensity already indicates that there are many contacts between two halves of the protein, which often involve hydrophobic residues. The two zinc-fingers are in contact through residues near Cys 11 and Ile 52. As shown in Figure 3 most long range NOEs were found between hydrophobic residues in or

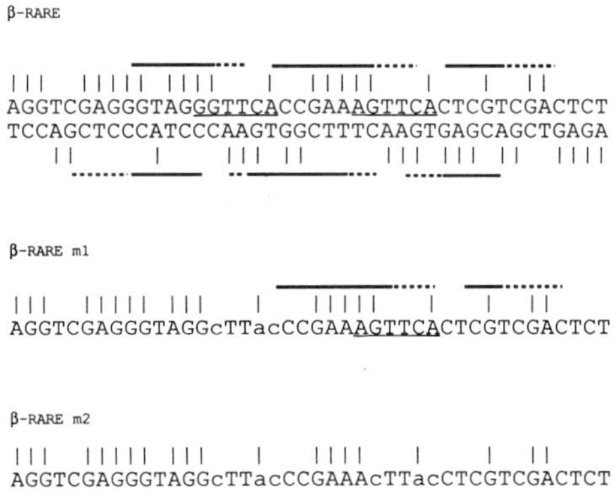

β-RARE

```
AGGTCGAGGGTAGGGTTCACCGAAAGTTCACTCGTCGACTCT
TCCAGCTCCCATCCCAAGTGGCTTTCAAGTGAGCAGCTGAGA
```

β-RARE m1

```
AGGTCGAGGGTAGGcTTacCCGAAAGTTCACTCGTCGACTCT
```

β-RARE m2

```
AGGTCGAGGGTAGGcTTacCCGAAAcTTacCTCGTCGACTCT
```

B. Mapping of protected areas on the sequences of the probes. The protection found in the footprinting experiments shown in A are mapped on the nucleotide sequences of the probes that were used. The purine residues in the labelled strands are indicated by vertical dashes over the sequence to facilitate easy reference to panel A. The RARE half sites are underlined; mutations in this sequence that abolish protein binding are indicated by using smallcaps instead of capitals in the sequence lettering. The protected areas are indicated by horizontal lines. Dotted lines are used wherever the exact ending of the protection could not be established.

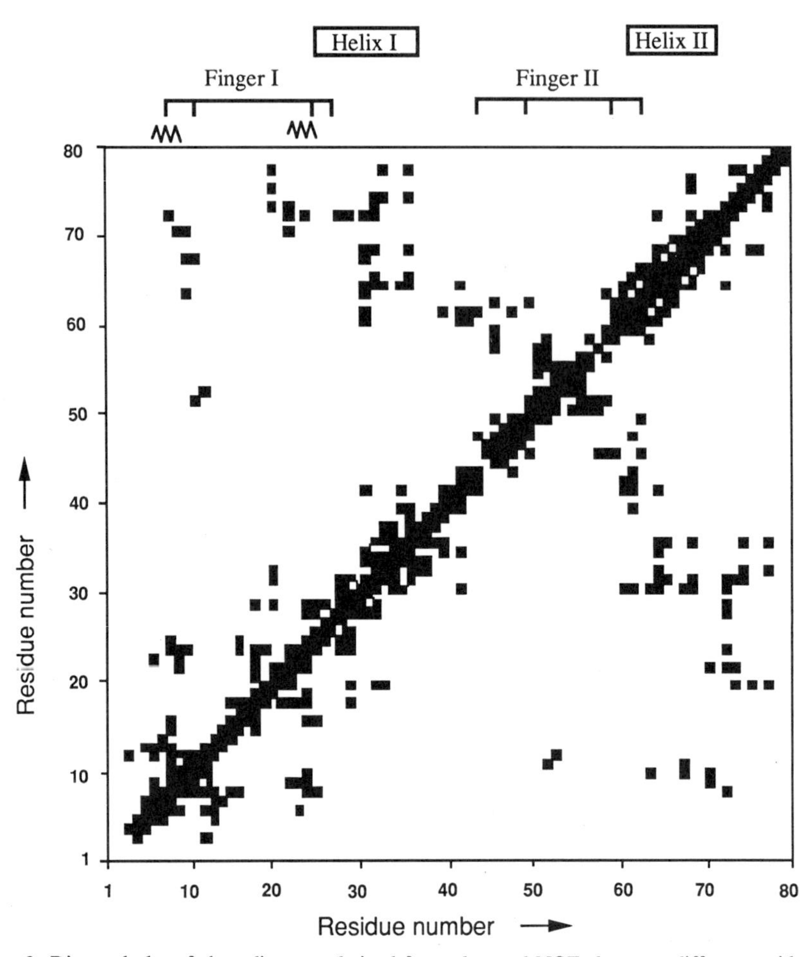

Figure 3. Diagonal plot of short distances derived from observed NOEs between different residues in the human retinoic acid receptor β DNA binding domain. Secondary structure elements and finger regions have been indicated.

close to the two α-helices and the β-sheet, whereas both zinc fingers show relatively few long range NOEs.

The NOE data were supplemented with hydrogen bond constraints from the previously identified helices (residues 27-37 and 61-70) and β-sheet regions (residues 7-9 and 23-25) (Katahira et al., 1992), restraining the distances between the carbonyl oxygen of residue (i) and the amide proton of residue (i+4) within a range of 1.85-2.30 Å. On the basis of homology with the GR DBD twelve sulphur-sulphur constraints on cysteines 8, 11, 25 and 28 in the first finger and 44, 50, 60 and 63 in the second finger were used, assuming a tetrahedral surrounding of the zinc and a 2.35 Å Zn-S bond distance. Structures generated without these constraints had similar zinc coordination and α-helical regions though less well determined backbone conformations.

With these distance constraints a family of 20 structures was generated using a metric matrix distance geometry algorithm and optimized using a conjugent gradient method against distances and chiralities. One structure was discarded because of wrong chiralities on the C-atoms, the other 19 were used as input for a distance-bound driven dynamics (DDD) (Scheek & Kaptein, 1988) calculation to obtain a better sampling of the allowed conformational space. After DDD, on average, five distance restraints are violated more than 0.2 Å with a maximum violation of 0.34 Å. Figure 4 shows the backbone trace of the 19 structures, superimposed on the helices, after DDD. From Figure 4 it is clear that the α-helical core of the protein is well determined with a backbone r.m.s.d. of 0.72 Å before and 0.79 Å after DDD. The r.m.s.d. for the backbone, excluding residues 1 through 4 for which no NOEs where observed, was 1.52 Å before and 1.93 Å after DDD. For the first finger (residues 8-28) the r.m.s.d. of the backbone was 1.08 Å before and 1.21 Å after DDD. For the second finger (residues 44-62) the r.m.s.d.'s were 1.66 Å and 1.85 Å, respectively. The fact that the second finger is less well determined by the distance restraints could imply that this region is more flexible in solution.

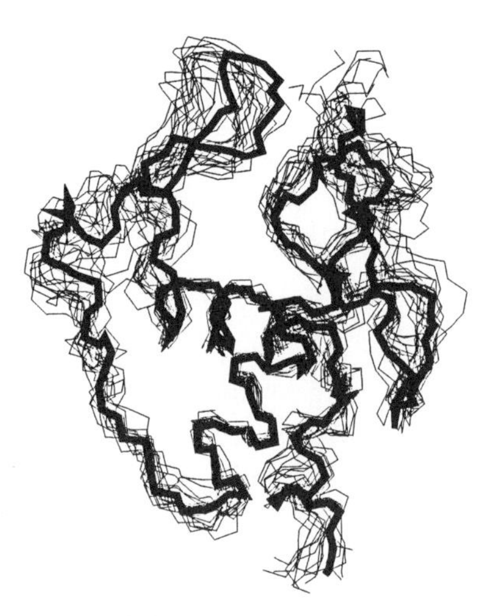

Figure 4. Polypeptide backbone traces of 19 distance geometry structures of the retinoic acid receptor DNA binding domain. The structures were obtained by distance geometry followed by distance-bound driven dynamics. Superpositioning was performed only on the helices. The structure with the lowest error function is drawn in thick line.

Figure 5 shows a more detailed view of the human RAR-β DBD. The two α-helices are perpendicular to each other, exposing hydrophilic residues to the solvent. They are connected to each other through a hydrophobic core made up of the aromatic and non-polar residues located in the two helices. Also near the short β-sheet region NOE contacts between hydrophobic residues are observed involving residues at opposite sides of the β-sheet and directly following the second helix. The two zinc atoms in the N- and C-terminal fingers have an S and R chirality, respectively, which was also found for the GR in the crystal (Luisi et al., 1991).

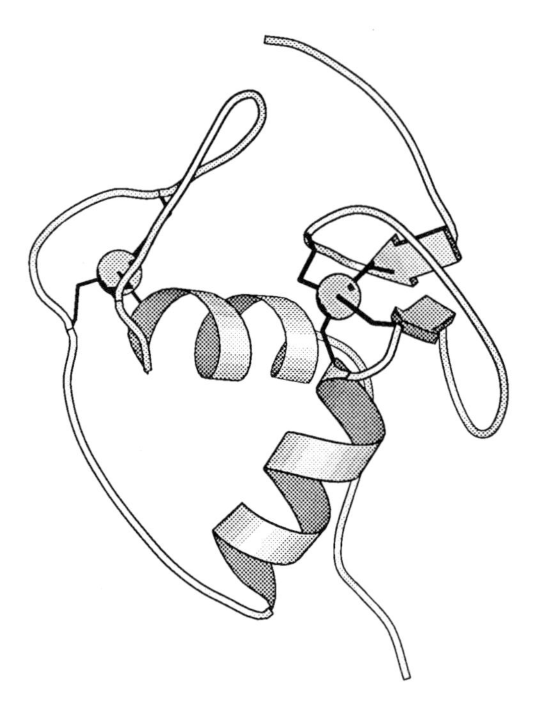

Figure 5. The best distance geometry structure of the retinoic acid receptor DNA binding domain. Both zinc ions, the two α-helices and the short strand of β-sheet are indicated schematically.

The availability of the tertiary structure of the RAR DBD allows us to explain some striking features observed in the NMR spectra. In 2D NOE spectra recorded in ^1H_2O a slowly exchanging hydroxyl proton was observed, that could be identified as the Thr 56 γ-hydroxyl group via a HOHAHA connectivity to the γ-methyl group of the same residue. Thr 56 is located in the second finger, which is not well defined by NOE restraints. This makes the identification of the hydrogen bond acceptor difficult. Both the Ile 52 and Asn 53 backbone carbonyl oxygens as well as the Asn 53 side-chain carbonyl oxygen could act as proton acceptors. It is interesting that a stable hydrogen bond is formed in a region that exhibits no regular secondary structure and appears to be disordered in the 19 DG structures.

The NH proton of Ser 23 is found to be extremely shifted to high field in the NMR spectra. This can be explained from the structure, where the Ser 23 amide proton is found above the plane of the aromatic ring of Phe 9 at a distance of roughly 2-3 Å. Also the aromatic ring of Tyr 18 is near, and its plane is oriented towards Ser 23. The orientation of the Phe 9 sidechain could well induce a ring current shift which causes the unusual shift of 5.28 ppm.

When the RAR DBD is compared to the solution and crystal structures of the GR DBD (Härd et al., 1990, Luisi et al., 1991) some differences are observed. In the RAR solution structure, the second helix, running from residue Gln 61 to Glu 70, is two residues shorter than in the GR (Härd et al., 1990, Katahira et al., 1992). In contrast to the GR, the ^1H resonances of the C-terminal extended chain directly following the second helix, could be assigned completely in the case of the RAR. Together with the observation of long range NOEs from Val 78 to residues in the first helix, this suggests that the C-terminus is folded against the protein surface and not directed into solution. This folding of the C-terminus of the RAR could account for the second α-helix being two residues shorter than that observed in the GR. Furthermore, in the Thr 43-Arg 49 segment of the RAR solution structure, two turns are observed which are approximately perpendicular. A similar folding of this region, which is responsible for protein-protein interaction in the complex, is also observed in the GR crystal structure (residues C476-C482) but in the GR solution structure it appears to be less structured with both turns orientated differently with respect to each other.

Overall, however, the RAR and GR DBD solution structures are highly similar. When the solution structures of the GR and RAR are superimposed on the helices, the r.m.s.d. of the N, C_α, C backbones of the helices is 0.7 Å. When the RAR is compared with the GR X-ray structure, however, the helix backbone r.m.s.d. is 1.3 Å. The difference in the relative orientation of the two helices could be explained by a change in the conformation of the protein on binding to the DNA, which involves the second helix directly, or a long range influence of the strong protein-protein contacts which occur during dimerization. The largest differences in conformation are observed in the second finger that contacts the DNA in the GR crystal structure. Here the overall fold of the finger in the RAR seems to be more similar to that in the GR crystal structure than to that in the GR solution structure but this region is not accurately determined in the NMR structures due to lack of NOEs. There is no short distorted α-helix present in the second finger of the RAR DBD, which was found in the GR DBD X-ray structure.

DNA-Binding

We shall now examine the consequences for DNA-binding by the RAR that can be deduced from the tertiary structure of the hRAR-β DBD. Since RAR and TR recognize very similar half sites in direct or inverted repeats, the TR will also be included in the discussion. According to recent findings the RAR and TR are thought to bind their response elements as heterodimers with RXRs. However, due to the 55-60% amino acid sequence homology and expected structural homology between the RAR and RXR DNA binding domains, the following discussion should be valid for homodimer as well as heterodimer DNA binding. For the RARs no biochemical data are available on amino-acid residues which are involved in specific recognition of the RA response element (RARE). Since the RAR DBD, however, is strongly homologous to that of the GR and ER it probably binds to DNA in a similar way. The DNA sequences recognized by RARs and TRs differ in the relative orientation and spacing in between of the two consensus half sites from those recognized by the GR and ER. The GRE and ERE consist of two palindromic DNA sequences of six basepairs which are separated by three arbitrary basepairs (Umesono & Evans, 1989) and two DBDs bind cooperatively with strong protein-protein interactions. For the RAR and the TR, two different types of response elements have been identified: an inverted repeat of the six basepair consensus half site without spacer and a direct repeat with a four (in case of the TR) or five (in case of the RAR) basepair spacing (Umesono & Evans, 1989, Umesono et al., 1991). The consensus sequences for the RAR and TR are AGGTCA and AGTTCA, respectively. For the RAR the sequence AGGTCA is functional in both inverted and direct repeats while the AGTTCA sequence appears to be only functional when arranged in a 5 bp spacing direct repeat (Vivanco Ruiz et al. 1991). The direct repeat TRE can be changed into a RARE by inserting a single basepair between the two half sites, since both RAR and TR recognize AGGTCA sequences. It appears, however, that the exact spacing between the direct repeats is less critical for TR then for RAR (Vivanco Ruiz et al. 1991). Palindromic TREs are able to bind both RAR and TR (Umesono et al., 1991). Furthermore, the TR is known to be unable to bind as a dimer to reponse elements existing of inverted repeats with a three basepair spacing similar to the GRE and ERE (B. Vennström, personal communication). The data therefore suggest that there are distinct differences between the protein-protein contacts in RAR and TR dimers complexed with DNA and those observed in the GR-DBD dimer (Luisi et al., 1991).

To examine the possibilities for protein-protein interactions between two RAR DBD monomers, which might determine the preference for the inverted and direct repeat RAREs, we modelled them bound to response elements with various spacings and directionalities. The RAR DBD is positioned in the major groove of the DNA similar to the GR DBD in the crystal structure (Luisi et al., 1991) while we assumed that the same conserved residues near the two C-terminal cysteines are involved in specific DNA recognition. In all cases a regular B-DNA conformation was assumed. In our model the 9-residue loop (51-59) in the second zinc finger is in close proximity to the phosphate backbone of the DNA with especially the conserved Arg 57 residue. A similar fold was observed in the crystalized GR-GRE complex (Luisi et al., 1991).

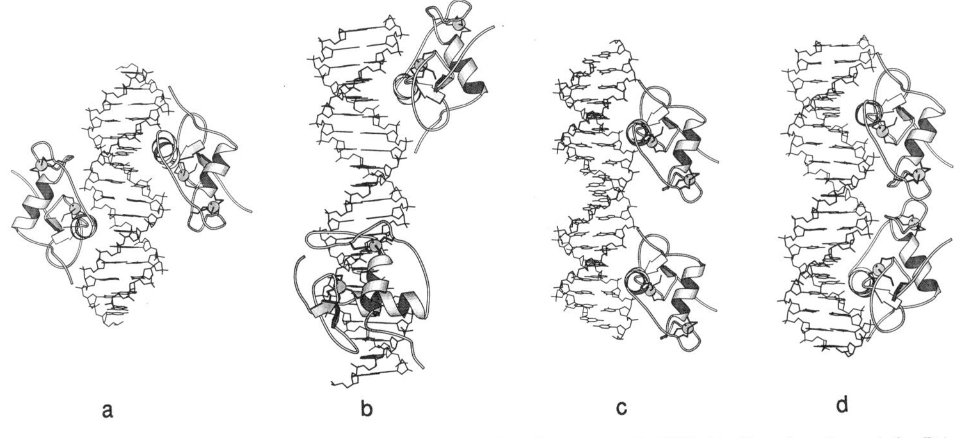

Figure 6. Models of dimeric complexes of the retinoic acid receptor-β DNA binding domain and the RA response element (RARE). (a) the zero basepair spacing inverted repeat RARE, (b) the five basepair spacing direct repeat RARE, (c) a hypothetical three basepair spacing direct repeat RARE and (d) a hypothetical three basepair spacing inverted repeat RARE.

Figure 6A shows a model for the complex of two hRAR-β DBDs and the inverted repeat TRE. Without spacer basepairs, the DNA makes roughly half a turn between the centers of the two half sites, thus placing both RAR DBD monomers on opposite sides of the DNA. In this configuration they are too far apart to make protein-protein interactions possible, unlike the GR DBD which is known to exhibit strong protein-protein interactions in the complex (Luisi et al., 1991). Our findings are in agreement with data for the intact TR which was found to bind with negative cooperativity to the palindromic TRE (B. Vennström, personal communication). In case of the inverted repeat TRE inter-monomer contacts might involve the hormone binding domain which was found to be crucial for TR/RAR heterodimer formation and DNA binding (Glass et al., 1989).

Also in the case of a direct repeat RARE with a five bp spacing, both DBD monomers are too far apart to allow for any protein-protein contacts between them, as is shown in Figure 6B. Based on this model no cooperativity of binding of RAR DBD to the direct repeat RARE is expected. Regarding the geometries of the different RAR/RARE complex models it can be deduced that the preference of the RAR for dimeric binding to REs with zero or five bp spacings does not reside in the DNA binding domain but is either due to other parts of the protein or to interactions with additional proteins of the transcriptional machinery which may interact with the RAR (Glass et al., 1989, Yang et al., 1991).

Figure 6C and 6D show the RAR DBD modelled on hypothetical RAREs with a three basepair spacer in a direct and an indirect repeat. The importance of interactions between other parts of the complete RAR proteins can also be deduced from Figure 6C where we modelled two RAR DBDs on a direct repeat RARE with a three bp spacer. Since the DBD extends more in one direction than the other, the monomers still appear not to be in contact. Thus, such a mode of binding for two DBDs could be a possibility although this DNA sequence is not functional as a RARE (Umesono et al., 1991). As shown in Figure 6D the failure of the TR (and possibly the RAR) to bind to inverted repeats of the consensus RE with a three basepair spacing (as in the GRE and ERE) may be due to unfavourable steric interactions in the D-box region and other parts of the second zinc-finger. By contrast, in the GR and ER the amino sequence in this region allows a good complementarity between the monomers leading to a strong cooperativity in DNA binding.

CONCLUSIONS

The 79 residue DNA binding C domain of the human retinoic acid receptor β is still functionally intact as was shown by DNase I protection assays. Overexpression of the RAR DBD in *E. coli* made determination of the solution structure by NMR possible. The solution structure of the hRAR-β DBD is highly similar to those of the GR and ER DBDs, with two perpendicular α-helices at the C-termini of both zinc fingers and a short stretch of β-sheet. Some differences are observed between the GR DBD and the RAR DBD. The second helix in the RAR DBD is two residues shorter and no helical structure was observed in the second zinc finger as was in the case of the GR DBD crystal structure. The second zinc finger, which is less well defined indicating flexibility in this region, appears to be folded similar as to what is observed in the ER solution structure and the GR-GRE crystal structure where positively charged residues contact the phosphate backbone.

The availability of structural data on the RAR DNA binding domain provides more insight into the factors determining the recognition of different response elements by nuclear hormone receptors. In contrast to the GR DBD, the specificity of the RAR DBD for a 5 basepair spacing direct repeat sequence seems not to be determined by protein-protein contacts between single DBDs, but is probably related to the structure of the whole RAR protein and its interactions with other proteins.

EPERIMENTAL PROCEDURES

Bacterial Expression and Production of the hRAR-β DNA Binding Domain

The DNA binding region of the human retinoic acid receptor β was obtained with a bacterial protein expression system using an inducible gene coding for the T7 RNA polymerase, in combination with transcription of the target gene under control of a strong T7 promotor (Studier & Moffat, 1986). A pET3 vector was used, which has a *Nde*I site overlapping the translation start site, and a *Bam*HI site downstream of it. Using oligodirected mutagenesis (Kunkel et al., 1987), a *Nde*I site in the hRAR-β cDNA (De Thé et al., 1987) with the ATG codon in frame directly upstream of codon 75 (Pro) was introduced and 154 (Asp) was changed into a stop codon, followed by a *Bam*HI site. By inserting the 270 bp *NdeI/Bam*HI fragment of the human RARβ cDNA into the pPET3 vector, all T7 coding sequences were eliminated from the vector, except for the initiator ATG codon, and a plasmid was created that directed the synthesis of a 80 amino acid protein fragment (Figure 1) containing the complete DNA binding region of hRAR-β. The hRAR-β DBD has a calculated molecular weight of 9439 daltons. BL21(DE3)pLysS bacteria transformed with the plasmid construct under control of the lacUV promoter, synthesized high levels of a protein with the expected mobility, as estimated by SDS-PAGE (results not shown), upon isopropyl β-D-thiogalactopyranoside (IPTG) induction. Bacteria were harvested by centrifugation at 3 hours after induction by IPTG (final concentration 0.1 mM).

Purification of Bacterially Expressed hRAR-β-DBD

The hRAR-β DBD was isolated using an adapted version of the protocol described for the glucocorticoid receptor (Freedman et al., 1988), as follows. Bacterial pellets were resuspended in lysis buffer (50 mM TrisHCl, pH=8.0; 10% glycerol; 4 mM $CaCl_2$; 40 mM $MgCl_2$; 0.5 mM PMSF; 1 mM EDTA; 5mM DTT) containing 250 mM NaCl at 5% of the culture volume. Cells were lysed by freezing them at -80℃ and thawing them quickly in a 37℃ waterbath. DNase I was added to the lysate to a final concentration of 10 µg/ml and the lysate was incubated at room temperature for 15 minutes. Bacterial debris was pelleted by centrifugation and discarded. Polymin P was then added over a 10 min. period at 4℃ to 0.2% under constant stirring, and the precipitate removed by centrifugation. $(NH_4)_2SO_4$ was added to the supernatant to 30% saturation, and the precipitated bacterial proteins were pelleted by centrifugation. The $(NH_4)_2SO_4$ concentration in the supernatant was adjusted to 80% saturation, and the precipitated protein was collected by centrifugation. After discarding the supernatant, the pellet was resuspended in $TGEDZ_{50}$ (50 mM TrisHCl; pH=8.0; 10% glycerol; 1 mM EDTA; 5 mM DTT; 50 µM $ZnCl_2$) containing 50 mM NaCl and dialysed overnight against the same buffer. The dialysate was loaded onto an Accel Plus CM cation exchange column (Millipore), column dimensions 0.75 x 18 cm. The column was eluted using a gradient of 50 mM to 1 M NaCl in $TGEDZ_{50}$ over 60 minutes. The RAR-β DBD protein eluted around 700 mM NaCl, and was over 95 % pure, as verified by SDS-PAGE (results not shown). The peak fractions were pooled and for NMR experiments, the $TGEDZ_{50}$ buffer was replaced with 200 mM NaCl, 0.1 mM DTT and 5 %

^2H_2O, and the protein concentrated to 2-4 mM final concentration, using an Amicon YM-3 filter membrane. The presence of zinc in the hRARβ DBD was confirmed by atomic absorption spectroscopy.

DNase I Protection Assays

A synthetic, double-stranded DNA fragment containing the sequence of interest (see text and Figure legend) was radioactively labelled at one end using Klenow enzyme and α-^{32}P labelled dCTP (3000 Ci/mmol, Amersham) to a specific activity of app. $3x10^3$ dpm/ng. Purified RAR-DBD protein was mixed with the radioactive probe in binding buffer (4% Ficoll 400; 15 mM KCl; 5 mM Hepes, pH=7.9; 0.5 mg/ml BSA; 10 ng/ml pdIdC; 1mM DTT; 0.1 mM EDTA; 0.5 mM $MgCl_2$; 5 μM $ZnCl_2$) in a total volume of 10 μl and incubated for one hour at room temperature. To this mix the specified amount (see Figure 2 legend) of DNaseI (Boehringer Mannheim) was added. The mixture was incubated for 30" at 25°C. The reaction was stopped by adding one volume of 5 mM EDTA, 0.1% SDS and cooling to 0°C. The mixture was phenol extracted, DNA subsequently precipitated and was analysed by electrophoresis on a 12% polyacrylamide gel under denaturing conditions. Next to the protection assays on the gel an equivalent amount of a purine specific sequence reaction (Maxam & Gilbert, 1980) of the same fragment was run as sequence specific size markers.

NMR Experiments

NMR samples of 2-4 mM were dissolved in 95/5 % $^1H_2O/^2H_2O$ mixtures containing 200 mM NaCl and 0.1 mM DTT.

^1H NMR spectra were recorded at 288-305K on Bruker AM500, AMX500 and AM600 spectrometers, operating at 500 and 600 MHz respectively. 2D NOE spectra (Jeener et al., 1982, States et al. 1982) with mixing times of 50, 100, 150 and 200 ms and 2D HOHAHA spectra (Griesinger et al., 1988) with a 40 ms mixing time in ^1H_2O were recorded in a phase sensitive mode using TPPI (Marion & Wüthrich, 1983). The water signal was suppressed by irradiation in the relaxation delay and the mixing time or by the SCUBA technique (Brown et al., 1988). For the 2D NOE and HOHAHA spectra, 400 FID's of 2K data points, 128 scans each, were recorded with a spectral width of 6250 Hz (at 500 MHz) or 7463Hz (at 600 MHz) for both domains. All data were processed with the "Triton" package to give real 1024 x 1024 real points data matrices. For both time domains sinebell windows shifted by $\pi/3$ or $\pi/4$ and fourth order polynomial baseline corrections were used. NOE cross-peaks were integrated using the "Triton" package and initial buildup rates were obtained by fitting to exponential or linear functions.

Structure Determination

NOE intensities were converted to upperbound distance restraints with the use of the initial buildup of NOE intensities. Tyrosine 2,3 and 5,6 NOE buildup rates representing a 2.48 Å distance were used for calibration after division by two. An error of 10% was added to all upperbounds to account for experimental and integration errors. Pseudo-atom corrections were applied to restraints involving methylene or methyl groups or aromatic ring protons (Wüthrich et al., 1986).

A total of 20 structures was generated with distance geometry (Havel et al., 1983) and optimized against distances and chiralities in 500 steps using a conjugate gradient method. Out of 20 structures, 19 structures with the correct chirality at the C-atoms were selected and subdued to a distance-bound driven dynamics calculation (DDD) (Scheek & Kaptein, 1988) of 500 steps at 300 K and 700 steps at 1 K. The time step for the DDD was 20 ps and a temperature coupling constant of 200 ps was applied. The constraint force constant was 10^4 kJ $mole^{-1}$ nm^{-4}. All DG and DDD calculations were performed on Silicon Graphics 4D/35 and 4D GTX workstations. Structures were examined and protein-DNA complex models were build on Silicon Graphics IRIS workstations using BIOSYMs Insight II software. The DNA sequences used for model building were 5'-AGGTCATGACCT-3' for the zero spacing inverted repeat, 5'-AGTTCACCCAGTTCA-3' for the three spacing direct repeat, 5'-AGTTCACCCCCAGTTCA-3' for the five spacing direct repeat and 5'-AGTTCACCCTGAACT-3' for the three spacing inverted repeat (Umesono et al., 1991). In all cases two GC basepairs were added to both ends of the strand. The RAR monomers were positioned in the major groove of the DNA so that residue Gly 30 was close to the central adenine-thymine basepair in the consensus operator and the first α-helix was directed analogously to that of the GR in the crystal structure of the GR-ER complex (Luisi et al., 1991). Diagrams were generated with MOLSCRIPT (Kraulis, P. J., 1991).

ACKNOWLEDGEMENTS

We thank Dr. P. Chambon (Strassbourg) for the hRAR-β cDNA, Dr. H. G. Stunnenberg (EMBL, Heidelberg) for making his manuscript available to us before publication and Dr. M.L. Ganadu for performing the atomic absorption experiments. This research was supported by the Netherlands Organization for Chemical Research (SON) and the Netherlands Organization of Scientific Research (NWO). The 600-MHz ^{1}H NMR spectra were recorded at the National Dutch HF-NMR facility in Nijmegen with the assistance of Dr. S. Wijminga and Mr. J. Joordens. M. K. was supported by a grant from the Human Frontier Science Program for Long-Term Fellowships. J. G. S. was supported by a fellowship of the Dutch Cancer Society. G.E.F. is also supported by the Dutch Cancer Society.

REFERENCES

Beato, M. (1989). Gene regulation by steroid hormones. *Cell* 56, 335-344.

Benbrook, D., Lernhardt, E. & Pfahl, M. (1988). A new retinoic acid receptor identified from a hepatocellular carcinoma. *Nature* 333, 669-672.

Berg, J.M. (1989). DNA binding specificity of steroid receptors. *Cell* 57, 1065-1069.

Brand, N.J., Petkovich, M., Krust, A.,Chambon, P., de Thé, H., Marchio, A., Tiollas, P. & Dejean, A. (1988). Identification of a second human retinoic acid receptor. *Nature* 332, 850-853.

Brown, S.C., Weber, P.L. & Mueller, L. (1988). Toward complete ^{1}H NMR spectra of proteins. *J. Magn. Res.* 77, 166-169.

Danielsen, M., Hinck, L. & Ringold, G.M. (1989). Two amino acids within the knuckle of the first zinc finger specify DNA response element activation by the glucocorticoid receptor. *Cell* 57, 1131-1138.

De Thé, H., Marchio, A., Tiollais, P. and Dejean, A. (1987). A novel steroid hormone recpetor-related gene inappropriately expressed. in human hepato cellular carcinoma. *Nature* 330, 667-670.

De Thé, H., del Mar Vivanco-Ruiz, M., Tiollais, P., Stunnenberg, H. & Dejean, A. (1990). Identification of a retinoic acid response element in the retinoic acid receptor β gene. *Nature* 343, 177-180.

Evans, R.M. (1988). The steroid and thyroid hormone receptor superfamily. *Science* 240, 889-895.

Freedman, L.P., Luisi, B.F., Korszun, Z.R., Basavappa, R., Sigler, P.B. and Yamamoto, K.R. (1988). The function and structure of metal coordination sites within the glucocorticoid receptor DNA binding domain. *Nature* 334, 543-546.

Giguere, V., Ong, E.S., Sequi, P. & Evans, R.M. (1987). Identification of a receptor for the morphogen retinoic acid. *Nature* 330, 624-627.

Green, L. M. & Berg, J. M. (1989). A retroviral Cys-Xaa_2-Cys-Xaa_4-His-Xaa_4-Cys peptide binds metal ions: spectroscopic studies and a proposed three-dimensional structure. *Proc. Natl. Acad. Sci. U.S.A.* 86, 4047-4051.

Griesinger, C., Otting, G., Wüthrich, K. & Ernst, R.R. (1988). Clean TOCSY for proton spin system identification in macromolecules. *J. Am. Chem. Soc.* 110, 7870-7872.

Härd, T., Kellenbach, E., Boelens, R., Maler, B. A., Dahlman, K., Freedman, L. P., Carlstedt- Duke, J., Yamamoto, K. R., Gustafsson, J. & Kaptein, R. (1990). Solution structure of the glucocorticoid receptor DNA binding domain. *Science* 249, 157-160.

Glass, C. K., Lipkin, S. M., Devary, O. V. & Rosenfeld, M. G. (1989). Positive and negative regulation of gene transcription by a retinoic acid-thyroid hormone recptor hetero dimer.. *Cell* 59, 697-708.

Havel, T.F., Kuntz, I.D. & Crippen, G.M. (1983). The theory and practice of distance geometry. *Bull. Math. Biol.* 45, 665-720.

Jeener, J.Meier, B.H., Bachmann, P. & Ernst, R.R. (1982). Investigation of exchange processes by two-dimensional NMR. *J. Chem. Phys.*, 71, 4546-4553.

Kliewer, S.A., Umesono, K., Mangelsdorf, D.J. & Evans, R.M. (1992). Retinoid X receptor interacts with nuclear receptors in retinoic, thyroid hormone and vitamin D_3 signalling. *Nature* 355, 446-449.

Katahira, M., Knegtel, R.M.A., Boelens, R., Eib, D., Schilthuis, J.G., van der Saag, P.T. and Kaptein, R. (1992). Homo- and heteronuclear NMR studies of the human retinoic acid receptor β DNA binding domain: sequential assignments and identification of secondary structure elements. *Biochemistry* 31, 6474-6480.

Kraulis, P. J. (1991). MOLSCRIPT: a program to produce both detailed and schematic plots of protein structures, *J. Appl. Cryst.* 24, 946-950.

Krust, A., Kastner, P., Petkovich, M., Zelent, A. & Chambon, P. (1989). A third human retinoic acid receptor, hRAR-γ. *Proc. Natl. Acad. Sci. USA* 86, 5310-5314.

Kunkel, T. A., Roberts, J. D. & Zakour, R. A. (1987). Rapid and efficient site-specific mutagenesis without phenotypic selection. *Meth. Enzymol.* 154, 367-382.

Leid, M., Kastner, P., Lyons, R., Nakshatri, H., Saunders, M., Zacharewski, T., Chen, J.-Y., Staub, A., Garnier, J.-M., Mader, S. & Chambon, P. (1992). Purification, cloning, and RXR identity of the HeLa cell factor with which RAR or TR heterodimerizes to bind target sequences efficiently. *Cell* 68, 377-395.

Luisi, B.F., Xu, W. X., Otwinowski, Z., Freedman, L.P., Yamamoto, K.R. & Sigler, P.B. (1991). Crystallographic analysis of the interaction of the glucocorticoid receptor with DNA. *Nature* 352, 497-505.

Mader, S., Kumar, V., de Verneuil, H. & Chambon, P. (1989). Three amino acids of the estrogen receptor are essential to its ability to distinguish an estrogen from a glucocorticoid responsive element. *Nature* 338, 271-274.

Marion, D. & Wüthrich, K. (1983). Application of phase sensitive two-dimensional correlated spectroscopy (COSY) for measurement of ^{1}H-^{1}H spin-spin coupling constants in proteins. *Biochem. Biophys. Res. Commun.* 113, 967-974.

Maxam, A. & Gilbert, W. (1980). Sequencing end-labelled DNA with base specific chemical cleavages. *Meth. Enzymol.* 65, 499-560.

Miller, J., McLachlan, A. D. & Klug, A. (1985). Repetitive zinc-binding domains in the protein transcription factor IIIA from Xenopus oocytes. *EMBO J.* 4, 1609-1614.

Petkovich, M., Brand, N.J., Krust, A. & Chambon, P. (1987). A human retinoic acid receptor which belongs to the family of nuclear receptors. *Nature* 330, 444-450.

Scheek, R.M. & Kaptein, R. (1988). Molecular dynamics techniques for determination of molecular structure from nuclear magnetic resonance data. in Methods in enzymology (Oppenheimer, N.J. & James, J.L., Eds.) Academic Press, New York.

Schwabe, J. W. R., Neuhaus, D. & Rhodes, D. (1990). Solution structure of the DNA-binding domain of the estrogen receptor. *Nature* 348, 458-461.

States, D.J., Haberkorn, R.A. & Ruben, D.J. (1982). A two-dimensional nuclear Overhauser experiment with pure phase in four quadrants. *J. Magn. Reson.* 48, 286-297.

Studier, F.W. & Moffat, B.A. (1986). Use of bacteriophage T7 RNA polymerase to direct high selective high-level expression of cloned genes. *J. Mol. Biol.* 189, 113-130.

Suck, D. & Oefner, C. (1986). Structure of DNase I at 2.0 Å resolution suggests a mechanism for binding to and cutting DNA. *Nature* 321, 620-625.

Thaller, C. & Eichele, G. (1987). Identification and spatial distribution of retinoids in the developing chick limb bud. *Nature* 327, 625-628.

Umesono, K. & Evans, R. M. (1989). Determination of target gene specificity for steroid/thyroid hormone receptors. *Cell* 57, 1131-1146.

Umesono, K., Murakami, K. K., Thompson, C. C. & Evans, R.M. (1991). Direct repeats as selective response elements for the thyroid hormone, retinoic acid and vitamin D_3 receptors., *Cell* 65, 1255-1266.

Vivanco Ruiz, M. d. M., Bugge, T. M., Hirschmann, P. & Stunnenberg, H. G. (1991). Functional characterization of a natural retinoic acid response element., *EMBO J.* 10, 3829-3838.

Wüthrich, K. (1986) in *NMR of Proteins and Nucleic Acids*, Wiley New York.

Yang, N, Schüle, R., Mangelsdorf, D.J., & Evans, R.M. (1991). Characterization of DNA binding and retinoic acid binding of retinoic acid receptor. *Proc. Natl. Acad. Sci. USA* 88, 3559-3563.

Yu, V.C., Delsert, C., Andersen, B., Holloway, J.M., Devarey, O. V., Näär, A.M., Kim, S.Y., Boutin, J.-M., Glass, C.K. & Rosenfeld M.G. (1991). RXRβ: A coregulator that enhaces binding of retinoic acid, thyroid hormone, and vitamin D receptors to their cognate response elements. *Cell* 67, 1251-1266.

Zhang, X., Hoffmann, B., Tran, P. B.-V., Graupner, G. & Pfahl, M. (1992). Retinoid X receptor is an auxiliary protein for thyroid hormone and retinoic acid receptors. *Nature* 355, 441-446.

MOLECULAR GENETICS OF PLASMA LIPID TRANSFER PROTEINS

Alan Tall

Division of Molecular Medicine
Department of Medicine
Columbia University, New York, NY 10032

The plasma lipoproteins are continuously modified during their metabolism in plasma, as result of the action of lipid modifying enzymes (lipases and lecithin: cholesterol acyltransferase, LCAT) and lipid transfer processes. Whereas the transfer of cholesterol between lipoproteins or between lipoproteins and cells is governed by simple diffusion, the transfer of phospholipids and neutral lipids (cholesteryl esters and triglycerides) between lipoproteins is mediated by plasma lipid transfer activities[1]. In plasma the cholesteryl ester transfer protein (CETP, also known as lipid transfer protein-1, LTP-1) mediates the transfer of cholesteryl esters and triglycerides between lipoproteins; the CETP can also facilitate phospholipid transfer. In addition, a separate protein or proteins without neutral lipid transfer activity, can also mediate phospholipid transfer between lipoproteins. Although not well characterized, these protein(s) are called the phospholipid transfer protein (PTP) or lipid transfer protein-2 (LTP-2). The CETP plays a particularly important role in the metabolism of HDL lipids[2]. In humans the HDL lipids turn over much more rapidly than the structural proteins of HDL, reflecting the continuous remodeling of HDL by lipases, LCAT and lipid transfer processes[2].

The importance of CETP in HDL metabolism has been demonstrated by the discovery of a human CETP deficiency state[3]. Individuals homozygous for CETP deficiency have extremely high levels of HDL cholesteryl esters (3-5 times normal levels), as well as reduced levels of cholesteryl esters in VLDL, IDL and LDL. CETP deficiency in these individuals is caused by a point mutation in the first nucleotide of an intron, a change which prevents the normal removal of intronic sequences during splicing of the primary gene transcript. The phenotype of this mutation shows that the normal function of CETP is to remove cholesteryl esters from HDL by transferring cholesteryl esters to the other plasma lipoproteins in exchange for triglyceride. The phenotype of genetic human CETP (high HDL, low levels of cholesterol in the apoB-containing lipoproteins) has

New Developments in Lipid-Protein Interactions and Receptor Function
Edited by K.W.A. Wirtz *et al.*, Plenum Press, 1993

strong anti-atherogenic potential, and has excited interest in the potential development of CETP inhibitory drugs.

The mechanism by which CETP mediates lipid transfer between lipoproteins is poorly understood. Kinetic experiments have been interpreted to support both carrier-mediated and collisional mechanisms of lipid transfer by CETP[1]. The CETP has binding sites for neutral lipids and phospholipids, which rapidly exchange these lipids with lipoprotein lipids, and which appear to be essential to CETP function. The existence of these lipid binding sites suggests that CETP acts as a lipid carrier, even if the lipid transfer event occurs during formation of a ternary complex involving donor and acceptor lipoprotein and CETP. A monoclonal antibody which neutralizes cholesteryl ester and triglyceride transfer activities of CETP has been useful in elucidating CETP function[4]. The epitope of this antibody has initially mapped to a region within the C-terminal 26 amino acids of CETP.

Recent experiments using a site-directed mutagenesis approach have further localized amino acids involved in the epitope and have shown that deletion of these sequences abolishes neutral lipid but not phospholipid transfer activities of CETP Point mutants suggest that the segment essential for neutral lipid transfer is an amphipathic helix. The neutralizing mAb binds to the polar face of this helix, but it is the hydrophobic face which is directly involved in neutral lipid transfer.

Thus, point mutants which abolish binding of the neutralizing mAb involve primarily charged or polar amino acids; these mutants have normal cholesteryl ester transfer activity. In contrast, mutants of the non-polar amino acids display essentially normal binding of the mAb TP2, but have impaired cholesteryl ester transfer activity. Low activity mutants in this region all involve the amino acids leucine and phenylalanine i.e. bulky hydrophobic amino acids. However, all of these mutants show only partial reductions of activity. Thus, the involvement of these amino acid residues does not appear to result from highly specific side-chain interactions, but rather reflects the general hydrophobic character of amino acids in this region. It appears that the hydrophobic face of the C-terminal amphipathic helix, as well as a contiguous region of hydrophobic amino acids comprise a fairly extensive hydrophobic surface, which is directly involved in the mechanism of cholesteryl ester transfer.

There are several potential reasons why this C-terminal hydrophobic region of CETP is necessary for neutral lipid transfer. Detailed digestions of C-terminal deletion mutants with proteases indicate the C-terminal helix can be removed from CETP without detriment to the overall folding pattern. Thus, the C-terminal helix appears to constitute a flexible tail. Movement of this tail may be involved in the neutral lipid transfer process. This could occur when the CETP becomes bound to the lipoprotein surface, exposing the hydrophobic side of the helix, as well as contiguous hydrophobic amino acids. This hydrophobic surface could directly bind neutral lipid molecules. Also, the C-terminal helix has some of the character of a fusion helix, i.e. it becomes increasingly hydrophobic along the axis of the helix, proceeding in a C- to N-terminal direction. This could allow the C-terminal helix to become tilted in the surface of the lipoprotein, disorganizing the lipid packing and allowing deeper penetration into the lipoprotein core. Other functions that are being considered for the C-terminal helix are that it directly mediates fusion between lipoprotein surfaces, or that it forms part of a dimerization domain. The leucine residues in the C-terminal helix may form a short "leucine zipper" motif, allowing dimer formation. Dimerization of CETP molecules initially bound to different lipoproteins could produce a ternary complex in which lipid transfer occurred.

The mouse is being increasingly used as a model of lipoprotein metabolism and early atherosclerosis. This reflects the availability of many strains of mice and the possibility of using genetics to understand factors regulating lipoproteins and atherosclerosis, and also the recent ability to express transgenes in mice or to knock out genes by homologous recombinations. These approaches allow causal analyses of complex in vivo events, and transcend the approach of correlational analysis.

The mouse strain C57Bl6 has been found to be susceptible to the development of fatty streak lesions (an early form of atherosclerosis) when placed on a high fat, high cholesterol diet. Other strains of mice show varying degrees of susceptibility to atherosclerosis, and some strains are completely resistant. Genetic analysis has shown that the susceptibility to atherosclerosis segregates with low HDL levels and an atherosclerosis susceptibility gene has been mapped to an unidentified locus called the Ath-1 gene on mouse chromosome 1. This locus may encode an unknown gene influencing HDL responses in mice. Strong evidence to support the role of HDL in mouse fatty streak lesions has been obtained by overexpressing the human apoA-I gene in mice[5]. Overexpression of human apoA-I in mice results in increased levels of HDL, human-like HDL speciation, and prevents the development of fatty streaks in C57BL6 mice.

Although the investigations are at an early stage, it has been possible to reconstitute many parts of the human lipoprotein system in mice. To date most of transgenic animals have been analyzed for single gene effects. For example, it has been discovered that transgenic mice overexpressing human apoC-III have marked hypertriglyceridemia, suggesting that human hypertriglyceridemia may sometimes result from overexpression of apoC-III[6]. Mice normally lack cholesteryl ester transfer activity. Introduction of the human CETP gene into mice results in a reduction in HDL cholesterol levels[7]. More surprisingly, when human apoA-I transgenic mice were crossed with human CETP transgenic mice, there was a much more profound reduction in HDL cholesterol, compared to animals only bearing the CETP transgene[8]. These experiments have uncovered a previously unsuspected specificity in the interaction between human CETP and HDL containing human apoA-I. Mice expressing three different transgenes (human apoC-III, apoA-I and CETP) have been de eloped by cross-breeding. These mice are hypertriglyceridemic and also have profound reductions in HDL cholesterol. This lipoprotein phenotype resembles a very common set of lipoprotein abnormalities often found in patients with coronary artery disease. Eventually, it may be possible by gene ablation techniques, combined with trangenesis, to reconstitute major portions of the human lipoprotein system in a mouse.

Transgenesis approaches have also been extremely important for investigating the regulation of gene expression. For example, a downstream regulatory element (about 14 Kbp 3' of the apoE gene promoter) was discovered to be necessary for liver expression of the human apoE and ApoC-I genes[9]. Similarly, an element for downstream of the human apoA-I gene seems to be necessary for its expression in the small intestine, one of the major sites of apoA-I synthesis[10]. Plasma cholesteryl ester transfer activity and CETP mass are increased in animals when placed on a high cholesterol diet. Transgenic animals expressing the human CETP gene with its natural flanking sequences have shown that the increase in plasma CETP in response to increased dietary cholesterol is due to increased hepatic gene transcription, an effect which depends on the presence of the natural flanking sequences of the CETP gene. These experiments indicate that a novel element present in the flanking sequences of the human CETP gene mediates a marked increase in gene transcription in response to a high cholesterol diet. The nature of this element will be defined by further transgenesis experiments combined with an analysis of promoter-reported gene activities in cultured cells. In each of the cases cited above, the nature of the tissue-specific or cholesterol-responsive elements was unsuspected on the basis of tissue culture experiments and only become apparent after transgenic mice were developed.

The CETP gene consists of 16 exons and comprises about 25 Kbp of genomic DNA and is located on chromosome 16 near the LCAT locus[11]. The major source of the CETP mRNA in humans in the liver. However, there are also significant amounts of the CETP mRNA in adipose tissue, small intestine and spleen[12]. In different species, the most conserved expression of CETP is in adipose tissue. The function of CETP in adipose tissue is unknown but it could be related to the fact the CETP activity is augmented during lipolysis. Thus, there may be local concentrations of CETP within blood vessels in adipose tissue, which help to recycle lipids transferred from triglyceride-rich lipoproteins into HDL during lipolysis.

There is an interesting variant of the CETP mRNA, in which sequences corresponding to exon 9 have been deleted[13]. The alternatively spliced form of the mRNA is found in all tissues containing the CETP mRNA, but is most abundant in human spleen where it represents 40 to 70% of total CETP mRNA. Analysis of the cell type expressing the CETP mRNA in the spleen indicates that it present predominantly in lymphocytes, particularly in B-lymphocytes. B-lymphocytes freshly isolated from human spleen contain relatively abundant CETP mRNA, with 40-60% in the alternatively spliced form, but the mRNA is not found in significant amounts in peripheral blood lymphocytes. Since exon 9 is within the coding sequence of the CETP cDNA, this results in a shortened form of the CETP protein. Expression of the exon 9 deleted form of CETP in Cos cells or CHO cells (by transfection of the exon 9 deleted cDNA) results in synthesis of a protein which is poorly secreted by the cells. Analysis with endoH indicates that the exon 9 deleted protein remains within the endoplasmic reticulum. Lipid transfer assays of cell lysates show that the form of CETP retained within the endoplasmic reticulum is not active, at least in standard lipid transfer assays. Thus, there are two possible functions for the alternative splicing of the CETP mRNA. First, it might carry out a novel function in specific cell types such as B-lymphocytes, where it may act intracellularly. Second, it might serve as a regulatory mechanism to limit the secretion of active CETP.

References

1. Tall, A.R., Plasma lipid transfer proteins, J. Lipid Res. 27:359-365 (1986).
2. Tall, A.R., Plasma high density lipoproteins - metabolism and relationship to atherogenesis, J. Clin. Invest. 86:379-384 (1990).
3. Brown, M.L., A. Inazu, C.B. Hesler, L.B. Agellon, C. Mann, M.E. Whitlock, Y.L. Marcel, R.W. Milne, J. Koisumi, H. Mabuchi, R. Takea, and A.R. Tall, Molecular basis of lipid transfer protein deficiency in a family with increased high-density lipoproteins, Nature 342:448-451 (1989).
4. Hesler, C.B., R.W. Milne, T.L. Swenson, P.K. Weech, Y.L. Marcel, and A.R. Tall, Monoclonal antibodies to the Mr 74,000 cholesteryl ester transfer protein neutralize all of the cholesteryl ester and triglyceride transfer activities in human plasma, J. Biol. Chem. 263:5020-5023 (1988).
5. Rubin, E.M., R.M. Krauss, E.A. Spangler, J.G. Verstuyft, and S.M. Clift, Inhibition of early atherogenesis in transgenic mice by human apolipoprotein AI, Nature 353:265-267 (1991).
6. Ito, Y., N. Azrolan, A. O'Connell, A. Walsh, and J.L. Breslow, Hypertriglyceridemia as a result of human apoCIII gene expression in transgenic mice, Science 249:790-793 (1990).
7. Agellon, L.B., A. Walsh, T. Hayek, P. Moulin, X.C. Jiang, S.A. Shelanski, J.L. Breslow, and A.R. Tall, Reduced high density lipoprotein cholesterol in human cholesteryl ester transfer protein transgenic mice, J. Biol. Chem. 266:10796-10801 (1991).

8. Hayek, T., T. Chajek-Shaul, A. Walsh, L.B. Agellon, P. Moulin, A.R. Tall, and J.L. Breslow, An interaction between the human CETP and apoA-I genes in transgenic mice results in a profound CETP mediated depression of HDL cholesterol levels, J. Clin. Invest. (1991), In press.
9. Simonet, W.S., N. Bucay, R.E. Pitas, S.J. Lauer, and J.M. Taylor, Multiple tissue-specific elements control the apolipoprotein E/C-I gene locus in transgenic mice, J. Biol. Chem. 266:8651-8654 (1991).
10. Walsh, A., Y. Ito, and J.L. Breslow, High levels of human apolipoprotein A-I in transgenic mice result in increased plasma levels of small high density lipoprotein (HDL) particles comparable to human HDL, J. Biol. Chem. 264:6488-6494 (1989).
11. Agellon, L.B., E. Quinet, T. Gillette, D. Drayna, M.E. Brown, and A.R. Tall, Organization of the human cholesteryl ester transfer protein gene, Biochemistry 29:1372-1376 (1990).
12. Jiang, X.C., P. Moulin, E.M. Quinet, I.J. Goldberg, L.K. Yacoub, L.B. Agellon, D. Compton, R.Polokoff, and A.R. Tall, Mammalian adipose tissue and muscle are major sources of lipid transfer protein mRNA, J. Biol. Chem. 266:4631-4639, 1991.
13. Inazu, A., E.M. Quinet, S. Wang., M.L. Brown, S. Stevenson, M.L. Barr, P. Moulin, and A.R. Tall, Alternative splicing of the mRNA encoding the human cholesteryl ester transfer protein, Biochemistry 31:2352-2358 (1992).

TARGETING SIGNALS AND MECHANISMS OF PROTEIN INSERTION INTO MEMBRANES

Gunnar von Heijne

Department of Molecular Biology
Karolinska Institute Center for Structural Biochemistry
S-141 57 Huddinge, Sweden

INTRODUCTION

Membrane proteins are critically involved in a host of important cellular functions, including such diverse processes as cell-cell adhesion, hormone responses, antigen presentation, nerve conduction, light detection, photosynthesis, transport of nutrients and ions, and drug resistance. Despite this impressive functional variablitiy, there are basic similarities in the way membrane proteins insert into membranes and in the structures they finally attain. A full understanding of any particular membrane protein thus includes an understanding of its biosynthesis and membrane assembly. In this chapter, I will try to trace some general principles of membrane protein biogenesis and structure, as they apply to proteins from a range of different membrane systems.

INSERTION OF PROTEINS INTO BACTERIAL MEMBRANES

Gram-negative bacteria have both an outer an inner membrane, and their respective proteins have very different characteristics. Gram-positive bacteria have only one membrane, corresponding to the inner membrane of their Gram-negative relatives.

The Sec-Pathway

To understand membrane protein biogenesis, one must first understand protein secretion. From this point of view, membrane proteins can be seen as incompletely secreted proteins: one or more parts have been translocated across the membrane, whereas other parts remain non-translocated. In *E. coli*, protein secretion is handled by the so-called *sec*-machinery (Schatz and Beckwith, 1990; Wickner et al., 1991). This includes the SecA, SecB, SecD, SecE, SecF, SecY, Band-1, signal peptidase I, and signal peptidase II proteins; for export across the outer membrane, an additional set of proteins is required (Pugsley et al., 1990).

Secretory proteins are initially made with an N-terminal extension, a signal peptide, that serves both to target the protein to the *sec*-machinery, and to retard the folding of the nascent chain into export-incompetent conformations (Hardy and Randall, 1991). During or shortly after synthesis, the preprotein is bound to the chaperone SecB, which prevents premature folding and aggregation. The polypeptide is subsequently handed over to the translocase complex proper, composed of the peripheral membrane protein SecA and the integral membrane proteins SecE, SecY, and Band-1 (Brundage et al., 1992). SecA apparently binds both to the signal peptide and to sites in other parts of the protein. Preprotein binding activates an ATP-binding site on SecA, and translocation across the membrane is initiated. The details of the translocation process are largely unknown, but it is clear that both ATP and the transmembrane electrochemical potential are needed. The role of the SecD and SecF proteins has not been clarified. Finally, the signal peptide is removed from the translocated protein by either signal peptidase I or II, and the mature molecule is released from the membrane.

Inner Membrane Proteins

All known integral inner membrane proteins have one or more transmembrane hydrophobic stretch of 15-30 residues. Simple energetic considerations suggest that these transmembrane stretches must be α-helical, a prediction that is borne out by the three-dimensional structures of the photosynthetic reaction center (Deisenhofer et al., 1985) and bacteriorhodopsin (Henderson et al., 1990). Multi-spanning inner membrane proteins thus form bundles of hydrophobic α-helices in the membrane.

It is likely that such helical bundles are assembled in two largely independent steps (Popot and Engelman, 1990). The nascent polypeptide chain is first inserted into the membrane with concomitant formation of the transmembrane helices, and these pre-formed helices then assemble into the final tertiary structure. The paucity of known three-dimensional structures has hampered the analysis of helix-helix interactions in a membrane environment, and no reliable methods for 3D-structure predictions based only on sequence data are available. On the other hand, our understanding of the insertion process is already quite good.

Both statistical and experimental studies have shown that the topology of bacterial inner membrane proteins is to a good approximation controlled by the distribution of positively charged amino acids in the polar regions that flank the apolar transmembrane segments; see Boyd and Beckwith (1990), Dalbey (1990), and von Heijne and Manoil (1990) for reviews. The "positive inside-rule" (von Heijne and Gavel, 1988) states that regions rich in lysine and arginine tend to remain on the cytoplasmic side of the membrane, whereas regions containing few such residues tend to be periplasmic. Thus, the introduction or removal of positively charged residues can be used to manipulate the topology of a protein (Nilsson and von Heijne, 1990). An analysis of the distribution of positively charged residues relative to the hydrophobic regions can also be used to significantly improve the prediction of the transmembrane topology of inner membrane proteins (von Heijne, 1992).

Some inner membrane proteins appear not to require a functional *sec*-machinery for proper assembly. Most of the available data suggests that the length of the translocated region determines its *sec*-dependence (Andersson and von Heijne, 1992), though this is still somewhat controversial (McGovern and Beckwith, 1991). In any case, a low content of positively charged residues is only found in periplasmic regions that are shorter than ~60-70 residues; longer segments have an amino acid composition similar to that of fully exported, *sec*-dependent periplasmic proteins (von Heijne and Gavel, 1988).

Outer Membrane Proteins

The basic architecture of the known outer membrane proteins is very different from the hydrophobic helix-bundles typical of the inner membrane proteins. Instead, the polypeptide chain folds into a large anti parallel ß-barrel with hydrophobic amino acids facing outwards towards the lipid environment (Weiss et al., 1991; Welte et al., 1991). If the ß-barrel is large enough, this creates a water-filled pore through the membrane that can serve as a passive (but selective) diffusion channel for small solutes.

Outer membrane proteins are initially synthesized with an amino-terminal signal peptide, and depend on the *sec*-machinery for translocation across the inner membrane. It is thought that they exist transiently as a soluble periplasmic intermediate before integration into the outer membrane. The integration process is poorly understood, but must conceivably be critically dependent on the formation of the entire ß-barrel structure either before or during insertion (Fourel et al., 1992).

INSERTION OF PROTEINS INTO THE ER MEMBRANE

All membrane proteins that are found along the secretory pathway of eukaryotic cells - from the ER to the plasma membrane - are initially synthesized by ER-bound ribosomes and inserted co-translationally into the ER membrane. Certain aspects of their biosynthesis are reminiscent of bacterial inner membrane proteins, but there are also a number of differences that make it risky to generalize from one system to the other.

The SRP Pathway

The biochemistry of protein translocation across the ER membrane is rather different from that of the *sec*-machinery (Rapoport, 1991). As it emerges from the ribosome, the signal peptide is initially recognized by a large protein-RNA complex, the signal recognition particle (SRP), that ensures the correct targeting of the pre-protein to translocation sites in the ER. The SRP-ribosome complex binds to the SRP-receptor protein, an integral, GTP-binding protein of the ER membrane. The actual mechanism of transfer of the nascent chain across the membrane is unclear, though a number of proteins have been shown by chemical crosslinking to be close to the translocating chain. The existence of transient ion-conducting channels in ER-derived membranes that seem to be "plugged" by nascent polypeptide chains has been taken as an indication that translocation takes place through a proteinaceous pore (Simon and Blobel, 1991).

Signal Peptides and Signal-Anchor Sequences

Signal peptides look very similar in prokaryotes and eukaryotes. Three regions can be easily delineated: a positively charged N-terminal region, a central hydrophobic region, and a C-terminal region that serves as a recognition site for the signal peptidase enzyme (von Heijne, 1990). If the hydrophobic region is ~15 residues or longer, or if no signal peptidase site is present, the signal peptide will be permanently attached to the protein and serve both as a translocation signal and an N-terminal membrane anchor (Sakaguchi et al., 1992). The orientation of an N-terminal signal-anchor sequence depends at least in part on the distribution of positively charged residues relative to the hydrophobic segment, and certain proteins with an "inverted" charge distribution insert with their N-terminus facing the ER lumen and the bulk of their mass in the cytoplasm (Beltzer et al., 1991; Parks and Lamb, 1991; Sato et al., 1990).

Stop-Transfer Sequences

Stop-transfer sequences are long (~20 residues) stretches of hydrophobic residues placed downstream of a signal peptide or a signal-anchor sequence. Such sequences will abort the translocation process, resulting in a transmembrane topology. Apparently, any sufficiently hydrophobic segment will have stop-transfer function; positively charged residues placed immediately downstream of a hydrophobic segment also contributes to the stop-transfer activity (Kuroiwa et al., 1991).

Multi-Spanning Proteins

Multi-spanning membrane proteins can be constructed from a succession of signal or signal-anchor sequences alternating with stop-transfer sequences (Lipp et al., 1989; Wessels and Spiess, 1988). In many cases, the topology of multi-spanning proteins can be predicted from the net charge difference between the N- and C-terminal flanks of the most N-terminal membrane-spanning segment (Hartmann et al., 1989), suggesting that membrane insertion is co-translational and that successive transmembrane segments simply insert in alternating orientations.

INSERTION OF PROTEINS INTO MITOCHONDRIAL MEMBRANES

Mitochondrial membrane proteins are either encoded in the mitochondrial genome and synthesized inside the organelle, or encoded in the nuclear genome and imported from the cytoplasm. In higher eukaryotes, the mitochondrial genome contains almost exclusively genes for very hydrophobic, multi-spanning membrane proteins, and the imported membrane proteins tend to have fewer and less hydrophobic transmembrane segments (de Vitry and Popot, 1989; von Heijne, 1986).

Mitochondrial Protein Import

A number of excellent reviews on the general mechanism of mitochondrial protein import have been written in recent years (Glick and Schatz, 1991; Hartl et al., 1989; Neupert et al., 1990; Pfanner and Neupert, 1990; Verner and Schatz, 1988). In short, proteins are targeted to mitochondria by N-terminal pre-sequences or targeting peptides that are rich in positively charged amino acids and have a potential to form amphiphilic α-helices. The targeting peptide is recognized by receptors in the outer mitochondrial membrane, and delivered to translocation-sites in areas of close contact between the outer and inner membranes. Upon entry into the matrix, the nascent chain is bound to chaperones of the hsp70 family, and subsequently transferred to an hsp60 chaperone that catalyzes the final folding. Inner membrane and intermembrane space proteins are either diverted from the import pathway after passage through the outer membrane, or are first imported into the matrix and then re-exported across the inner membrane.

Insertion Into the Inner Membrane

A recent statistical study has shown that mitochondrially encoded inner membrane proteins follow the positive-inside rule (more Arg and Lys in matrix-facing segments), whereas the situation is less clear for imported proteins (Gavel and von Heijne, 1992). It is thus possible that the mechanism of insertion of proteins delivered from the matrix side of the inner membrane is similar to that of *E. coli*, and that analogous topological rules apply. Whether the insertion of imported inner membrane proteins proceeds via a "stop-transfer" mechanism where hydrophobic segments halt the initial translocation into the matrix (Glick et al., 1992), or via a "conservative sorting" mechanism where the protein is first fully imported into the matrix and subsequently inserted into the inner membrane from the matrix side (Mahlke et al., 1990), is controversial. It is thus at present impossible to predict the orientation of imported inner membrane proteins directly from their amino acid sequence.

Insertion Into the Outer Membrane

The membrane insertion of only two proteins of the outer membrane, porin and a major 70 kDa protein (MAS70), has been reasonably well studied. Porin is believed to form a transmembrane ß-barrel similar to the bacterial outer membrane porins. Although lacking a cleavable N-terminal targeting peptide, it is nevertheless targeted to the same outer membrane receptor used by the matrix proteins, but then apparently insert directly into the outer membrane (Pfanner and Neupert, 1990).

The 70 kDa protein has an interesting targeting sequence with a short N-terminal, positively charged segment followed by a long, fairly hydrophobic stretch that serves to anchor the molecule in the outer membrane. When the hydrophobic region is deleted, the protein ends up in the matrix (Nakai et al., 1989); when the positively charged segment is replaced with a normal matrix-targeting pre-sequence the protein apparently inserts into the outer membrane in an "inverted" orientation with the N-terminus exposed outside the mitochondrion and the bulk of the chain inside the outer membrane (Li and Shore, 1992).

INSERTION OF PROTEINS INTO CHLOROPLAST MEMBRANES

The chloroplast has three different membrane systems: the outer and inner envelope membranes, and the thylakoid membranes. Only a handful of envelope membrane proteins have been cloned, and little is known about their membrane insertion (Li et al., 1991). In contrast, many thylakoid membrane proteins have been extensively characterized in terms of their topology and assembly mechanisms. Genes encoding thylakoid membrane proteins are found both in the plastid and nuclear genomes, and nuclearly encoded proteins must thus first be imported into the chloroplast before membrane insertion.

Chloroplast Protein Import

For reviews on chloroplast protein import, see deBoer and Weisbeek (1991) and Keegstra (1989). Briefly, proteins are targeted for import into the stromal compartment by an N-terminal transit peptide rich in hydroxylated amino acids (von Heijne et al., 1989) but with no discernible potential to form a regular secondary structure (Theg and Geske, 1992; van't Hof et al., 1991; von Heijne and Nishikawa, 1991). A weakly conserved cleavage-site motif is found in about one-third of all known transit peptides (Gavel and von Heijne, 1990).

Insertion Into the Thylakoid Membrane

After delivery to the stroma, insertion into the thylakoid membrane depends on hydrophobic amino acid segments (Auchincloss et al., 1992); in some cases, an N-terminal signal peptide-like sequence serves as the thylakoid targeting signal (von Heijne, et al., 1989). Stromal proteins of the hsp70 family have also been implicated in the insertion process (Yalovsky et al., 1992). Thylakoid membrane proteins follow the positive inside-rule (Gavel et al., 1991), and it is thus likely that many aspects of their membrane assembly are similar to the corresponding bacterial process. A proton gradient across the thylakoid membrane can drive insertion (Klosgen et al., 1992). The requirement for ATP is different for different proteins, and may thus not be universal (Cline et al., 1992).

CONCLUSION

Over the past few years, much progress has been made towards a complete understanding of both the signals and the biochemical machineries that guide proteins into membranes. Two structural families have been recognized among the integral membrane proteins: the helical bundle proteins and the ß-barrel proteins.

Membrane assembly of the helical bundle proteins can generally be understood in terms of the *sec*-pathway in prokaryotes and the SRP-pathway in eukaryotes; in addition, the positive inside-rule is a strong unifying principle underlying the determination of the topology of helical bundle proteins from a variety of membrane systems.

The membrane assembly of ß-barrel proteins appears to be more dependent on global features of the nascent chain, and the formation of the barrel presumably proceeds in concert with insertion into the membrane.

REFERENCES

Andersson, H., and G. von Heijne. 1992. *Sec*-dependent and *sec*-independent assembly of *E. coli* inner membrane proteins: the topological rules depend on chain-length. *submitted.*

Auchincloss, A. H., A. Alexander, and B. D. Kohorn. 1992. Requirement for 3 Membrane-Spanning alpha-Helices in the Post-Translational Insertion of a Thylakoid Membrane Protein. *J Biol Chem.* 267:10439-10446.

Beltzer, J. P., K. Fiedler, C. Fuhrer, I. Geffen, C. Handschin, H. P. Wessels, and M. Spiess. 1991. Charged Residues Are Major Determinants of the Transmembrane Orientation of a Signal-Anchor Sequence. *J Biol Chem.* 266:973-978.

Boyd, D., and J. Beckwith. 1990. The Role of Charged Amino Acids in the Localization of Secreted and Membrane Proteins. *Cell.* 62:1031-1033.

Brundage, L., C. J. Fimmel, S. Mizushima, and W. Wickner. 1992. SecY, SecE, and Band-1 Form the Membrane-Embedded Domain of *Escherichia coli* Preprotein Translocase. *J Biol Chem.* 267:4166-4170.

Cline, K., W. F. Ettinger, and S. M. Theg. 1992. Protein-Specific Energy Requirements for Protein Transport Across or into Thylakoid Membranes - Two Lumenal Proteins Are Transported in the Absence of ATP. *J Biol Chem.* 267:2688-2696.
Dalbey, R. E. 1990. Positively Charged Residues Are Important Determinants of Membrane Protein Topology. *Trends Biochem Sci.* 15:253-257.
de Vitry, C., and J.-L. Popot. 1989. Le micro-assemblage des protéines membranaires mitochondriales et chlorplastique. *C.R.Acad.Sci.Paris III.* 309:709-714.
deBoer, A. D., and P. J. Weisbeek. 1991. Chloroplast Protein Topogenesis - Import, Sorting and Assembly. *Biochim Biophys Acta.* 1071:221-253.
Deisenhofer, J., O. Epp, K. Miki, R. Huber, and H. Michel. 1985. Structure of the protein subunits in the photosynthetic reaction centre of *Rhodopseudomonas viridis* at 3Å resolution. *Nature.* 318:618-624.
Fourel, D., S. Mizushima, and J. M. Pagès. 1992. Dynamics of the Exposure of Epitopes on OmpF, an Outer Membrane Protein of Escherichia-Coli. *Eur J Biochem.* 206:109-114.
Gavel, Y., J. Steppuhn, R. Herrmann, and G. von Heijne. 1991. The Positive-Inside Rule Applies to Thylakoid Membrane Proteins. *FEBS Lett.* 282:41-46.
Gavel, Y., and G. von Heijne. 1990. A Conserved Cleavage-Site Motif in Chloroplast Transit Peptides. *FEBS Lett.* 261:455-458.
Gavel, Y., and G. von Heijne. 1992. The Distribution of Charged Amino Acids in Mitochondrial Inner Membrane Proteins Suggests Different Modes of Membrane Integration for Nuclearly and Mitochondrially Encoded Proteins. *Eur J Biochem.* 205:1207-1215.
Glick, B., and G. Schatz. 1991. Import of Proteins into Mitochondria. *Annu Rev Genet.* 25:21-44.
Glick, B. S., A. Brandt, K. Cunningham, S. Muller, R. L. Hallberg, and G. Schatz. 1992. Cytochromes-c1 and Cytochromes-b2 Are Sorted to the Intermembrane Space of Yeast Mitochondria by a Stop-Transfer Mechanism. *Cell.* 69:809-822.
Hardy, S. J. S., and L. L. Randall. 1991. A Kinetic Partitioning Model of Selective Binding of Nonnative Proteins by the Bacterial Chaperone SecB. *Science.* 251:439-443.
Hartl, F. U., N. Pfanner, D. W. Nicholson, and W. Neupert. 1989. Mitochondrial protein import. *Biochim Biophys Acta.* 988:1-45.
Hartmann, E., T. A. Rapoport, and H. F. Lodish. 1989. Predicting the orientation of eukaryotic membrane proteins. *Proc.Natl.Acad.Sci.USA.* 86:5786-90.
Henderson, R., J. M. Baldwin, T. A. Ceska, F. Zemlin, E. Beckmann, and K. H. Downing. 1990. A Model for the Structure of Bacteriorhodopsin Based on High Resolution Electron Cryo-Microscopy. *J Mol Biol.* 213:899-929.
Keegstra, K. 1989. Transport and routing of proteins into chloroplasts. *Cell.* 56:247-53.
Klosgen, R. B., I. W. Brock, R. G. Herrmann, and C. Robinson. 1992. Proton Gradient-Driven Import of the 16 kDa Oxygen-Evolving Complex Protein as the Full Precursor Protein by Isolated Thylakoids. *Plant Mol Biol.* 18:1031-1034.
Kuroiwa, T., M. Sakaguchi, K. Mihara, and T. Omura. 1991. Systematic Analysis of Stop-Transfer Sequence for Microsomal Membrane. *J Biol Chem.* 266:9251-9255.
Li, H. M., T. Moore, and K. Keegstra. 1991. Targeting of Proteins to the Outer Envelope Membrane Uses a Different Pathway Than Transport into Chloroplasts. *Plant Cell.* 3:709-717.
Li, J.-M., and G. C. Shore. 1992. Reversal of the Orientation of an Integral Protein of the Mitochondrial Outer Membrane. *Science.* 256:1815-1817.
Lipp, J., N. Flint, M.-T. Haeuptle, and B. Dobberstein. 1989. Structural requirements for membrane assembly of proteins spanning the membrane several times. *J.Cell Biol.* 109:2013-22.
Mahlke, K., N. Pfanner, J. Martin, A. L. Horwich, F. U. Hartl, and W. Neupert. 1990. Sorting Pathways of Mitochondrial Inner Membrane Proteins. *Eur J Biochem.* 192:551-555.
McGovern, K., and J. Beckwith. 1991. Membrane Insertion of the *Escherichia coli* MalF Protein in Cells with Impaired Secretion Machinery. *J Biol Chem.* 266:20870-20876.
Nakai, M., T. Hase, and H. Matsubara. 1989. Precise determination of the mitochondrial import signal contained in a 70 kDa protein of yeast mitochondrial outer membrane. *J Biochem (Tokyo).* 105:513-519.
Neupert, W., F. U. Hartl, E. A. Craig, and N. Pfanner. 1990. How Do Polypeptides Cross the Mitochondrial Membranes. *Cell.* 63:447-450.
Nilsson, I. M., and G. von Heijne. 1990. Fine-tuning the Topology of a Polytopic Membrane Protein. Role of Positively and Negatively Charged Residues. *Cell.* 62:1135-1141.
Parks, G. D., and R. A. Lamb. 1991. Topology of Eukaryotic Type-II Membrane Proteins - Importance of N-Terminal Positively Charged Residues Flanking the Hydrophobic Domain. *Cell.* 64:777-787.
Pfanner, N., and W. Neupert. 1990. The Mitochondrial Protein Import Apparatus. *Annu Rev Biochem.* 59:331-353.

Popot, J. L., and D. M. Engelman. 1990. Membrane Protein Folding and Oligomerization - The 2-Stage Model. *Biochemistry*. 29:4031-4037.
Pugsley, A. P., C. Denfert, I. Reyss, and M. G. Kornacker. 1990. Genetics of Extracellular Protein Secretion by Gram-Negative Bacteria. *Annu Rev Genet*. 24:67-90.
Rapoport, T. A. 1991. Protein Transport Across the Endoplasmic Reticulum Membrane - Facts, Models, Mysteries. *FASEB J*. 5:2792-2798.
Sakaguchi, M., R. Tomiyoshi, T. Kuroiwa, K. Mihara, and T. Omura. 1992. Functions of Signal and Signal-Anchor Sequences Are Determined by the Balance Between the Hydrophobic Segment and the N-Terminal Charge. *Proc Natl Acad Sci USA*. 89:16-19.
Sato, T., M. Sakaguchi, K. Mihara, and T. Omura. 1990. The Amino-Terminal Structures That Determine Topological Orientation of Cytochrome-P-450 in Microsomal Membrane. *EMBO J*. 9:2391-2397.
Schatz, P. J., and J. Beckwith. 1990. Genetic Analysis of Protein Export in *Escherichia coli*. *Annu Rev Genet*. 24:215-248.
Simon, S. M., and G. Blobel. 1991. A Protein-Conducting Channel in the Endoplasmic Reticulum. *Cell*. 65:371-380.
Theg, S. M., and F. J. Geske. 1992. Biophysical Characterization of a Transit Peptide Directing Chloroplast Protein Import. *Biochemistry*. 31:5053-5060.
van't Hof, R., R. A. Demel, K. Keegstra, and B. de Kruijff. 1991. Lipid Peptide Interactions Between Fragments of the Transit Peptide of Ribulose-1,5-Bisphosphate Carboxylase Oxygenase and Chloroplast Membrane Lipids. *FEBS Lett*. 291:350-354.
Verner, K., and G. Schatz. 1988. Protein translocation across membranes. *Science*. 241:1307-13.
von Heijne, G. 1986. Why mitochondria need a genome. *Febs Lett*. 198:1-4.
von Heijne, G. 1990. The signal peptide. *J Membr Biol*. 115:195-201.
von Heijne, G. 1992. Membrane Protein Structure Prediction: Hydrophobicity Analysis and the 'Positive Inside' Rule. *J Mol Biol*. 225:487-494.
von Heijne, G., and Y. Gavel. 1988. Topogenic signals in integral membrane proteins. *Eur J Biochem*. 174:671-8.
von Heijne, G., and C. Manoil. 1990. Membrane Proteins - From Sequence to Structure. *Protein Eng*. 4:109-112.
von Heijne, G., and K. Nishikawa. 1991. Chloroplast Transit Peptides - The Perfect Random Coil? *FEBS Lett*. 278:1-3.
von Heijne, G., J. Steppuhn, and R. G. Herrmann. 1989. Domain structure of mitochondrial and chloroplast targeting peptides. *Eur.J.Biochem*. 180:535-545.
Weiss, M. S., A. Kreusch, E. Schiltz, U. Nestel, W. Welte, J. Weckesser, and G. E. Schulz. 1991. The structure of porin from *Rhodobacter capsulata* at 1.8Å resolution. *FEBS Lett*. 280:379-382.
Welte, W., M. S. Weiss, U. Nestel, J. Weckesser, E. Schiltz, and G. E. Schulz. 1991. Prediction of the General Structure of OmpF and PhoE from the Sequence and Structure of Porin from Rhodobacter-Capsulatus - Orientation of Porin in the Membrane. *Biochim Biophys Acta*. 1080:271-274.
Wessels, H. P., and M. Spiess. 1988. Insertion of a multispanning membrane protein occurs sequentially and requires only one signal sequence. *Cell*. 55:61-70.
Wickner, W., A. J. M. Driessen, and F. U. Hartl. 1991. The Enzymology of Protein Translocation Across the *Escherichia coli* Plasma Membrane. *Annu Rev Biochem*. 60:101-124.
Yalovsky, S., H. Paulsen, D. Michaeli, P. R. Chitnis, and R. Nechushtai. 1992. Involvement of a Chloroplast HSP70 Heat Shock Protein in the Integration of a Protein (Light-Harvesting Complex Protein Precursor) into the Thylakoid Membrane. *Proc Natl Acad Sci USA*. 89:5616-5619.

RECOGNITION OF PRECURSOR PROTEINS BY THE MITOCHONDRIAL PROTEIN IMPORT APPARATUS

Karin Becker and Walter Neupert

Institute of Physiological Chemistry
University of Munich
Goethestrasse 33
D-W-8000 München 2
Germany

1. BIOGENESIS OF MITOCHONDRIA

Eukaryotic cells are functionally and morphologically subdivided into a number of membrane-bounded compartments. Each of the subcompartments, or organelles, has a specific protein composition according to its function in the cell. Mitochondria and, in plant cells, chloroplasts are unique in that they have their own genetic systems including DNA, ribosomes, tRNAs. Both organelles are surrounded by a double membrane and have extended internal membrane systems. According to the endosymbiotic hypothesis of the origin of mitochondria and chloroplasts both organelles are derived from prokaryotic endosymbionts in the ancestor of the eukaryotic cell (Margulis, 1970; Schwartz and Dayhoff, 1978). A large number of the original prokaryotic features are retained in the organelles. In the course of evolution, both organelles have lost most of their genetic information, which has been transferred to the nucleus (Schwartz and Dayhoff, 1978).

Mitochondria fulfil several important functions in the metabolism of a cell. The most prominent are catabolic reactions yielding NADH or $FADH_2$, namely the tricarboxylic acid cycle and the ß-oxidation of fatty acids. NADH and $FADH_2$ are funneled into the electron transport chain which supplies the cell with ATP via oxidative phosphorylation thereby providing the energy required for metabolic processes. The enzymes of the respiratory chain reside in the inner mitochondrial membrane, predominantly in the cristae, which are the invaginated regions of the inner membrane. The genes of the mitochondrial genome code for some proteins of the respiratory chain complexes and mitochondrial tRNAs and rRNAs (Attardi, 1981).

The vast majority of mitochondrial proteins, however, are encoded by nuclear genes. Their translation takes place in the cytoplasm and subsequently they are imported into the organelle. This raises the question of how these precursors are correctly sorted. It is generally recognized that mitochondrial precursors carry specific targeting signals and these are recognized by the mitochondrial protein import apparatus. As a result of this, translocation into the organelle and sorting to the various mitochondrial subcompartments takes place. The specific targeting to and translocation into mitochondria has been analysed in some detail and will be reviewed in the following.

2. GENERAL PRINCIPLES OF MITOCHONDRIAL PROTEIN IMPORT

Most of the characteristics of mitochondrial protein import described in the following were analysed *in vivo* and *in vitro* with the fungi *Neurospora crassa* and *Saccharomyces cerevisiae* (Attardi and Schatz, 1988; Hartl et al., 1989; Wienhues et al., 1992). A general model of the mitochondrial protein import is shown in figure 1.

Protein translocation into mitochondria is independent of translational elongation, it can occur post-translationally *in vitro* (Korb and Neupert, 1978; Maccecchini et al., 1979) and *in vivo* (Hallermayer et al., 1977; Schatz, 1979; Wienhues et al., 1991).

Proteins destined to the mitochondria contain specific targeting sequences which are recognized by the mitochondrial protein import apparatus. After binding to a receptor in the mitochondrial outer membrane, precursor proteins are inserted into this membrane. Translocation proceeds at the contact sites where the two mitochondrial membranes are in close apposition. In the presence of a membrane potential the precursors are then transported across the inner membrane into the matrix space. This translocation step is dependent on mitochondrial hsp70 which directly interacts with the precursor proteins in transit.

Upon arrival in the matrix space, the presequence is cleaved by the matrix localized mitochondrial processing peptidase (MPP) (Arretz et al., 1991; Hawlitschek et al., 1988). Some precursors have to undergo a second processing step by the mitochondrial intermediate peptidase (MIP) (Kalousek et al., 1988) or by MPP (subunit 9 of the F_1-F_0-ATPase) to be processed to their mature size. Folding of the proteins and assembly into oligomeric structures are mediated by interaction with mitochondrial hsp60 in the matrix (Ostermann et al., 1989). After import, sorting to other intramitochondrial compartments (matrix space, inner membrane, intermembrane space) takes place. Precursors destined to the inner membrane are inserted into this membrane or translocated across it (Mahlke et al., 1990). Some proteins destined to the intermembrane space have bipartite signal sequences, which consist of the matrix targeting sequence and a sorting signal to the intermembrane space. The sorting pathway of the Rieske iron-sulfur protein includes import into the matrix followed by retranslocation to the intermembrane space (Hartl et al., 1986). In the case of the cytochromes b_2 and c_1, two mechanisms have been proposed for their sorting to the intermembrane space: The "stop-transfer model" is based on the assumption that the second part of the bipartite presequence functions as a stop-transfer signal by which proteins get arrested at the inner membrane and then laterally diffuse into the intermembrane space (Glick et al., 1992; Van Loon and Schatz, 1987). According to the "conservative sorting model", proteins are first imported into the matrix space and then retranslocated to the intermembrane space by an export process resembling the general pathway of protein secretion in bacteria. This model takes into account the prokaryotic origin of mitochondria and the similarity between the intermembrane space targeting sequence and bacterial export signals (Hartl and Neupert, 1990; Koll et al., 1992).

There are a few exceptions to the above outlined general pathway of protein import into mitochondria. Some proteins use only part of the components by either deviating from the general pathway at an early stage (cytochrome c heme lyase, which uses only the receptor complex and GIP) (Lill et al., 1992b) or else getting inserted at a later stage (subunit Va of cytochrome oxidase, which is imported independently of receptors and inserted into the outer membrane presumably at the GIP-stage) (Miller and Cumsky, 1991).

Targeting of the intermembrane space protein cytochrome c is independent of any of the components of the general protein import apparatus: it is inserted into the outer membrane without the aid of protease-sensitive surface receptors and directly reaches the intermembrane space (Lill et al., 1992a; Stuart et al., 1990).

3. STEPS IN MITOCHONDRIAL PROTEIN IMPORT

3.1. Targeting Sequences Direct Precursor Proteins to Mitochondria

Proteins destined to the mitochondria carry targeting sequences which are recognized by the mitochondrial protein import apparatus. For most mitochondrial proteins the targeting sequence consists of an amino-terminal presequence which is cleaved upon arrival of the protein in the mitochondrial matrix. Mitochondrial presequences are characterized by being positively charged and having a tendency to form amphipathic helices in a hydrophobic environment (Von Heijne, 1986). Some mitochondrial preproteins, however, contain internal, sometimes multiple, noncleaved targeting sequences which have not yet been defined in detail (Pfanner et al., 1987).

The targeting function of mitochondrial presequences has been demonstrated by fusing mitochondrial targeting sequences to cytosolic proteins and showing that the fusion proteins are imported into mitochondria (Hurt et al., 1984; Rassow et al., 1990).

3.2. Precursors Are Recognized by Mitochondrial Import Receptors

An initial approach to the identification of mitochondrial protein receptors consisted in raising antisera to the various proteins of the outer membrane of *Neurospora crassa* mitochondria. IgGs and Fab fragments purified from these monospecific antisera were prebound to mitochondria, and inhibition of import of precursor proteins was tested.

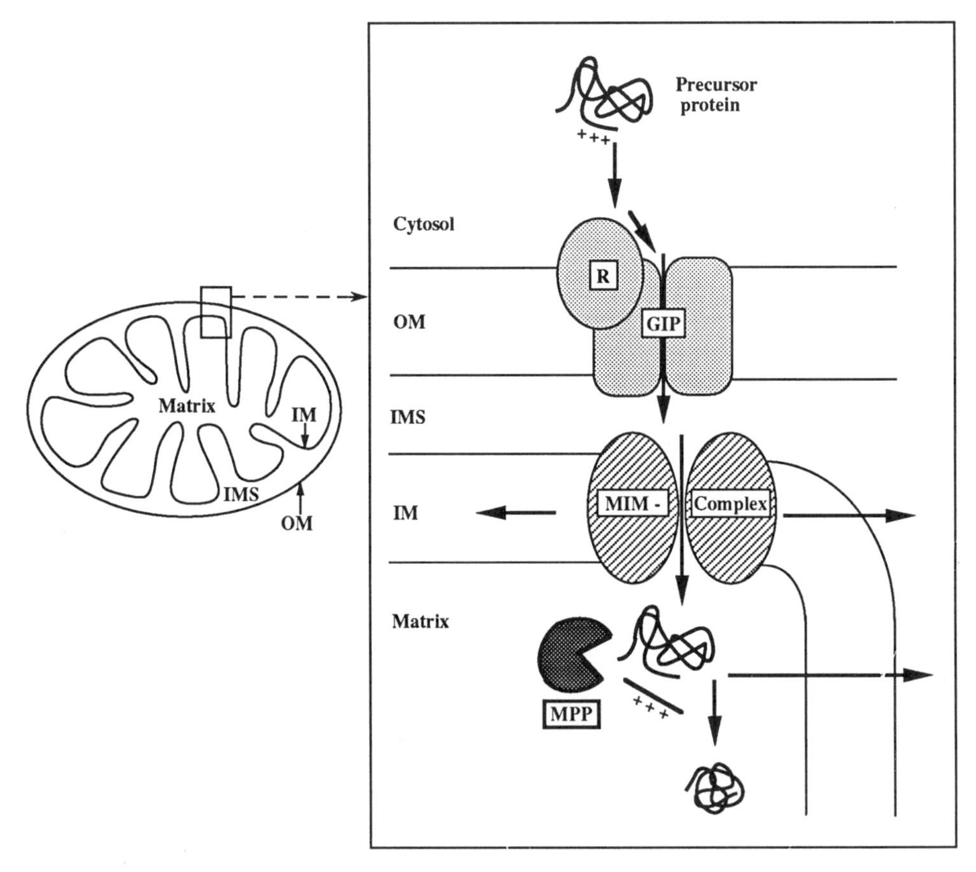

Fig. 1. Working model for the import of proteins into mitochondria
OM: Outer Membrane, IM: Inner Membrane, IMS: Intermembrane Space,
R: Receptor, GIP: General Insertion Pore, MIM-Complex: Inner Membrane Complex,
MPP: Mitochondrial Processing Peptidase

Antibodies directed against two outer membrane proteins of 19kDa and 72kDa, termed MOM19 and MOM72, were effective in inhibiting import. Antibodies to MOM19 inhibited import of most mitochondrial precursor proteins including all those containing cleavable presequences (Söllner et al., 1989). In contrast to this, antibodies to MOM72 strongly inhibited import of the AAC and only weakly inhibited import of other precursors (Söllner et al., 1990). These results suggested that MOM19 and MOM72 are mitochondrial protein import receptors of *Neurospora crassa.*

In yeast only one mitochondrial import receptor has been identified so far, named MAS70 or yeast-MOM72. Preincubation of yeast mitochondria with antibodies to this protein resulted in a decrease of the import of some precursor proteins (Hines et al., 1990). Since the MOM72 proteins of yeast and *Neurospora crassa* are homologous (32% identity, 46% similarity), they are supposed to fulfil similar functions in both organisms (Steger et al., 1990). A homolog of MOM19 in yeast has not yet been identified.

Following binding to the mitochondrial receptors MOM19 or MOM72, a precursor protein gets inserted into the outer membrane (Pfaller et al., 1988). At this step the pathways coming from the two different receptor proteins converge. This was demonstrated in competition experiments using several different precursor proteins. Precursors with different receptor specificity, like porin (MOM19) and AAC (MOM72), did not compete for binding to these receptors, as expected. But they were found to compete at a later stage of import, the embedding into the outer membrane (Pfaller et al., 1988). The corresponding activity was named General Insertion Pore (GIP). Beyond this step, import pathways were shown to diverge again depending on the final intramitochondrial localisation of the respective proteins.

3.3. A Protein Complex in the Mitochondrial Outer Membrane Contains Receptors and the General Insertion Pore

In an approach to identify components involved in binding and insertion of precursor proteins into the mitochondrial outer membrane, mitochondria from *Neurospora crassa* were lysed with the mild detergent digitonin and analysed for a protein complex containing the mitochondrial protein import receptors. A high molecular weight protein complex was found after fractionation by gel chromatography and subsequent coimmunoprecipitation with antibodies to MOM19. It contained the outer membrane proteins MOM19, MOM22, MOM38, and MOM72 (Kiebler et al., 1990). In another approach mitochondria were lysed with digitonin and directly subjected to immunoprecipitation using antibodies monospecific for MOM19. Several proteins were coimmunoprecipitated which were named according to their molecular weight: MOM7, MOM8, MOM19, MOM22, MOM30, MOM38, and MOM72 (Söllner et al., 1992) (see also figure 2). The proteins MOM19, MOM22, and MOM72 were very sensitive to treatment of mitochondria with proteases indicating that they were exposed at the mitochondrial surface and they have therefore been implicated in the receptor function. MOM38 is resistant to protease treatment and behaves exactly as predicted for a GIP-component. Therefore MOM38 has been tentatively assigned to the GIP function (Kiebler et al., 1990).

The characterisation of the numerous components of the mitochondrial import apparatus was greatly facilitated by establishing techniques to generate translocation intermediates at various stages of the import pathway of precursor proteins. This was most extensively studied with the ADP-ATP-carrier (AAC), a protein of the mitochondrial inner membrane (Pfanner and Neupert, 1987).

Upon ATP-depletion and dissipation of the membrane potential, the AAC was not completely imported but instead remained bound to the surface of the mitochondria. If a membrane potential was regenerated and ATP added to the reaction, this bound precursor could be chased to complete its import and assembly into the inner membrane, showing that the arrested precursor was on the correct import pathway. The AAC arrested at this stage was crosslinked to MOM19 and MOM72 (Söllner et al., 1992), again indicating that MOM19 and MOM72 are mitochondrial import receptors. Additionally, the arrested AAC could be coimmunoprecipitated with antibodies to MOM72 showing a direct interaction between these two proteins.

In attempts to characterize the GIP-activity, the AAC was stalled at this stage by dissipating the membrane potential in the presence of ATP prior to addition of the protein. It could be crosslinked to several outer membrane proteins: MOM7, MOM8, MOM19, MOM30, MOM38 (Söllner et al., 1992). The crosslink products corresponding to MOM7, MOM8, and MOM30 were not generated with the AAC arrested at the receptor-bound stage, suggesting that they are related to the GIP-function. MOM7 and MOM8 were crosslinked with very high efficiency suggesting that they might be located very closely to the translocating polypeptide chain. Since, on the other hand, these two proteins are extractable at pH 11.5, indicating that they are associated with the outer membrane through interaction with other membrane proteins, they have been implied in being part of the putative translocation channel (Söllner et al., 1992).

A similar protein complex was found in yeast mitochondria (Moczko et al., 1992). It contains proteins tentatively named MOM7, MOM8, MOM19, MOM22, MOM30, MOM38, and MOM72 according to their putative *Neurospora*-counterparts. The MOM38 of yeast mitochondria (also termed ISP42 for Import Site Protein of 42kDa) was shown to be part of the translocation machinery of yeast mitochondria (Vestweber et al., 1989). A yeast mutant depleted of ISP42 is not viable indicating that this protein has a crucial function in mitochondrial protein import (Baker et al., 1990).

To summarize these data, the several components of the mitochondrial receptor complex can be assigned to two activities: receptor activity and GIP activity. MOM19 and MOM72 have been shown to be mitochondrial import receptors (Söllner et al., 1989; Söllner et al., 1990). MOM38 is believed to be a main component of the GIP function since it fulfils the criteria defined for the GIP-stage of protein import (Kiebler et al., 1990). MOM7 and

MOM8 may be part of the translocation channel in the outer membrane which would be functionally identical to GIP (Söllner et al., 1992). The roles of MOM22 and MOM30 are still being investigated. A tentative model of the mitochondrial receptor complex is shown in figure 2.

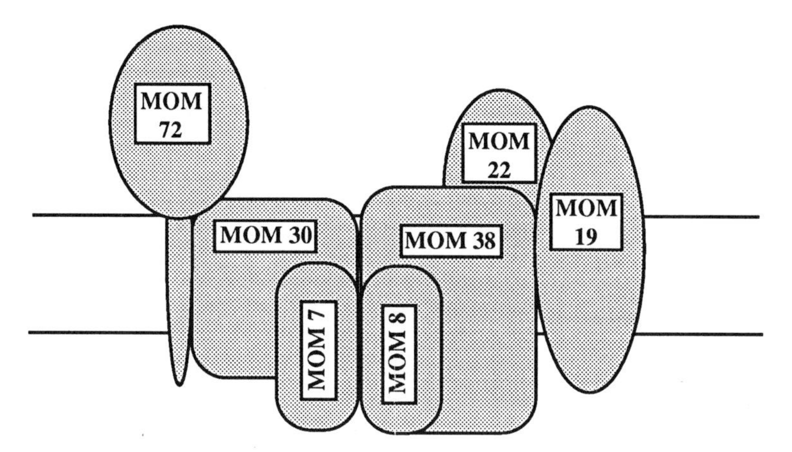

Fig. 2. Hypothetical working model for the mitochondrial receptor complex

3.4. The Inner Membrane Contains Independent Import Sites

Mitochondria from which the receptors have been removed by treatment with trypsin can import proteins only at a very low efficiency (Pfaller et al., 1989). Import into these mitochondria could be restored by removing the outer membrane (Ohba and Schatz, 1987). This showed that the mitochondrial inner membrane contains an independent translocation machinery. This inner membrane translocation machinery seems to be present in considerably higher numbers than the outer membrane machinery. This has been demonstrated by saturating the outer membrane import sites and the contact sites with two-membrane-spanning translocation intermediates which led to a complete block of the import of precursors into intact mitochondria. If, however, the outer membrane of these mitochondria was disrupted, import was restored (Hwang et al., 1989). This demonstrated that the inner membrane contains a fully functional import system which can operate independently of the outer membrane translocation machinery.

The existence of independent import sites in the inner membrane was also suggested by the following observations. In intact mitochondria, translocation intermediates were characterized which were localized in such a way that the aminoterminus was already protruding into the matrix while the carboxyterminus was found to be exposed to the intermembrane space (Hwang et al., 1991; Rassow and Pfanner, 1991). The precursors arrested in this position could be chased to the fully imported forms showing that they were on the correct import pathway and that import could be accomplished by the inner membrane machinery alone. These experimental data strongly indicate that protein translocation is accomplished by two distinct yet cooperating protein translocation machineries in the outer and inner mitochondrial membrane rather than a continuous channel across both mitochondrial membranes (Glick et al., 1991; Pfanner et al., 1992).

This notion was also supported by *in vitro* experiments with membrane vesicles derived from the mitochondrial membranes. Inner membrane vesicles virtually free of outer membrane components imported several mitochondrial precursor proteins in an ATP- and membrane potential dependent manner (Hwang et al., 1989). Outer membrane vesicles which contained no detectable amount of inner membrane proteins were able to import proteins such as cytochrome c heme lyase, porin, MOM19, and MOM22 (A. Mayer, R. Lill, W. Neupert, in preparation). This again showed that both membranes contain protein translocation complexes which can function independently of each other.

3.5. Translocation Contact Sites are Dynamic Structures

Electron microscopic studies revealed that protein import takes place at the so-called contact sites (Kellems et al., 1975; Pfanner et al., 1990; Schwaiger et al., 1987). These are the regions of the mitochondrial envelope where the two membranes are in close proximity being only 2-8 nm apart. The parts of the inner membrane contained within these regions have been termed inner boundary membrane as opposed to the cristae which protrude into the matrix space. Around 90% of the outer membrane is adjacent to the inner boundary membrane in intact mitochondria (Pfanner et al., 1992; Van der Klei et al., submitted). Stable contacts between the two membranes have been demonstrated with mitochondria in which the matrix space was condensed due to hyperosmotic treatment (Hackenbrock, 1968). This caused a retraction of the inner membrane which stayed attached to the outer membrane only at these stable morphological contact sites which then comprised about 5-10% of the mitochondrial surface (Rassow et al., 1989).

Precursor proteins arrested during translocation in a two-membrane-spanning manner were demonstrated to be localized at or close to the morphological contact sites by immuno electron microscopy (Schleyer and Neupert, 1985; Schwaiger et al., 1987).

In apparent contrast, the receptor proteins were found to be distributed all over the mitochondrial outer membrane, although MOM72 was concentrated at morphological contact sites (Söllner et al., 1989). This suggests that the protein translocation machinery of mitochondria is dynamic in that the components are able to move laterally within the membranes. The receptors and general insertion protein can assemble and disassemble again indicating a dynamic nature of the complexes (Kiebler et al., 1990). The translocation of proteins is facilitated by the proximity of the two membranes at the contact regions which ensures that a precursor protein can directly bind to an inner membrane complex upon emerging from the outer membrane import site. This is further facilitated by the higher number of the inner membrane complexes as compared to the outer membrane receptor complexes (Hwang et al., 1989; Pfanner et al., 1992).

In summary, protein import into the mitochondrial matrix appears to occur by successive translocation across the two mitochondrial membranes each of which contains a distinct import apparatus. Translocation can occur simultaneously across both membranes since an incoming polypeptide chain can directly be handed over to the inner membrane machinery upon emerging in the intermembrane space. This may be facilitated by the close proximity of the membranes and the higher number of inner membrane translocation complexes.

3.6. Mitochondrial Hsp70 is Essential for Translocation

Protein translocation into the matrix not only requires the electrical membrane potential, it is also dependent on mitochondrial hsp70, a heat shock protein of 70 kDa in the mitochondrial matrix. In a yeast mutant deficient of hsp70, import of precursor proteins into mitochondria was inhibited. In the mutant, precursors were arrested during translocation in a membrane-spanning manner (Kang et al., 1990). The block could be overcome by artificially unfolding the precursor protein in urea prior to import. The protein then was completely imported into the matrix but did not attain its native conformation. This suggested a dual role for hsp70 in import and folding of precursor proteins.

Hsp70 was shown to bind to incoming precursor proteins (Scherer et al., 1990). This binding keeps precursor proteins in an unfolded conformation. Release of the protein from hsp70 requires hydrolysis of ATP. Therefore, interaction with hsp70 was proposed to provide the driving force of mitochondrial protein import (Hartl et al., 1992). In addition, binding of hsp70 to segments of the precursor appearing on the matrix side may facilitate unfolding on the cytosolic side of the outer membrane (Neupert et al., 1990).

4. CYTOSOLIC FACTORS IN MITOCHONDRIAL PROTEIN IMPORT

How are precursor proteins targeted to the mitochondrial receptor complex? Precursors can only be recognized by the translocation machinery if the targeting sequence is exposed. But are the precursors, or their targeting sequences, directly recognized by the receptor proteins or are they targeted to the receptors by cytosolic factors analogous to the signal recognition particle (SRP) of the endoplasmic reticulum (ER) or the SecB protein of bacteria?

The signal sequences of proteins destined to the ER are recognized by the SRP as soon as they emerge from the ribosome. Translation then may slow down and the precursor-SRP-complex is targeted to the ER membrane via recognition of the SRP receptor (docking protein) on the ER surface and donates the precursor protein to a translocation channel in the ER membrane whereas SRP is recycled (Walter et al., 1984; Meyer, 1991).

In bacteria such as *E. coli*, proteins have been identified that are able to keep precursor proteins in an import competent conformation *in vitro*, SecB, DnaK, and GroEL (Lecker et al., 1989). Proteins that are to be exported are recognized by SecB, a soluble cytoplasmic protein, or other components which target the precursor to the SecA protein, a peripheral protein of the export machinery of the plasma membrane. Translocation then involves the membrane proteins SecY, SecE, other membrane proteins, and lipids (Hartl et al., 1990; Wickner et al., 1991).

In mitochondria, however, little is known about the role of cytosolic factors in the import of precursor proteins. A number of cytosolic factors have been identified which stimulated import into isolated mitochondria, including cytosolic 70kDa-heat-shock proteins (hsp70s) (Deshaies et al., 1988b; Sheffield et al., 1990; Murakami et al., 1988a), a "presequence binding factor" (PBF) (Murakami et al., 1988b; Murakami and Mori, 1990) and a "cytosolic factor" (Ono and Tuboi, 1988). Two main functions for a putative cytosolic factor can be distinguished: a chaperone-like function which consists in preserving the import-competence of precursor proteins by preventing misfolding, and a targeting function which would direct the precursor protein to the target membrane, in this case the mitochondrial membrane. It is well established that cytosolic hsp70 proteins play an important role in the translocation of preproteins into mitochondria (and the ER) (Deshaies et al., 1988a; Murakami et al., 1988a). Much less is known about the targeting of precursor proteins to the mitochondria: are precursors directly recognized by the mitochondrial protein import receptors, or do the receptors recognize a "targeting factor" which binds to the precursors?

4.1. Unfolding of Precursors is Required for Import

With a number of protein translocation systems studied it was shown that unfolding is a prerequisite for translocation of most proteins (Randall and Hardy, 1986; Schleyer and Neupert, 1985; for a review see: Langer and Neupert, 1991).

Protein translocation into mitochondria also requires unfolding of the precursor proteins (Eilers and Schatz, 1986; Schleyer and Neupert, 1985). The unfolded conformation is stabilized by interaction with the cytosolic hsp70 protein and probably other factors, which probably binds to the precursor directly after synthesis (Nelson et al., 1992).

Conversely, a stable tertiary structure inhibits import of proteins into mitochondria. This was demonstrated *in vitro* by inducing a tightly folded tertiary structure to a precursor protein and showing that this precursor cannot be imported afterwards (Eilers and Schatz, 1986; Rassow et al., 1989). This is not due to an artificial situation in the *in vitro* system. When in intact yeast cells a protein containing DHFR (dihydrofolate reductase) was expressed in the presence of a folate analog, aminopterine (which induced tight folding of the protein) import was inhibited. The fusion protein was accumulated in a two-membrane-spanning manner. Upon removal of the aminopterine, the protein was unfolded and subsequently import was completed (Wienhues et al., 1991).

4.2. Targeting Sequences are Recognized by Mitochondrial Receptors

It has been stated before that the mitochondrial outer membrane protein MOM19 is a receptor for most mitochondrial precursor proteins, including all those carrying amino-terminal cleavable targeting sequences. This suggests that MOM19 might recognize the targeting sequence of the precursor proteins. To test this hypothesis we constructed a fusion protein consisting of the matrix targeting sequence of cytochrome b_2 fused to DHFR, a cytosolic protein. The construct was imported into mitochondria, as expected. Upon preincubation of the mitochondria with antibodies to MOM19, import of the fusion protein was almost completely inhibited showing that import occurred via MOM19 (Becker et al., 1992). This demonstrated that a cleavable presequence was sufficient for directing import of a fusion protein via MOM19 and suggests that MOM19 indeed recognizes the presequence of mitochondrial proteins.

4.3. Purification of Mitochondrial Precursor Proteins

For most of the experiments described above, precursor proteins were synthesized *in vitro* in a rabbit reticulocyte lysate in the presence of ^{35}S-methionine (Pelham and Jackson, 1976). These radiolabelled precursors were added to isolated mitochondria and their import into the organelles was analysed. Most characteristics of the mitochondrial import pathways known to date were established using this experimental system.

To be able to analyse if cytosolic targeting factors are essential for mitochondrial protein import we needed purified precursor proteins which were devoid of the cytosolic factors present in reticulocyte lysate. To achieve this we overexpressed in *E. coli* and purified several mitochondrial proteins.The proteins were produced as insoluble inclusion bodies in the bacteria. They were purified by several washing steps in the presence of detergent and solubilized in urea-containing buffer. The proteins were radiolabelled by adding ^{35}S-sulfate to the media for the bacteria during induction (Wienhues et al., 1992). A fusion protein consisting of the first 148 amino acids of cytochrome b_2 -including the complete matrix targeting sequence- fused to the first half of the mature ß-subunit of the F_1-F_0-ATPase (pb_2(148)*) was chosen for the experiments described in the following section. This protein was produced in high amounts in the bacteria. It was purified to apparent homogeneity and solubilized in urea-containing buffer (Becker et al., 1992).

4.4. Import of a Purified Preprotein into Mitochondria is Independent of the Addition of a Cytosolic Signal Recognition Factor

The purified protein pb_2(148)* was imported into mitochondria independently of cytosolic hsp70 since it was present in an unfolded conformation due to its denaturation in urea. Import of the protein was dependent on the receptor MOM19 since antibodies to MOM19 inhibited import of the protein (Becker et al., 1992). Therefore this protein was used to test for the requirement for cytosolic targeting factors in mitochondrial protein import.

Mitochondria were washed in buffers containing up to 2M KCl to remove any attached cytosolic factors which might have been copurified during isolation of the mitochondria. These salt-washed mitochondria showed normal import activity.

The purified precursor protein pb_2(148)* was imported into the salt-washed mitochondria and showed the usual import characteristics: It was correctly processed in the matrix and the processed form was completely resistant to protease K added from outside showing that it had been completely imported. Preincubation of mitochondria with antibodies to MOM19 strongly inhibited the import of the protein showing that also in this case the import occurred via MOM19.

These results taken together suggest that MOM19 directly recognizes mitochondrial targeting sequences and that mitochondrial protein import *in vitro* is independent of the addition of cytosolic signal recognition factors (Becker et al., 1992). This, however, does not exclude that *in vivo* protein import into mitochondria can be mediated or enhanced by cytosolic factors which might increase the efficiency of targeting and translocation or prevent aggregation of precursor proteins.

5. ACKNOWLEDGMENTS

Work in the authors´ laboratory was supported by the Fonds der Chemischen Industrie and the Deutsche Forschungsgemeinschaft (Sonderforschungsbereich 184).

6. REFERENCES

Arretz, M., Schneider, H., Wienhues, U. , and Neupert, W., 1991, Processing of mitochondrial precursor proteins, *Biomed. Biochim. Acta* 50: 403-412.

Attardi, G., 1981, Organisation and expression of the mammalian mitochondrial genome: a lesson in economy, *Trends Biochem. Sci.* 6:86-89, 100-103.
Attardi, G. and Schatz, G., 1988, Biogenesis of mitochondria, *Ann. Rev. Cell. Biol.* 4:289-333.
Baker, K.P., Schaniel, A., Vestweber, D. , and Schatz, G., 1990, A yeast mitochondrial outer membrane protein essential for protein import and cell viability, *Nature* 348:605-609.
Becker, K., Guiard, B., Rassow, J., Söllner, T. , and Pfanner, N., 1992, Targeting of a chemically pure preprotein to mitochondria does not require the addition of a cytosolic signal recognition factor, *J. Biol. Chem.* 267:5637-5643.
Deshaies, R.J., Koch, B.D. , and Schekman, R., 1988a, The role of stress proteins in membrane biogenesis, *Trends Biochem. Sci.* 13:384-388.
Deshaies, R.J., Koch, B.D., Werner-Washburne, M., Craig, E.A. , and Schekman, R., 1988b, A subfamily of stress proteins facilitates translocation of secretory and mitochondrial precursor polypeptides, *Nature* 332:800-805.
Eilers, M. and Schatz, G., 1986, Binding of a specific ligand inhibits import of a purified precursor protein into mitocondria, *Nature* 322:228-232.
Glick, B., Wachter, C. , and Schatz, G., 1991, Protein import into mitochondria: two systems acting in tandem?, *Trends Cell Biol.* 1:99-103.
Glick, B., Brandt, A., Cunningham, K., Müller, S., Hallberg, R.L. , and Schatz, G., 1992, Cytochromes c1 and b2 are sorted to the intermembrane space of yeast mitochondria by a stop-transfer mechanism, *Cell* 69:809-822.
Hackenbrock, C.R., 1968, Chemical and physical fixation of isolated mitochondria in low-energy and high-energy states, *Proc. Natl. Acad. Sci. USA* 61:598-605.
Hallermayer, G., Zimmermann, R. , and Neupert, W., 1977, Kinetic studies on the transport of cytoplasmically synthesized proteins into the mitochondria in intact cells of *Neurospora crassa*, *Eur. J. Biochem.* 81:523-532.
Hartl, F.-U., Schmidt, B., Wachter, E., Weiss, H. , and Neupert, W., 1986, Transport into mitochondria and intramitochondrial sorting of the Fe/S protein of ubiquinol-cytochrome c reductase, *Cell* 47:939-951.
Hartl, F.-U., Pfanner, N., Nicholson, D.W. , and Neupert, W., 1989, Mitochondrial protein import, *Biochim. Biophys. Acta* 988:1-45.
Hartl, F.-U. and Neupert, W., 1990, Protein sorting to mitochondria: evolutionary conservations of folding and assembly, *Science* 247:930-938.
Hartl, F.-U., Lecker, S., Schiebel, E., Hendrick, J.P. , and Wickner, W., 1990, The binding cascade of SecB to SecA to SecY/E mediates preprotein targeting to the E. coli plasma membrane, *Cell* 63:269-279.
Hartl, F.-U., Martin, J. , and Neupert, W., 1992, Protein folding in the cell: the role of molecular chaperones Hsp70 and Hsp60, *Ann. Rev. Biophys.* 21:293-322.
Hawlitschek, G., Schneider, H., Schmidt, B., Tropschug, M., Hartl, F.-U. , and Neupert, W., 1988, Mitochondrial protein import: identification of processing peptidase and of PEP, a processing enhancing protein, *Cell* 53:795-806.
Hines, V., Brandt, A., Griffiths, G., Horstmann, H., Brütsch, H. , and Schatz, G., 1990, Protein import into yeast mitochondria is accelerated by the outer membrane protein MAS70, *EMBO J.* 9:3191-3200.
Hurt, E.C., Pesold-Hurt, B. , and Schatz, G., 1984, The amino-terminal region of an imported mitochondrial precursor polypeptide can direct cytoplasmic dihydrofolate reductase into the mitochondrial matrix, *EMBO J.* 3:3149-3156.
Hwang, S., Jascur, T., Vestweber, D., Pon, L. , and Schatz, G., 1989, Disrupted yeast mitochondria can import precursor proteins directly through their inner membrane, *J. Cell Biol.* 109:487-493.
Hwang, S.T., Wachter, C. , and Schatz, G., 1991, Protein import into the yeast mitochondrial matrix: a new translocational intermediate between the two mitochondrial membranes, *J. Biol. Chem.* 266:21083-21089.
Kalousek, F., Hendrick, J.P. , and Rosenberg, L.E., 1988, Two mitochondrial matrix proteases act sequentially in the processing of mammalian matrix enzymes, *Proc. Natl. Acad. Sci. USA* 85:7536-7540.
Kang, P.J., Ostermann, J., Shilling, J., Neupert, W., Craig, E.A. , and Pfanner, N., 1990, Requirement for hsp70 in the mitochondrial matrix for translocation and folding or precursor proteins, *Nature* 348:137-143.
Kellems, R.E., Allison, V.F. , and Butow, R.A., 1975, Cytoplasmic type 80S ribosomes associated with yeast mitochondria, *J. Cell Biol.* 65:1-14.
Kiebler, M., Pfaller, R., Söllner, T., Griffiths, G., Horstmann, H., Pfanner, N. , and Neupert, W., 1990, Identification of a mitochondrial receptor complex required for recognition and membrane insertion of precursor proteins, *Nature* 348:610-616.

Koll, H., Guiard, B., J., R., Ostermann, J., Horwich, A.L., Neupert, W. , and Hartl, F.-U., 1992, Antifolding activity of hsp60 couples protein import into the mitochondrial matrix with export to the intermembrane space, *Cell* 68:1163-1175.

Korb, H. and Neupert, W., 1978, Biogenesis of cytochrome c in *Neurospora crassa*, *Eur. J. Biochem.* 91:609-620.

Langer, T. and Neupert, W., 1991, Heatshock proteins hsp60 and hsp70: their roles in folding, assembly and membrane translocation of proteins, *Curr. Top. Microbiol. Immunol.* 160:3-30.

Lecker, S., Lill, R., Ziegelhoffer, T., Bassford, J.P.J., Kumamoto, C.A. , and Wickner, W., 1989, Three pure chaperone proteins of Escherichia coli - SecB, trigger factor and GroEL - form soluble complexes with precursor proteins in vitro, *EMBO J.* 8:2703-2709.

Lill, R., Hergersberg, C., Schneider, H., Söllner, T., Stuart, R. , and Neupert, W., 1992a, General and exceptional pathways of protein import into the sub-mitochondrial compartments, *in:* Membrane biogenesis and protein targeting, W. Neupert and R. Lill, ed., Elsevier Science Publishers.

Lill, R., Stuart, R.A., Drygas, M.E., Nargang, F.E. , and Neupert, W., 1992b, Import of cytochrome c heme lyase into mitochondria: a novel pathway into the intermembrane space, *EMBO J.* 11:449-456.

Maccecchini, M.-L., Rudin, Y. , and Schatz, G., 1979, Transport of proteins across the mitochondrial outer membrane: a precursor form of the cytoplasmically made intermembrane enzyme cytochrome c peroxidase, *J. Biol. Chem.* 254:7468-7471.

Mahlke, K., Pfanner, N., Martin, J., Horwich, A.L., Hartl, F.-U. , and Neupert, W., 1990, Sorting pathways of mitochondrial inner membrane proteins, *Eur. J. Biochem.* 192:551-555.

Margulis, L., 1970, Aerobiosis and the mitochondrion, *in:* Origin of eukaryotic cells, Yale University Press, New Haven and London.

Meyer, D.I., 1991, Protein translocation into the endoplasmic reticulum: a light at the end of the tunnel, *Trends Cell Biol.* 1:154-158.

Miller, B.R. and Cumsky, M.G., 1991, An unusual mitochondrial import pathway for the precursor to yeast cytochrome c oxidase subunit Va, *J. Cell. Biol.* 112:833-841.

Moczko, M., Dietmeier, K., Söllner, T., Segui, B., Steger, H.F., Neupert, W. , and Pfanner, N., 1992, Identification of the mitochondrial receptor complex in *Saccharomyces cerevisiae*, *FEBS Lett.* 310:265-268.

Murakami, H., Pain, D. , and Blobel, G., 1988a, 70-kD heat shock-related protein is one of at least two distinct cytosolic factors stimulating protein import into mitochondria, *J. Cell. Biol.* 107:2051-2057.

Murakami, K., Amaya, Y., Takigushi, M., Ebina, Y. , and Mori, M., 1988b, Reconstitution of mitochondrial protein transport with purified ornithine carbamoyltransferase precursor expressed in *Escherichia coli*, *J. Biol. Chem.* 263:18437-18442.

Murakami, K. and Mori, M., 1990, Purified presequence binding factor (PBF) forms an import-competent complex with a purified mitochondrial precursor protein, *EMBO J.* 9:3201-3208.

Nelson, R.J., Ziegelhoffer, T., Nicolet, C., Werner-Washburne, M. , and Craig, E.A., 1992, The translation machinery and 70kd heat schock protein cooperate in protein synthesis, *Cell* 71:97-105.

Neupert, W., Hartl, F.-U., Craig, E.A. , and Pfanner, N., 1990, How do polypeptides cross mitochondrial membranes?, *Cell* 63:447-450.

Ohba, M. and Schatz, G., 1987, Disruption of the outer membrane restores protein import to trypsin-treated yeast mitochondria, *EMBO J.* 6:2117-2122.

Ono, H. and Tuboi, S., 1988, The cytosolic factor required for import of precursors of mitochondrial proteins into mitochondria, *J. Biol. Chem.* 263:3188-3193.

Ostermann, J., Horwich, A.L., Neupert, W. , and Hartl, F.-U., 1989, Protein folding in mitochondria requires complex formation with hsp60 and ATP hydrolysis, *Nature* 341:125-130.

Pelham, H.R.B. and Jackson, R.J., 1976, An efficient mRNA-dependent translation system from reticulocyte lysates, *Eur. J. Biochem.* 67:247-256.

Pfaller, R., Steger, H.F., Rassow, J., Pfanner, N. , and Neupert, W., 1988, Import pathways of precursor proteins into mitochondria: multiple receptor sites are followed by a common insertion site, *J. Cell. Biol.* 107:2483-2490.

Pfaller, R., Pfanner, N. , and Neupert, W., 1989, Mitochondrial protein import: bypass of proteinaceous surface receptors can occur with low specificity and efficiency, *J. Biol. Chem.* 264:34-39.

Pfanner, N., Hoeben, P., Tropschug, M. , and Neupert, W., 1987, The carboxyterminal two-thirds of the ADP/ATP carrier polypeptide contains sufficient information to direct translocation into mitochondria, *J. Biol. Chem.* 262:14851-14854.

Pfanner, N. and Neupert, W., 1987, Distinct steps in the import of the ADP/ATP carrier into mitochondria, *J. Biol. Chem.* 262:7528-7536.

Pfanner, N., Rassow, J., Wienhues, U., Hergersberg, C., Söllner, T., Becker, K. , and Neupert, W., 1990, Contact sites between inner and outer membranes: structure and role in protein translocation into mitochondria, *Biochim. Biophys. Acta* 1018:239-242.

Pfanner, N., Rassow, J., van der Klei, I. , and Neupert, W., 1992, A dynamic model of the mitochondrial protein import machinery, *Cell* 68:999-1002.

Randall, L.L. and Hardy, S.J.S., 1986, Correlation of competence for export with lack of tertiary structure of the mature species: a study in vivo of maltose binding protein in E. coli, *Cell* 46:921-928.

Rassow, J., Guiard, B., Wienhues, U., Herzog, V., Hartl, F.-U. , and Neupert, W., 1989, Translocation arrest by reversible folding of a precursor protein imported into mitochondria. A means to quantitate translocation contact sites., *J. Cell. Biol.* 109:1421-1428.

Rassow, J., Hartl, F.-U., Guiard, B., Pfanner, N. , and Neupert, W., 1990, Polypetides traverse the mitochondrial envelope in an extended state, *FEBS Lett.* 275:190-194.

Rassow, J. and Pfanner, N., 1991, Mitochondrial preproteins en route from the outer membrane to the inner membrane are exposed to the intermembrane space, *FEBS Lett.* 293:85-88.

Schatz, G., 1979, How mitochondria import proteins from the cytoplasm, *FEBS Lett.* 103:203-211.

Scherer, P.E., Krieg, U.C., Hwang, S.T., Vestweber, D. , and Schatz, G., 1990, A precursor protein partly translocated into yeast mitochondria is bound to a 70 kD mitochondrial stress protein, *EMBO J.* 9:4315-4322.

Schleyer, M. and Neupert, W., 1985, Transport of proteins into mitochondria: translocational intermediates spanning contact sites between outer and inner membranes, *Cell* 43:339-350.

Schwaiger, M., Herzog, V. , and Neupert, W., 1987, Characterization of translocation contact sites involved in the import of mitochondrial proteins, *J. Cell. Biol.* 105:235-246.

Schwartz, R.M. and Dayhoff, M.O., 1978, Origins of prokaryotes, eukaryotes, mitochondria, and chloroplasts, *Science* 199:395-403.

Sheffield, W.P., Shore, G.C. , and Randall, S.K., 1990, Mitochondrial precursor protein. Effects of 70-kD heat shock protein on polypeptide folding, aggregation, and import competence, *J. Biol. Chem.* 265:11069-11076.

Söllner, T., Griffiths, G., Pfaller, R., Pfanner, N. , and Neupert, W., 1989, MOM19, an import receptor for mitochondrial precursor proteins, *Cell* 59:1061-1070.

Söllner, T., Pfaller, R., Griffiths, G., Pfanner, N. , and Neupert, W., 1990, A mitochondrial import receptor for the ATP/ADP carrier, *Cell* 62:107-115.

Söllner, T., Rassow, J., Wiedmann, M., Schlossmann, J., Keil, P., Neupert, W. , and Pfanner, N., 1992, Mapping of the protein import machinery in the mitochondrial outer membrane by crosslinking of translocation intermediates, *Nature* 355:84-87.

Steger, H.F., Söllner, T., Kiebler, M., Dietmeier, K.A., Pfaller, R., Trülzsch, K.S., Tropschug, M., Neupert, W. , and Pfanner, N., 1990, Import of ADP/ATP-carrier into mitochondria: two receptors act in parallel, *J. Cell Biol.* 111:2353-2363.

Stuart, R.A., Nicholson, D.W. , and Neupert, W., 1990, Early steps in mitochodrial protein import: receptor functions can be substituted by the membrane insertion activity of apocytochrome c, *Cell* 60:31-49.

Van der Klei, I., Veenhuis, M. , and Neupert, W., submitted, A morphological view on mitochondrial protein targeting.

Van Loon, A.P.G.M. and Schatz, G., 1987, Transport of proteins to the inner membrane: the "sorting" domain of the cytochrome c1 presequence is a stop-transfer sequence for the mitochondrial inner membrane, *EMBO J.* 6:2441-2448.

Vestweber, D., Brunner, J., Baker, A. , and Schatz, G., 1989, A 42K outer-membrane protein is a component of the yeast mitochondrial protein import site, *Nature* 341:205-209.

Von Heijne, G., 1986, Mitochondrial targeting sequences may form amphiphilic helices, *EMBO J.* 5:1335-1342.

Walter, P., Gilmore, R. , and Blobel, G., 1984, Protein translocation across the endoplasmic reticulum, *Cell* 38:5-8.

Wickner, W., Driessen, A.J.W. , and Hartl, F.-U., 1991, The enzymology of protein translocation across the Escherichia coli plasma membrane, *Ann. Rev. Biochem.* 60:101-124.

Wienhues, U., Becker, K., Schleyer, M., Guiard, B., Tropschug, M., Horwich, A.L., Pfanner, N. , and Neupert, W., 1991, Protein folding causes an arrest of preprotein translocation into mitochondria in vivo, *J. Cell. Biol.* 115:1601-1609.

Wienhues, U., Koll, H., Becker, K., Guiard, B. , and Hartl, F.-U., 1992, Protein targeting to mitochondria, *in:* A Practical Approach to Protein Targeting, IRL (Oxford University Press).

A SHORT N-TERMINAL DOMAIN OF MITOCHONDRIAL CREATINE KINASE IS INVOLVED IN OCTAMER FORMATION BUT NOT IN MEMBRANE BINDING

Philipp Kaldis, Hans M. Eppenberger, and Theo Wallimann

Institute for Cell Biology

Swiss Federal Institute of Technology, ETH Hönggerberg

CH-8093 Zürich, Switzerland

INTRODUCTION

Creatine kinase (CK) isoenzymes catalyze the reversible transfer of the phosphoryl group of phosphocreatine to ADP (Kenyon and Reed, 1983). CK genes are expressed in several tissues with high energy demands, e.g. in skeletal and cardiac muscle, brain, photoreceptor cells, and spermatozoa. In most tissues, nuclear genes for cytosolic as well as mitochondrial CK isoenzymes are coexpressed. CK isoenzymes are similar in terms of activity, but differ in biochemical and structural aspects (Basson et al., 1985). Cytosolic CKs have an acidic pI, are dimeric molecules and exist either free in solution or associated with subcellular structures, e.g. the myofibrillar M-line or the sarcoplasmic reticulum where they are functionally coupled to the myosin ATPase (Wallimann et al., 1984) and the Ca^{2+}-pump (Rossi et al., 1990), respectively. Mitochondrial CK (Mi-CK) isoenzymes in most cases

New Developments in Lipid-Protein Interactions and Receptor Function
Edited by K.W.A. Wirtz *et al.*, Plenum Press, 1993

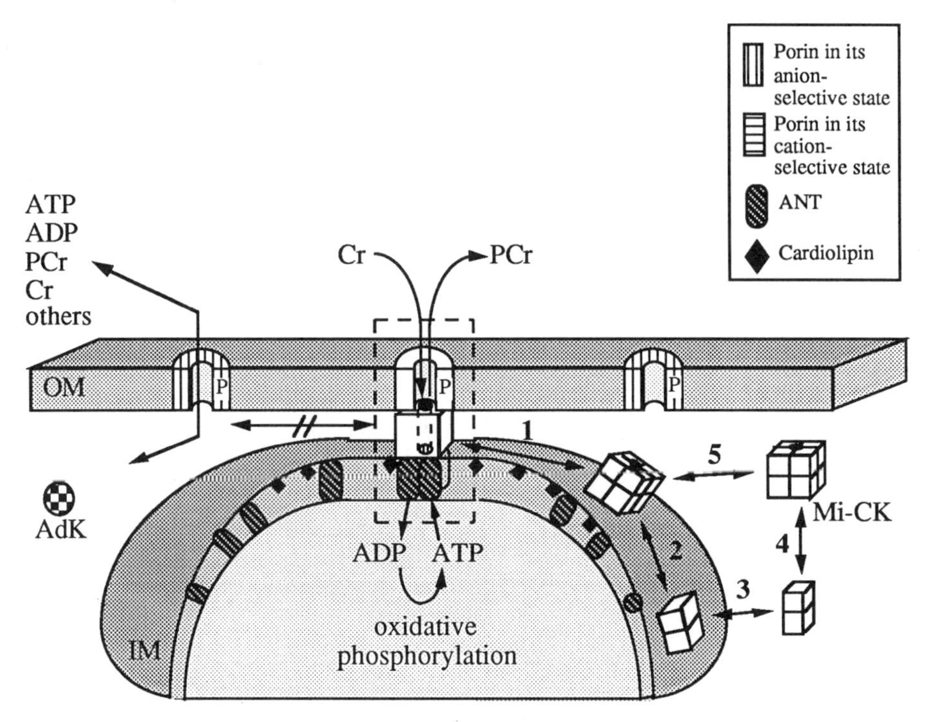

Figure 1

Structure and function of Mi-CK within the mitochondrial intermembrane space with special reference to the contact sites (CS).
Within CS (surrounded by a dashed line), exclusively octameric Mi-CK interacts with both the inner and the outer mitochondrial membrane and is thought to be functionally coupled to ANT of the inner and porin of the outer membrane. Below the contacts, Mi-CK may be dimeric or octameric, and it may be either bound to the inner membrane or free in the intermembrane space. In contrast to Mi-CK, adenylate kinase (AdK) is not firmly bound to the membranes. In the model, it is schematically indicated that two different microcompartments for CK substrates exist within and beyond the CS. Due to this microcompartmentation, PCr production by Mi-CK may still proceed even at a high cytosolic ATP/ADP ratio. Note that by the schematic representation it is *not* intended to suggest that Mi-CK is only active within the CS. Furthermore, it should not be inferred that below the CS, the two mitochondrial membranes are far from each other. In fact, electron microscopic analysis of in situ mitochondria revealed that the mitochondrial inner and outer membrane form a 5-layered structure with a width of only 12 nm, whereas a 7-layered structure would be expected for two separated membranes. Tetrameric ANT of the inner membrane, (tetrameric) porin of the outer membrane and octameric Mi-CK are thought to form a highly organized multienzyme complex and thereby create a microcompartment allowing efficient substrate channelling between the three enzymes. According to this model, PCr is produced within the central channel of the octameric Mi-CK molecule and is directly pulled out of the mitochondria by electrostatic repulsion through the superimposed porin molecule, thus allowing further PCr production even in the presence of a high cytosolic ATP/ADP ratio. For simplicity, the subunit "boundaries" were omitted in the Mi-CK octamer localized within the CS. Reprinted from Wyss et al. (1992), with copyright permission from Elsevier Sience Publishers, Amsterdam, The Netherlands.

have a basic pI, and form either dimeric or octameric molecules. Mi-CK is restricted to the mitochondrial intermembrane compartment and is located along the entire cristae membrane, and is probably accumulated at mitochondrial contact sites (Wegmann et al., 1991; Kottke et al., 1991). According to the phosphocreatine shuttle or circuit hypothesis, the CK/phosphocreatine system ensures the availability of ATP at sites of energy consumption and facilitates phosphocreatine synthesis at sites of ATP production (reviewed in [Wallimann et al., 1992]).

Several studies on isolated mitochondria have shown that mitochondrial oxidative phosphorylation and the Mi-CK reaction are functionally coupled: Mi-CK preferentially utilizes the ATP synthesized via oxidative phosphorylation for phosphocreatine synthesis (see Figure 1). Consequently, phosphocreatine, instead of ATP, is the main product of oxidative phosphorylation (for a review see [Wyss et al., 1992]). For mitochondrial export of high energy phosphates, substrates of the CK reaction have to be transported across the inner (ATP, ADP) and outer (creatine, phosphocreatine) mitochondrial membrane. Transport is mediated by the adenine nucleotide translocator (ANT) of the inner membrane and by mitochondrial porin of the outer membrane, respectively. Recently, it was shown that Mi-CK interacts similarly well with inner and outer mitochondrial membranes (Rojo et al., 1991a) and that Mi-CK mediates intermembrane adhesion between inner and outer mitochondrial membranes (Rojo et al., 1991b) (see Figure 2). In these studies, octameric Mi-CK proved to be the oligomeric form that preferentially interacts with membranes (see also [Schlegel et al., 1990]).

Although evidence is accumulating that, in vivo, Mi-CK exists mostly as octamer, it is not known whether or to what extent dimer/octamer interconversions are physiologically relevant (reviewed in [Wyss et al., 1992]). *In vitro*, the chicken heart Mi-CK octamer is stable at high protein concentrations and dissociates very slowly into dimers upon dilution. The octamer/dimer equilibrium is critically influenced by Mi-CK protein concentration and pH, as well as by the presence of CK substrates (Schlegel et al., 1990). Up to now, little is known about the factors governing the dissociation/reassociation of Mi-CK oligomers from and with mitochondrial membranes, except that by high pH phosphate buffer and nucleotides, Mi-CK can be released almost quantitatively from mitoplasts if the outer membrane has been removed by hypo-osmotic treatment before extraction (Schlegel et al., 1988, 1990; Wyss et al., 1990, 1992).

Thus, a better understanding of the factors involved in the oligomerization and membrane interaction of Mi-CK is relevant for gaining a deeper insight into the enzyme's function in active mitochondria *in vivo*. In addition, information on the domains of Mi-CK which are involved in oligomerization and membrane interaction may help to better understand the function of Mi-CK on a molecular level. In this work, an N-terminal deletion mutant of Mi-CK was generated. The membrane binding of wild-type and mutant Mi-CK to mitoplasts and liposomes was studied and compared. In addition, the role of the oligomeric structure, as well as the influence of the N-terminus of the enzyme with respect to membrane binding of Mi-CK, was investigated.

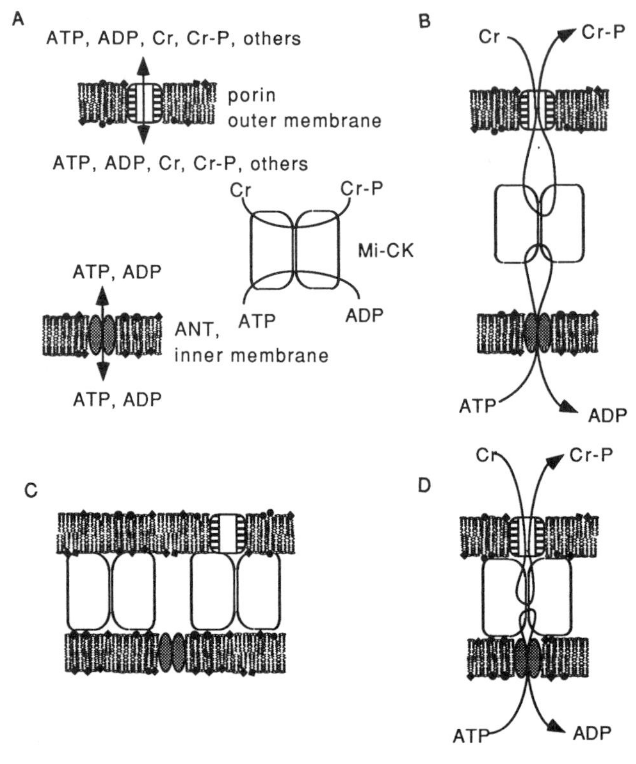

Figure 2

Model of a multienzyme complex that couples the creatine kinase reaction to oxidative phosphorylation. A: The enzymes involved in this process all catalyze reversible reactions. **B**: The coupled creatine kinase reaction utilizes ATP synthesized by oxidative phosphorylation in the mitochondrial matrix and creatine (Cr) from the cytosolic compartment and transports the reaction products ADP and phosphocreatine (P-Cr) back to the matrix and the cytosolic compartment, respectively. **C**: Model depicting the simultaneous interaction of Mi-CK with the mitochondrial inner and outer membrane. **D**: The hypothesis of this work is that a highly ordered multienzyme complex consisting of ANT, Mi-CK and porin is maintained by the membrane binding properties of Mi-CK and ensures vectoriality of the coupled creatine kinase reaction. (Courtesy of M. Rojo).

MATERIALS AND METHODS

Enzyme Assays, Electrophoresis

CK activity was determined in the direction of ATP synthesis by the pH-Stat method as described by Wallimann et al. (1984). 1 IU corresponds to 1 μmol of PCr transphosphorylated per min at pH 7.0 and 25°C. SDS-PAGE was performed according to Lämmli et al. (1970), and proteins were visualized by staining with Coomassie Brilliant Blue R-250.

E. coli Strains, Plasmids, DNA Manipulations

E. coli strains BL21(DE3)pLysS and expression vector pET-3b have been described by Studier et al. (1990), media and DNA manipulations by Furter et al. (1992). The expression plasmid PK-Δ5 was derived from the wild-type Mi-CK expression plasmid pRF23 (Furter et al., 1992) by deleting the 15 N-terminal nucleotides and by introducing a new initiating methionine codon. Thus, PK-Δ5 was designed to encode a mitochondrial CK lacking the first five amino acids of the mature wild-type enzyme.

Expression and Purification of wild-type and mutant Mi-CK

Wild-type Mi-CK and the N-terminally deleted derivative were expressed in E. coli from the plasmids pRF23 and PK-Δ5, respectively, and purified to homogeneity as previously described (Furter et al., 1992). N-terminal protein sequencing showed that the initiating Met residue was not removed in Δ5-Mi-CK:

NH_2-	T	V	H	E	K	R	L	wild-type Mi-CK
			NH_2-		M	R	L	Δ5-Mi-CK

Generation of Mi-CK Dimers and Octamers

The proportion of octamers and dimers is dependent on the protein concentration of Mi-CK, on the presence of substrates and to a lesser extent on the pH. Purified wild-type Mi-CK consists of approximately 85% octamers and 15% dimers. Octamers are stable over weeks, but can be dissociated into dimers within 30 min by adding 4 mM ADP, 5 mM Mg^{2+}, 20 mM creatine and 50 mM nitrate, thus forming a transition state-analogue complex (TSAC) with Mi-CK. Reassociation of dimers into octamers can be achieved by removing the substrates on a desalting column or by dialysis. In contrast, Δ5-Mi-CK was simply dimerized by dilution to 0.1 mg/ml (see results).

Binding of Mi-CK to Mitoplasts

Isolation of chicken heart mitochondria and extraction of Mi-CK from mitoplasts were performed as described by Schlegel et al. (1990). After isolation of the mitochondria, the outer membrane was removed by hypotonic treatment and the endogenous Mi-CK solubilized by incubation in sodium phosphate buffer at pH 9.0. With this procedure, approximately 87% of the endogenous Mi-CK was removed from mitoplasts. The remaining Mi-CK activity was determined in order to correct for endogenous activity in the subsequent binding experiments. For the binding of octamers, 1 ml of mitoplasts (about 5.5 μmol phospholipids) was mixed with 1.5 ml of binding buffer (10 mM NaH_2PO_4, 5 mM BME). Then, 15 μl of Δ5-Mi-CK (8 mg/ml, 8 IU, consisting of 90% octamers) were added. For the binding of Δ5-Mi-CK dimers, 1 ml of mitoplasts was mixed with 0.4 ml of binding buffer, followed by 1.1 ml of diluted Δ5-Mi-CK consisting exclusively of dimers (0.1 mg/ml, 6 IU, stored for 50 hours at 4°C). Finally, in the individual binding experiments, the desired pH was adjusted to the values indicated by adding diluted NaOH or HCl. After incubation of the samples for 15 min at 4°C on a shaker, mitoplasts were sedimented for 15 min at 100'000 g, the supernatants removed, and the pellets resuspended in the same volume of binding buffer. The CK activity was measured in both fractions to calculate the percentage of bound Mi-CK. In all experiments the ratio of phospholipids to CK activity was 700-900 nmol/1 IU.

Preparation of Liposomes

Lipids were extracted from chicken heart mitoplasts as described (Hovius et al., 1990), dissolved in $CHCl_3$/MeOH (1:1, v/v) to a concentration of 1-5 mM lipid phosphorus, and stored under N_2 at -20°C. Lipids were dried by evaporation, suspended in binding buffer and sonicated for 20 min with a Branson sonifier under N_2 and with good cooling in an ice bath. Liposomes were centrifuged for 65 min at 105'000 g. The pellet containing the liposomes was resuspended in binding buffer, and the content of lipid phosphorus was measured by the method of Fiske and SubbaRow (1925) after destruction of phospholipids with 70% perchloric acid.

Binding of Mi-CK to Liposomes

For the binding of Δ5-Mi-CK octamers, 175 μl of liposomes (about 1250 nmol phospholipids) were mixed with 297 μl of binding buffer. Then, 3 μl of Δ5-Mi-CK (8 mg/ml, 1.5 IU, consisting of 90% octamers) were added. For the Δ5-Mi-CK dimers, 175 μl of liposomes were mixed with 300 μl of Δ5-Mi-CK (0.1 mg/ml, 1.5 IU, consisting of 100% dimers). After adjustment of the pH with diluted NaOH or HCl, the samples were incubated for 15 min at 4°C on a shaker and afterwards centrifuged for 30 min in an Airfuge™ (Beckmann, 30° rotor, 178'000 g). The supernatants were removed and the pellets resuspended in the same volume of binding buffer. The CK activity in both fractions was measured to calculate the percentage of bound Mi-CK. In all experiments, the ratio of phospholipids to CK activity was 700-900 nmol/1 IU.

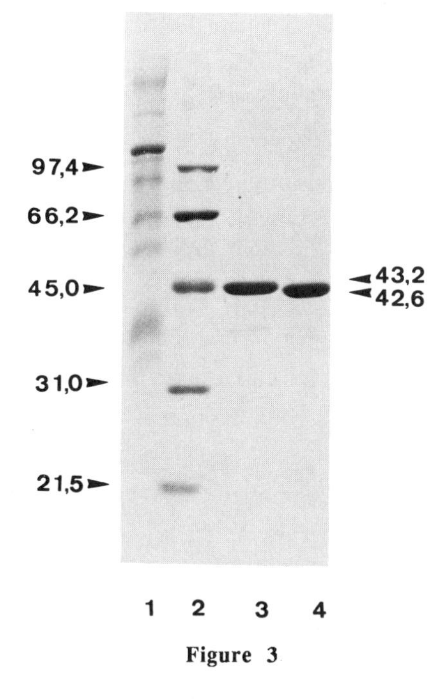

Figure 3

SDS-PAGE of wild-type and Δ5-Mi-CK expressed in E. coli
10% SDS-polyacrylamide gel showing in lanes 1 and 2 molecular weight markers. Wild-type Mi-CK, expressed in E. coli and purified to homogeneity, is shown in lane 3, Δ5-Mi-CK in lane 4. Δ5-Mi-CK has a slightly higher apparent mobility, consistent with the lack of five amino acids. The relative molecular mass of wild-type and Δ5-Mi-CK subunits are indicated in kDa.

RESULTS

In order to locate the regions in the Mi-CK molecule relevant for association of Mi-CK dimers into octamers, limited proteolytic digests with Mi-CK dimers and octamers were performed. A specific removal of a few amino acids at the N-terminus by proteinase Lys C led to a drastic decrease in the stability of the Mi-CK octamers (Ph. Kaldis, unpublished observation).

To obtain larger quantities of N-terminally deleted Mi-CK, a deletion mutant of the Mi-CK cDNA, called PK-Δ5, was generated. The protein expressed from this mutant cDNA was lacking the first five amino acids of the mature Mi-CK N-terminus, but retained the initiating methionine as the first N-terminal amino acid. Δ5-Mi-CK had a slightly higher apparent mobility on SDS-PAGE (Fig. 3), consistent with a difference in molecular mass of 600 Da deduced from the primary structure.

Δ5-Mi-CK displayed a specific activity and a K_m for PCr (80 IU/mg and 1.4 mM, respectively) similar to the wild-type enzyme, suggesting that Δ5-Mi-CK is correctly folded. While the catalytic properties of Δ5-Mi-CK were unchanged, its ability to form stable octamers was drastically decreased. Even at high protein concentration (>1mg/ml), only 70% of Δ5-Mi-CK existed as octamers. Upon dilution to a protein concentration of 0.1 mg/ml, Δ5-Mi-CK dissociated completely into dimers within hours. Wild-type Mi-CK, on the other hand, normally consists of ≥85% octamers which are stable after dilution for several weeks (see Materials and Methods). These findings show that the N-terminus of Mi-CK is important for the formation of stable octameric molecules. These properties make Δ5-Mi-CK an attractive alternative to study the membrane binding properties of dimeric Mi-CK. In the past, studies with the dimeric form of wild-type Mi-CK have been hampered by the fact that dimeric Mi-CK samples always contained a considerable amount (approximately 15%) of octameric Mi-CK and that substrates have to be used to induce dissociation into dimers. In addition, removal of the substrates or concentration of the sample led to fast reassociation into octamers (see Materials and Methods). Consequently, interpretation of experiments with wild-type Mi-CK dimers has often been complicated. With Δ5-Mi-CK at hand, experiments can now be designed where the properties of octamers and dimers can be compared under identical experimental conditions and where close to 100% dimers or octamers are present. An initial set of experiments with Δ5-Mi-CK was done to reinvestigate the binding properties of dimeric and octameric Mi-CK to mitoplasts and liposomes.

Binding of octameric and dimeric Δ5-Mi-CK to Mitoplasts

It was decided to measure the pH dependency of Mi-CK association to mitoplasts, since the binding of Mi-CK to membranes is thought to involve electrostatic interactions that are pH sensitive (Schlegel et al., 1990). Between pH 7.0 and pH 8.0 both Δ5-Mi-CK octamers as well as wild-type octamers, bound almost completely (about 80%) to Mi-CK-depleted chicken heart mitoplasts (Fig. 4). In contrast, Δ5-Mi-CK dimers were quantitatively bound only at pH 7.0. With increasing pH, the percentage of bound dimers decreased drastically, so that at pH 8.0 almost no Mi-CK was bound (Fig. 4). These results are in agreement with those of Schlegel et al. (1990) and corroborate that the interaction of dimeric Mi-CK with the inner mitochondrial membrane is much more pH dependent than that of octameric Mi-CK. Since the binding behavior of wild-type and Δ5-Mi-CK octamers was very similar, the membrane binding of Mi-CK seems not to be dependent on the N-terminus.

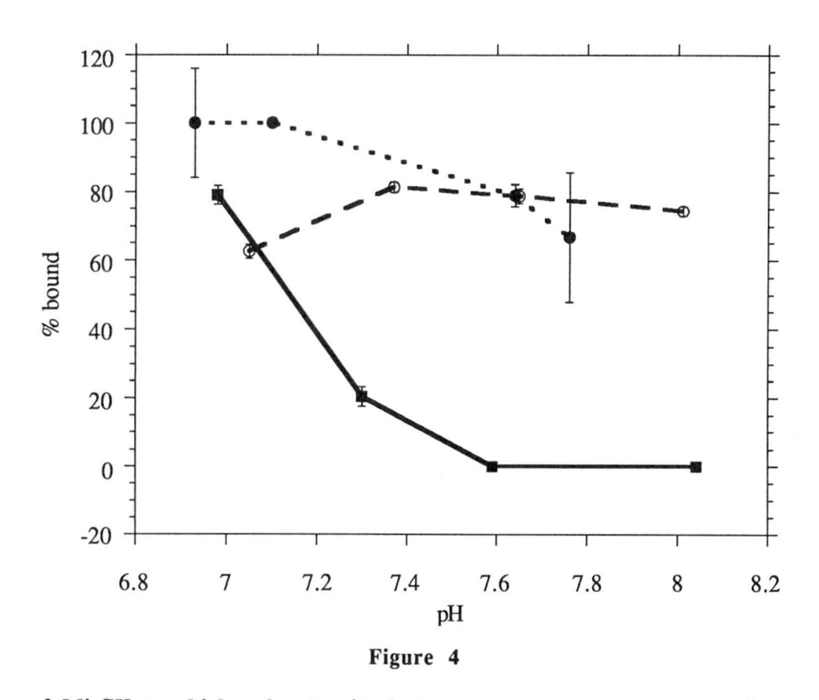

Figure 4

Binding of Mi-CK to chicken heart mitoplasts
Octameric wild-type Mi-CK (●), as well as octameric (○) and dimeric (■) Δ5-Mi-CK were incubated at different pH values with chicken heart mitoplasts. Mitoplasts were sedimented, the supernatants removed, and the pellets resuspended in the same volume of binding buffer. The CK activity in both fractions was measured to calculate the percentage of bound Mi-CK.

Binding of octameric and dimeric Δ5-Mi-CK to Liposomes

Mi-CK, a peripheral membrane protein, is also able to bind to membranes formed with lipid extracts of different subcellular membranes (Rojo et al., 1991a). When the interaction with pure lipid species was studied, a preferential interaction with anionic phospholipids, especially cardiolipin, was observed (Müller et al., 1985; Rojo et al., 1991a). Therefore, it was proposed that the in vivo membrane binding of Mi-CK is mediated by the lipid components of the membrane. It is an open question whether Mi-CK binding is due to an electrostatic interaction with membrane phospholipids or with a specific receptor protein. Since the mitoplast binding assay involves highly complex components, i. e. phospholipids and a variety of integral and peripheral inner membrane proteins, the observed effects can not be attributed to a single component. As a consequence, similar measurements were performed with liposomes formed from mitoplast phospholipids (see Materials and Methods).

As observed in the binding assay with mitoplasts, the binding of octamers to liposomes was almost independent of pH in the range between 7.0 and 8.0 (Fig. 5). Remarkably, the portion of bound Mi-CK was higher in the liposome than in the mitoplast system. For Δ5-Mi-CK dimers, a similar pH dependency of binding was observed in experiments done with both mitoplasts and liposomes.Whereas almost 100% of Δ5-Mi-CK octamers bound to liposomes in the pH range between 7.0 and 8.0, a qualitative binding of Δ5-Mi-CK dimers was observed only at pH 7.0. The proportion of bound Mi-CK dimers decreased with increasing pH, but in contrast to the mitoplasts experiments, a significant amount of Δ5-Mi-CK dimers still bound to liposomes at pH 8.0 (about 70%, see Fig. 5).

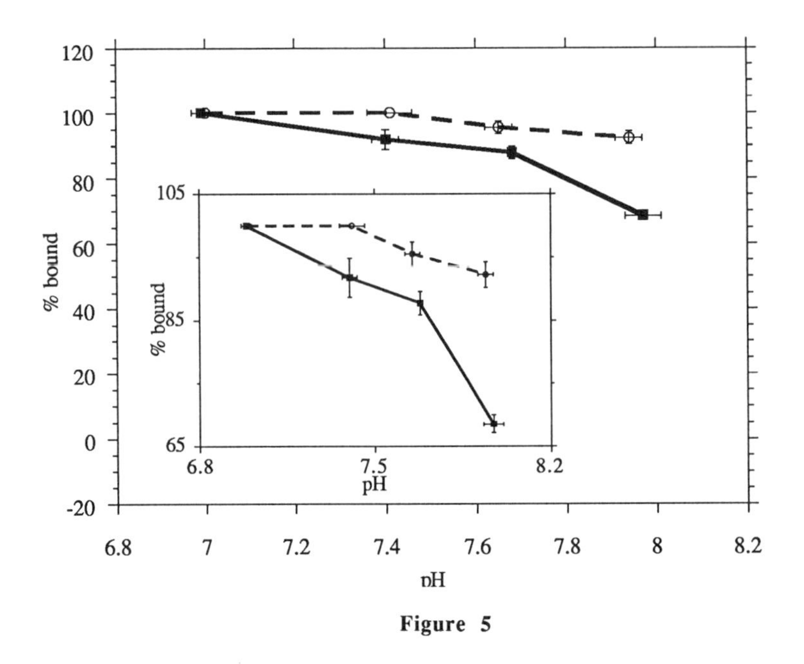

Figure 5

Binding of Mi-CK to liposomes
Octameric (O) and dimeric (■) Δ5-Mi-CK were incubated at different pH values with liposomes, prepared from mitoplast phospholipids. Liposomes were sedimented, the supernatants removed, and the pellets resuspended in the same volume of binding buffer. The CK activity in both fractions was measured to calculate the percentage of bound Mi-CK. The inset shows figure 5, drawn at a different scale.

DISCUSSION

Deletion of five amino acids at the N-terminus of Mi-CK did not affect its catalytic activity, suggesting that the N-terminus of the protein is not directly involved in catalysis and that the N-terminally deleted protein folds correctly. At high protein concentrations, Δ5-Mi-CK consisted predominantly of octamers and behaved like the wild-type enzyme. At low protein concentrations, Δ5-Mi-CK consisted only of dimers, even in the absence of the substrate mixture that is required for quantitative dissociation of wild-type Mi-CK into dimers. These properties of Δ5-Mi-CK can be extremely helpful to study several questions concerning the role of Mi-CK octamer/dimer ratios. The fact that the absence of the five N-terminal amino acids leads to the destabilization of the octameric form of Mi-CK strongly suggests that the N-terminus directly participates in the contacts between neighbouring dimers within an octamer. N-termini of all cytosolic isoforms of CK, which do not form octamers, are highly conserved among each other, but differ from the N-termini of the Mi-CK isoenzymes (Babbitt et al., 1986, Hossle et al., 1988). This difference in the N-termini of cytosolic and mitochondrial CKs also emphasizes their role as a functional domain necessary for octamer stabilization. However, this domain by itself is probably not sufficient for octamer formation, and it is likely that additional regions of the enzyme are also involved in this process.

Octameric Δ5-Mi-CK bound to mitoplasts in the same way as wild-type octameric Mi-CK. The proportion of both wild-type and Δ5-Mi-CK octamers bound to mitoplasts was much greater than the proportion of bound dimers. Since the binding behaviour of Δ5-Mi-CK octamers is nearly identical to that of the wild-type enzyme, it can be concluded that the five N-terminal amino acids, shown to be necessary for octamer stabilization, are not directly involved in membrane binding. However, an earlier study has shown that a larger N-terminal peptide of Mi-CK binds to cardiolipin-containing liposomes with a possible involvement of amino acid residues Arg_{19} Lys_{20} His_{21} (Cheneval and Carafoli, 1988).

Δ5-Mi-CK dimers showed a strong pH dependency in binding to both mitoplasts and liposomes. A similar pH dependency was observed in earlier studies (Schlegel et al., 1990) using wild-type Mi-CK dimers induced by a transition state analog complex. In the pH range tested, wild-type Mi-CK octamers and Δ5-Mi-CK octamers showed no significant pH dependency in binding. However, when mitoplasts are extracted with sodium phosphate buffer at pH 9.0, less than 15% of the endogenous, mostly octameric, Mi-CK remains bound. Therefore, it seems likely that octameric Mi-CK will also show reduced binding to mitoplasts and liposomes at pH values higher than 8.0.

The pH dependent binding of Δ5-Mi-CK dimers to mitoplasts and liposomes suggests that these interactions have either a pronounced electrostatic component or are entirely electrostatic. Since all the pK values of the phospholipids are higher than the pH range of the binding experiments, the observed pH dependency suggests that the binding of the dimeric Mi-CK molecule is determined primarily by the charges of the enzyme rather than by those of the lipids. It is probable that the same holds true for the octamers.

Δ5-Mi-CK octamers bound to a slightly higher extent, and Δ5-Mi-CK dimers bound to a significantly higher extent to liposomes as compared to mitoplasts. This result indicates that proteins of the inner mitochondrial membrane may not or weakly participate in the binding of Mi-CK to membranes and suggest that negative phospholipids are primarily involved in this binding. The quantitatively better binding of dimeric and octameric Mi-CK to liposomes versus mitoplasts was somewhat unexpected, since both types of binding experiments were performed at the same phospholipid to Mi-CK ratio and under identical conditions. However, several explanations have to be considered. First, it is conceivable that positively charged membrane proteins in mitoplasts diminish the overall available negative charge potential of mitoplast phospholipids and thus reduce Mi-CK binding. Second, due to the presence of membrane proteins in the mitoplasts, a certain proportion of mitoplast phospholipids interacting with these proteins may no longer be available for Mi-CK binding for steric reasons.

The findings that Mi-CK octamers are preferentially bound to membranes compared to dimers can be explained by the fact that octameric Mi-CK has a higher isoelectric point compared to dimeric Mi-CK (see Wyss et al., 1992) and thus, the former has more positive charges to interact with phospholipids. The stronger pH dependency of membrane binding of Mi-CK dimers compared to that of octamers in the range pH 7.0 to 8.0 may in part also be explained by the difference in isoelectric points between the two oligomeric species of Mi-CK. The different quaternary structures of Mi-CK dimers and octamers should also be considered. These different structures could lead to the involvement of a different number and/or different kinds of positive amino acids in the interaction of dimers and octamers with membranes. Additionally, according to the 'large-ligand' binding model of Stankowsky (1983), steric hindrance of binding would be more pronounced for four individual Mi-CK dimers compared to one intact Mi-CK octamer.

Based on the results, it seems likely that the interaction between Mi-CK and negative phospholipids is the underlying basis for to the binding of Mi-CK to mitochondrial membranes. It is further suggested that once bound to the membranes, Mi-CK is able, by lateral diffusion on the membrane, to form dynamic collision complexes with other membrane proteins. For example, a simultaneous interaction at the mitochondrial contact sites, where inner and outer membrane are in close vicinity, of Mi-CK with ANT of the inner and with porin of the outer mitochondrial membrane, may result in a multienzyme complex that is well suited to facilitate "metabolite channelling" of high-energy phosphates in mitochondria (see Fig. 1; Wyss and Wallimann, 1992).

Future studies will focus on the domains of Mi-CK involved in membrane binding, on the nature of this interaction and on additional parameters affecting the membrane binding behaviour of Mi-CK octamers and dimers.

Acknowledgements

E. Furter-Graves, R. Furter, M. Wyss and M.Rojo are gratefully acknowledged for experimental help, helpful discussion, proof-reading and for providing figures. We also thank E. Zanolla for technical assistance, R. Falchetto for protein sequence analysis and Prof. E. Carafoli for support. This work was supported by Swiss National Science Foundation grant No. 31-33907.92 and by the Swiss Foundation for Muscle Diseases (both to T.W.).

REFERENCES

Babbitt, P.C., Kenyon, G.L., Kuntz, I.D., Cohen, F.E., Baxter, J.D., Benfield, P.A., Buskin, J.D., Gilbert, W.A., Hauschka, S.D., Hossle, J.P., Ordahl, C.P., Pearson, M.L., Perriard, J.-C., Pickering, L.A., Putney, S.D., West, B.L., and Ziven, R.A., 1986, Comparison of creatine kinase primary structures, *J. Prot. Chem.* 5:1-13

Basson, C.T., Grace, A.M., and Roberts, R., 1985, Enzyme kinetics of a highly purified mitochondrial creatine kinase in comparison with cytosolic forms, *Mol. Cell. Biochem.* 67:151-159

Cheneval, D., and Carafoli, E., 1988, Identification and primary structure of the cardiolipin-binding domain of mitochondrial creatine kinase, *Eur. J. Biochem.* 171:1-9

Fiske, L.M., and SubbaRow, Y., 1925, *J. Biol. Chem.* 66:375-389

Furter, R., Kaldis, Ph., Furter-Graves, E.M., Schnyder, T., Eppenberger, H.M., and Wallimann, T., 1992, Expression of active, octameric chicken cardiac mitochondrial creatine kinase in Escherichia coli, *Biochem. J.* 288:771-775

Hossle, J.P., Schlegel, J., Wegmann, G., Wyss, M., Böhlen, P., Eppenberger, H.M., Wallimann, T., and Perriard, J.-C., 1988, Distinct tissue specific mitochondrial creatine kinases from chicken brain and striated muscle with a conserved CK framework, *Biochem. Biophys. Res. Commun.* 151:408-416

Hovius, R., Lambrechts, H., Nicolay, K., and de Kruijff, B., 1990, Improved methods to isolate and subfractionate rat liver mitochondria. Lipid composition of the inner and outer membrane, *Biochim. Biophys. Acta* 1021:217-226

Kenyon, G.L., and Reed, G.H., 1983, Creatine kinase: structure-activity relationships, *in:* Advances in Enzymology 26:367-426, A. Meister, ed., J. Wiley & Sons, New York

Kottke, M., Adams, V., Wallimann, T., Nalam, V.K., and Brdiczka, D., 1991, Location and regulation of octameric mitochondrial creatine kinase in the contact sites, *Biochim. Biophys. Acta,* 1061:215-225

Lämmli, U.K., 1970, Cleavage of structural proteins during the assembly of the head of bacteriophage T4, *Nature* 227:680-685

Müller, M., Moser, R., Cheneval, D., and Carafoli, E., 1985, Cardiolipin is the membrane receptor for mitochondrial creatine phosphokinase, *J. Biol. Chem.* 260:3839-3843

Rojo, M., Hovius, R., Demel, R.A., Nicolay, K., and Wallimann, T., 1991a, Mitochodrial creatine kinase mediates contact fromation between mitochodrial membranes, *J. Biol. Chem.* 266:20290-20295

Rojo, M., Hovius, R., Demel, R., Wallimann, T., Eppenberger, H.M., and Nicolay, K., 1991b, Interaction of mitochondrial creatine kinase with model membranes, *FEBS letters* 281:123-129

Rossi, A.M., Eppenberger, H.M., Volpe, P., Cotrufo, R., and Wallimann, T., 1990, Muscle-type MM creatine kinase is specifically bound to sarcoplasmatic reticulum and can support Ca^{2+} uptake and regulate local ATP/ADP ratios, *J. Biol. Chem.* 265:5258-5266

Schlegel, J., Zurbriggen, B., Wegmann, G., Wyss, M., Eppenberger, H.M., and Wallimann, T., 1988, Native mitochondrial creatine kinase forms octameric structures, *J. Biol. Chem.* 263:16942-16953

Schlegel, J., Wyss, M., Eppenberger, H.M., and Wallimann, T., 1990, Functional studies with the octameric and dimeric form of mitochondrial creatine kinase, *J. Biol. Chem.* 265:9221-9227

Stankowsky, S., 1983, Large-ligand adsorption to membranes II. Disk-like ligands and shape-dependence at low saturation, *Biochim. Biophys. Acta* 735:352-360

Studier, F.W., Rosenberg, A.H., Dunn, J.J., and Dubendorff, J.W., 1990, Use of T7 RNA polymerase to direct expression of cloned genes, *Methods Enzymol.* 185:60-89

Wallimann, T., Schlösser, T., and Eppenberger, H.M., 1984, Function of M-line-bound creatine kinase as intramyofibrillar ATP regenerator at the receiving end of the phosphorylcreatine shuttle in muscle, *J. Biol. Chem.* 259:5238-5246

Wallimann, T., Wyss, M., Brdiczka, D., Nicolay, K., and Eppenberger, H.M., 1992, Intracellular compartmentation, structure and function of creatine kinase isoenzymes in tissues with high and fluctuating energy demands: the 'phosphocreatine circuit' for cellular energy homeostasis, *Biochem. J.* 281:21-40

Wegmann, G., Huber, R., Zanolla, E., Eppenberger, H.M., and Wallimann, T., 1991, Differential expression and localization of brain-type and mitochondrial creatine kinase isoenzymes during development of the chicken retina: Mi-CK as a marker for differentiation of photoreceptor cells, *Differentiation* 46:77-87

Wyss, M., and Wallimann, T., 1992, Metabolite channelling in aerobic energy metabolism, *J. theor. Biol.* 158: 29-132

Wyss, M., Schlegel, J., James, P., Eppenberger, H.M., and Wallimann, T., 1990, Mitochondrial creatine kinase from chicken brain, *J. Biol. Chem.* 265:15900-15908

Wyss, M., Smeitink, J., Wevers, R.A., and Wallimann, T., 1992, Mitochondrial creatine kinase: a key enzyme of aerobic energy metabolism, *Biochim. Biophys. Acta* 1102:119-166

TRANSMEMBRANE LIPID ASYMMETRY IN EUKARYOTES

Philippe F. Devaux and Alain Zachowski

Institut de Biologie Physico-Chimique
13, rue Pierre et Marie Curie
F-75005 Paris, France

PHOSPHOLIPID ASYMMETRY IN BIOMEMBRANES

Following the pioneering work of Bretscher (1972), many researchers measured the transmembrane distribution of endogenous phospholipids in mammalian red cells using various techniques. These techniques comprise chemical labeling with non penetrating agents (for example with TNBS or fluorescamine), immunological methods, enzymatic assays with phosphatidylserine (PS)-dependent proteins, phospholipase digestion of membrane phospholipids, use of phospholipid exchange proteins, and physical methods such as X ray diffraction and NMR. Previous reviews have discussed the validity of the various approaches (Op den Kamp, 1979; Etemadi, 1980).

Figure 1 shows a scheme corresponding to the outer and inner distribution of the main phospholipids of the human red cell membrane. The data presented in Figure 1 should be completed by more recent information on the transmembrane distribution of plasmalogen phosphatidylethanolamine (PE), which, like the corresponding diacylglycerol phospholipid (Hullin et al., 1991), is found preferentially in the inner monolayer as well as of phosphatidylinositol (PI) derivatives. Reportedly, 20% of PI, PIP_2 and phosphatidic acid (PA), but no PIP, are located in the outer monolayer (Bütikofer et al., 1990; Gascard et al., 1991). In addition to the asymmetry of the head group distribution, the average fatty acid composition of PS and PE shows more unsaturation than that of PC and SM (Myher et al., 1989); furthermore, within the same class of phospholipids (SM and PE) , acyl chains from the outer monolayer differ from those of the inner monolayer (Boeghein et al., 1983; Hullin et al., 1991).

The compositional asymmetry in red cells is accompanied by the asymmetrical physical properties of this membrane (Seigneuret et al., 1984; Tanaka et Ohnishi, 1976; Williamson et al., 1982). Several laboratories have shown that the viscosity of the outer monolayer is greater than that of the inner monolayer as determined with fluorescent or paramagnetic probes.

Erythrocytes from other mammals have yielded similar results in spite of the different proportion of the four main phospholipids in different species. The situation is more complex

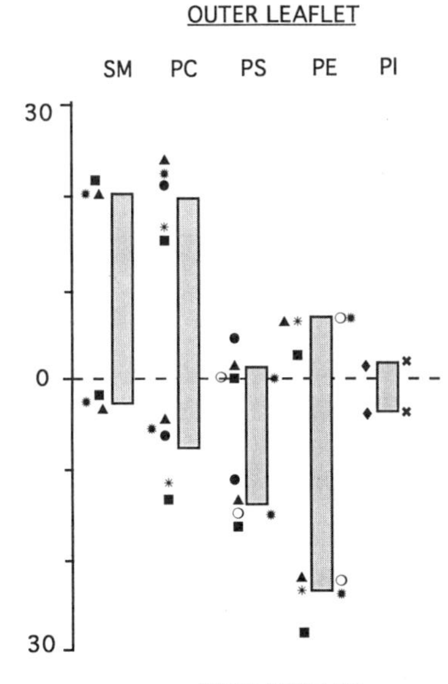

Figure 1. Transmembrane distribution of the main phospholipid species in human erythrocytes. The results are indicated in percentage of total phospholipids. The data are gathered from different publications and refer to non pathological cells. (star) Verkleij et al., (1973); (solid triangle) Van Meer et al. (1981); (asterisk) Williamson et al. (1982); (solid square) Bergmann et al. (1984); (solid circle) Dressler et al.(1984) ; (open circle) Gordesky et al.(1975); (cross) Bütikofer et al. (1990); (diamond) Gascard et al. (1991).

Table 1. Quantitative Results on the Phospholipid Asymmetry in the Plasma Membrane of Eukaryotic Cells

Cell type	Percent in outer layer					Reference
	PC	SM	PE	PS	PI	
red blood cell (man)			see figure 1			
platelet (man)	62		54	6	34	Wang et al. (1986)
	45	93	20	9	16	Perret et al. (1979)
platelet (pig)	40	91	34	6		Chap et al. (1977)
kidney brush border (rabbit)	34	80	23	15		Vénien et Le Grimellec (1988)
intestinal brush border (rabbit)	26		28			Barsukov et al. (1986)
intestinal brush border (trout)			50	32		Pelletier et al. (1987)
heart sarcolemna (rat)	43	93	25	0		Post et al. (1988)
embryo fibroblast (chick)			34	17		Sessions and Horwitz (1983)
embryo myoblast (chick)			66	46		Sessions and Horwitz (1983)
brain synaptosomes (mouse)			10-15	20		Fontaine et al. (1980)
LM fibroblast (mouse)			4-6	5		Fontaine et al. (1979)
hepatocytes (rat)						
bile canalicular surface	85	63	50	0	0	Higgins and Evans (1978)
contiguous surface	82	0	0	14	0	Higgins and Evans (1978)
sinusoidal surface	85	66	55	0	0	Higgins and Evans (1978)
Krebs II ascites (mouse)	51	46	45	20	30	Record et al. (1984)
yeast			30-25	10-20	30-25	Cerbon and Calderon (1991)

with other eukaryotic cells because the plasma membrane is only a small fraction of the total cell membranes. Nevertheless, many laboratories have analyzed the lipid topology in purified plasma membranes. The main results are collected in Table 1.

Table 1 clearly shows that, overall, the asymmetry of phospholipid distribution in human red cells is found in the plasma membrane of most eukaryotes. To what extent the quantitative variations reported for specific cell membranes are significant is a matter of debate. Indeed, purified plasma membrane may be contaminated by subcellular organelles. Moreover the purification of the membranes is certainly responsible for a partial scrambling of their lipids. We recently showed that hypotonic lysis of red cells induces a partial scrambling of their lipids (Schrier et al., 1992). This phenomenon is more pronounced if the lysis is carried out in the presence of Ca^{2+} (Williamson et al., 1985). Also, as discussed later in this review, in certain cells, such as platelets, a reorientation of a fraction of the lipids accompanies activation (Bevers et al., 1983). Thus, the population of cells examined in some instances may have contained a fraction of activated cells with a non representative lipid asymmetry.

As it is the case with erythrocytes, the phospholipid asymmetry of the plasma membrane of eukaryotes is accompanied by an asymmetrical transmembrane viscosity. Such asymmetries were reported for platelets (Kitagawa et al., 1991) and fibroblasts (Sandra and Pagano, 1978).

The labile character of the phospholipid transmembrane distribution and the difficulty to purify homogenous fractions are certainly also responsible for the discrepancies in the results obtained in cell organelles. In a living organism, lipids are continuously involved in the cellular traffic associated with exocytosis, endocytosis, lipid catabolism and lipid synthesis. Thus, the lipid distribution is a dynamic process and the state of an isolated organelle such as the endoplasmic reticulum or a mitochondria might depend drastically upon the conditions of isolation.

In conclusion, lipid asymmetry is probably a general feature in eukaryotic as well as in prokaryotic membranes. The model of the human red cell is probably representative of the topology of the plasma membrane of eukaryotes. General rules are more difficult to infer in subcellular organelles.

PROTEIN INVOLVEMENT IN TRANSMEMBRANE LIPID PASSAGE

The spontaneous passage of neutral lipids such as diacylglycerol (Ganong and Bell, 1984) or fatty esters and probably cholesterol is rapid, with characteristic times of the order of seconds. On the contrary, phospholipid transverse diffusion in a bilayer is a slow process with characteristic times of the order of hours or days depending upon the chain length, degree of unsaturation and the nature of the head groups (Middelkoop et al., 1986) as well as upon the overall composition of the membrane. Cholesterol appears to increase the transverse stability of phospholipids (Morrot et al., 1989), while detergent (Kramer et al., 1981) and/or addition of membrane proteins can accelerate lipid flip-flop in reconstituted systems (De Kruijff et al., 1978). Thus, the packing problem caused within the lipid lattice by the rough surface of any intrinsic protein inserted in the bilayer destabilizes the phospholipids at lipid-protein interfaces, and because the rate of lateral diffusion is fast (in-plane exchange rates on the order of 10^7-10^6 sec^{-1}), a single protein can accelerate the flip-flop of a large population of lipids. For example gramicidin enhances the transbilayer reorientation of lipids in erythrocytes membranes by inducing the locally hexagonal H_{II} phase (Classen et al., 1987; Tournois et al., 1987). Such extreme situations may exist at least temporarily in biological membranes.

Various proteins play specific roles in the establishment, maintenance, and cancellation of

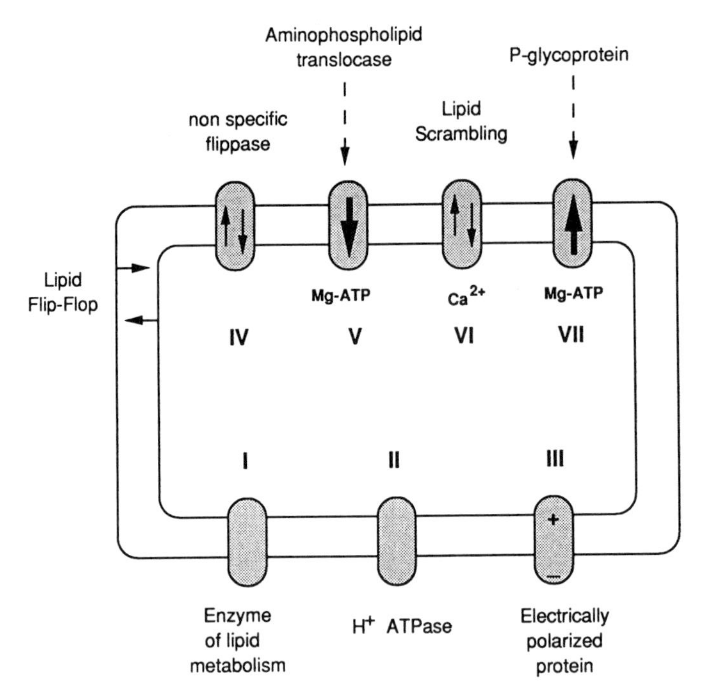

Figure 2. Protein involvement in transmembrane lipid passage.

phospholipid asymmetry in eukaryotic cells and probably also in prokaryotic cells. Figure 2 is a scheme that attempts to summarize the various classes of proteins involved. These proteins are in reality distributed in the plasma membrane and in subcellular organelles and may not coexist in the same membrane as schematically shown in this figure for the sake of simplicity. The first class of proteins, probably the most important one in relationship with lipid asymmetry, is constituted by membrane bound enzymes that are responsible for phospholipid synthesis. At least the last stages of lipid synthesis and lipid remodeling are membranous and, hence, vectorial by nature. The orientation of the enzymes in the endoplasmic reticulum (ER), in mitochondria, and in the Golgi complex where most lipids are synthesized, therefore plays a crucial role in their subsequent orientation once shuttled to their target membranes, e.g. the plasma membrane (for a review, see Bishop and Bell (1988) or Hjelmstod and Bell (1991)). Even erythrocytes that have a low metabolic activity contain enzymes that influence the transmembrane distribution. For example Hirata and Axelrod (1978) showed that the methylation of PE is accompanied by its transmembrane reorientation. More recently, Andrick et al. (1991) demonstrated the fast translocation of PC to the outer membrane leaflet of human erythrocytes after its synthesis at the inner membrane surface by reacylation of lyso-PC. Thus, unlike model systems, biomembranes are made asymmetrically, and apparently fusion events, which take place during cell life and cell division, do not totally scramble lipid orientation.

A second class of proteins (II) that can be indirectly responsible for lipid asymmetry is composed of the ATPases that generate potential gradients and/or pH gradients. The strong pH gradient found in lysosomes polarizes the membrane of these organelles. But even the small acidic character of erythrocyte cytoplasm can explain the accumulation of exogenously added free fatty acids on the outer monolayer of erythrocyte (this is accompanied by the formation of echinocytes - see below). Changes in the compositional asymmetry of phospholipids associated with the increment in the membrane surface potential have been reported in yeast (Cerbon and Calderon, 1991). However, one should keep in mind that the potential or pH does not suffice to force most charged lipids to flip. As reported by Cullis' group, only the neutral form of the lipids can traverse the bilayer; the pH gradient traps a certain distribution but is not the driving force (Cullis and de Kruijff, 1979).

Hubbell (1990) proposed an interesting model to explain the stable lipid asymmetry which supposedly exists in disc membranes. The model is based on the polarization through the membrane that results from fixed charged distributed asymmetrically on rhodopsin (III). This electric polarization would eventually segregate PS on the cytoplasmic side of the discs and indirectly influence PE distribution, because of preferential interaction between PE and PS molecules. However, this model requires free passage of lipids between both monolayers in order to reach equilibrium. Such a rapid flip-flop is not excluded in disc membranes that contain very long unsaturated lipids, though this observation is yet to be verified.

Molecules of classes IV, V, VI and VII correspond to a protein family sometimes called *phospholipid flippases,* following the terminology invented by M. Bretscher. This family includes very different proteins. Molecule IV is a flippase whose existence was originally demonstrated in rat liver endoplasmic reticulum (ER) by Bishop and Bell in 1985. Using a water soluble short chain phospholipid (dibutyroyl-PC), these authors demonstrated the rapid passage through purified ER of the PC analogue. The process was saturable and sensitive to proteases and chemical modifications and, according to these authors, specific for PC. A different transporter was postulated for lyso-PC (Kawashima and Bell, 1987). Using the same membrane preparation and spin-labeled lipids with one short chain and one long chain, we found also a rapid lipid flip-flop ($\tau_{1/2}$ ~ 10 min at 37°C) but with very little specificity (see fig. 3-bottom). Competition experiments demonstrated that lyso-PS and PC used the same pathway (Herrmann et al., 1990). Reconstitution experiments using a crude extract of microsomal proteins confirmed the existence of the "flippase" in ER (Baker and Dawidowicz, 1987). The same experiment using proteins from the red cell membrane failed. This process, which is a non energy-requiring catalysis of phospholipid flip-flop can be classified as a facilitated diffusion. Clearly this protein is indispensable in the membranes where lipids are being processed in order to allow a proper equilibration between lipids in both leaflets. One, therefore, may postulate the existence of similar protein(s) in bacterial membranes.

Protein V represents the aminophospholipid translocase. This protein originally discovered in the membrane of human erythrocytes (Seigneuret and Devaux, 1984) and later in the plasma membrane of many eukaryotic cells such as platelets (Sune et al., 1987), lymphocytes (Zachowski et al., 1987), fibroblasts (Martin and Pagano, 1987), and synaptosomes (Zachowski and Morot Gaudry-Talarmain, 1990) was also found in chromaffin granules, i.e. in a subcellular organelle (Zachowski et al., 1989). This protein transports selectively aminophospholipids (PS and PE) from the outer to the inner monolayer of plasma membrane. Contrary to the "flippase" in endoplasmic reticulum (protein IV), the lipid selectivity is very high as indicated in figure 3-top where the initial rates of outside-inside movements of different lipids have been plotted. The movement is vectorial, requires cytosolic Mg-ATP (K_m ~ 1 mM), works against a chemical gradient and, thus, corresponds to an active transport. The mechanism is inhibited by N-ethyl maleimide (K_i ~ 0.3 mM), cytosolic vanadate (~ 50 μM) or AlF_4 (~ 50 μM) as well as by cytosolic Ca^{2+} (~ 1 μM).

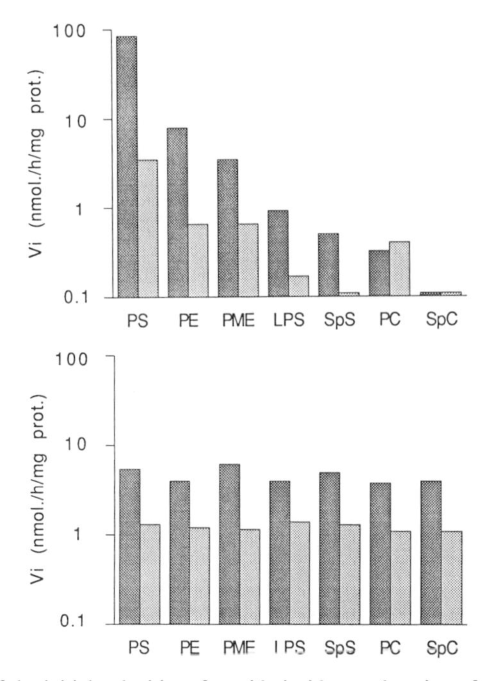

Figure 3. Comparison of the initial velocities of outside-inside translocation of spin-labeled phospholipids, at 37°C. (Top) human red blood cells; (bottom) rat liver microsomes. The dashed bars correspond to experiments carried out after incubation with 1 mM N-ethyl-maleimide. Abbreviations are :PS, phosphatidylserine; PE, phosphatidylethanolamine; PC, phosphatidylcholine; PME, monomethyl-phosphatidylethanolamine; LPS, lyso-phosphatidylserine; SpS, sphingosyl-phosphorylserine; SpC, sphingomyelin. (Adapted from Herrmann et al. (1990) and Morrot et al. (1989)).

The transmembrane equilibrium distribution of exogenously added lipids appears remarkably similar to that of endogenous lipids. For example, spin-labeled PS, PE, PC and SM reach an equilibrium corresponding respectively to 95%, 80%, 30% and 5% of the lipid analogues in the inner monolayer. At 37°C, the plateau is attained in approximately 10 min with PS, 1 hr with PE and more than 10 hrs are necessary to estimate the final distribution of the other lipids (PC and SM) (Morrot et al., 1989). Similar results were obtained in various laboratories with radioactive lipids (Tilley et al., 1986), fluorescent lipids (Connor and Schroit, 1987) and short chain lipids (Daleke and Huestis, 1985, 1989). Note that in all these experiments, the labeled lipids incorporated in the outer monolayer corresponded to a small fraction of the total lipids. To date, no studies have demonstrated that a totally scrambled membrane from red cells can recover its native lipid asymmetry by the action of the aminophospholipid translocase. This aspect will be discussed later in the review.

The aminophospholipid translocase has not yet been unambiguously identified. Connor and Schroit have proposed, based on experiments with radioactive photoactivable PS, a 32 kDa polypeptide as the putative protein (Connor and Schroit, 1988, 1990). We and other groups have selected a vanadate sensitive Mg^{+}ATPase of molecular weight 120 kDa (Morrot et al., 1990; Daleke et al., 1991).

Protein VI is a very speculative protein of the "flippase" family. The evidence is strong that under physiological conditions, in some cells, the lipid asymmetry can be suddenly scrambled, thereby exposing PS on the outer layer of the plasma membrane. Platelets are the best documented system in this regard (Bevers et al., 1983, 1989). Depending on the stimulus, phospholipid transbilayer asymmetry of the plasma membrane is more or less lost during platelet activation. Increase of cytosolic Ca^{2+} by addition of Ca^{2+} and ionophore is a powerful stimulation of lipid scrambling which is accompanied by vesicle shedding (Comfurius et al., 1990). Similar observations were made with erythrocytes although the lipid scrambling triggered by cytosolic Ca^{2+} is less efficient (Chandra et al., 1987; Comfurius et al., 1990; Connor et al., 1990; Henseleit et al., 1990). Importantly, not only PS but PE also flops to the outer monolayer while PC flips inside. However, the efficiency of Ca^{2+}-induced reorientation of SM is weak (Williamson et al., 1992). The time scale of this lipid reorientation precludes a simple inhibition of the aminophospholipid translocase by Ca^{2+}. Indeed when the translocase is inhibited, the randomization of lipid distribution takes normally hours or days through the spontaneous diffusion pathway. Ca^{2+} alone, although it interacts with PS or PA to form rigid domains, does not suffice to destroy the bilayer. Thus, a specific protein could be responsible for this Ca^{2+}-induced randomization in platelets and possibly in other cells. One mechanism which could account for the observed phenomenon would involve the formation of non bilayer structures following the phospholipase induced production of diacylglycerol. Indeed, researchers have shown in model systems that physiological levels of diacylglycerol can destabilize lipid bilayers (Siegel et al., 1989). Alternatively, partial disturbances of the bilayer structure could occur during fusion of the granules which accompanies in some instances platelet stimulation or during fusion of the plasma membrane that take place upon shedding of membrane microvesicles (Comfurius et al., 1990).

Recently we have discovered that the simutaneous addition of phosphatidyl inositol-diphosphate ($PI\text{-}P_2$) and Ca^{2+} to the outer monolayer of human red blood cells trigger important lipid scrambling in this membrane (J.C. Sulpice, A. Zachowski, P. Devaux, F. Giraud, unpublished). This is yet another plausible mechanism by which the lipid asymmetry can be destroyed. However, in conclusion the physiological mechanism of fast lipid scrambing is not known.

Finally protein VII is another candidate flippase, acting on amphiphilic drugs. Higgins and Gottesman have point out that the protein responsible for multidrug resistance in cancer cells is in fact a "flippase" since it selectively transports hydrophobic molecules from the inner monolayer to the outer monolayer (Higgins and Gottesman, 1992).

GENERATION AND MAINTENANCE OF LIPID ASYMMETRY. MODEL SYSTEMS AND BIOMEMBRANES

Could the aminophospholipid translocase alone generate the normal lipid asymmetry in a totally randomized erythrocyte membrane ? This question pertains to the role of the aminophospholipid translocase on the transmembrane distribution of PC and SM. In several laboratories, labeled lipids, including PC and SM, were added in trace amounts to the outer monolayer of erythrocytes. With variable kinetics, the phospholipid analogues eventually reached a transmembrane distribution similar to that of the corresponding naturally occuring lipids. In fact, the similarity between the steady-state distribution of the labeled lipids and that of unlabeled lipids is striking. Thus, one may conclude that the activity of the amino-lipid pump suffices to sort aminolipids from choline-containing lipids. To explain the asymmetrical distribution of PC and SM, one would not need to postulate the existence of a PC (or SM) outward pump but simply that the occupancy of the inner lipid state by the aminophospholipids forces PC and SM to eventually accumulate in the outer monolayer. In case of scrambling, the mere passive diffusion of PC and SM concomitantly to the active inward transport of PS and PE would establish the segregation. But this view may in fact be erroneous. The model, as it is, is derived from the model explaining the formation of ion gradients through membranes but it may not apply directly to lipids. Indeed, the inward movement of lipids without **simultaneously** outward movement of another lipid (that is a "flip" without a "flop") means that the two monolayers acquire progressively unequal surface

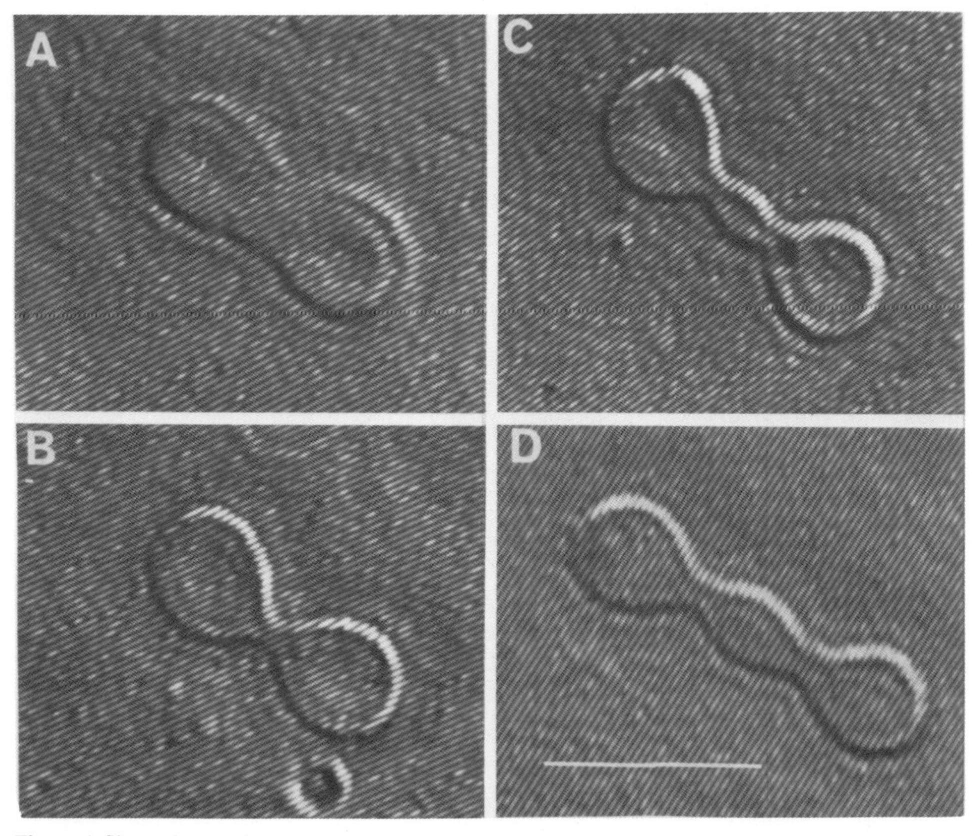

Figure 4. Shape change of a giant liposome induced by the reorientation of ~ 1% of the phospholipids. The liposome contains 99% PC and 1% PG. The latter lipids accumulate on the outer monolayer when the external pH is brought to 9.5. The bar corresponds to 20 μm. (Farge and Devaux, 1992).

areas and can no longer fit to each other. This process leads to membrane curvature (Figure 4) and, if lipid transfer is continued, to vesiculation and shedding. Thus, generating a lipid asymmetry from a totally scrambled lipid distribution appears to be impossible if the two mechanisms of "flip" and "flop" have very different kinetics, which would be the case in erythrocytes where the aminophospholipid translocase activity is only counterbalanced by the very slow diffusion of PC and SM.

However, if only a small fraction of the lipids are involved in a transmembrane redistribution, as in the articles cited above, the difference in kinetics may be acceptable. In fact Daleke and Huestis (1985, 1989) used shape changes to monitor the transmembrane passage of lipids in red cells. Recent work showed that the reorientation of less than 1% of the phospholipids of a giant liposome, triggered by a change in pH, is accompanied by important shape changes (Farge and Devaux, 1992).

In conclusion the aminophospholipid translocase fulfils an important function in controling the aminophospholipid asymmetry, particularly in sequestering PS from the outer leaflet of plasma membranes.Probably because of the lower affinity of the enzyme for PE, under steady-state conditions the equilibrium distribution of PE is only ~ 70-80%. But the aminophospholipid translocase is just one of the several enzymes responsible for the lipid asymmetry. Thus, if scrambling of a large fraction of the phospholipids can be triggered physiologically by a Ca^{2+} burst in some cells (such as platelets) we infer that return to normal asymmetry, if it ever occurs, cannot be achieved without partial vesiculation of the membranes.

BIOLOGICAL FUNCTIONS OF PHOSPHOLIPID FLIPPASES

The four types of protein included here in the "flippase family" (IV, V, VI and VII in figure 2) play very different roles in the lipid transmembrane traffic. As already mentioned the nonspecific phospholipid flippase (class IV) probably does not coexist with classes V and VI in the plasma membrane of eukaryotes. This protein was discovered in ER membranes and most likely functions in the redistribution of newly synthesized lipids on both sides of the ER.

The association of the aminophospholipid translocase (protein V) with the putative Ca^{2+}-triggered flippase (protein VI) can be compared to the association of an ATP dependent ion pump and an ionophore triggered by a specific ligand. However, as discussed above the reversibility of a lipid scrambling is questionable. So the real question is : what is the function of a transmembrane lipid segregation ?

We envision two types of functions. The first type of function deals with specific interactions of a subclass of lipids, particularly PS, that, if present, can be recognized on particular cell surface or can modulate enzyme activity. A fraction of the cell membranes are therefore tagged not only by proteins but also by the lipids. This may be important in cell sorting within the plasma as well as vesicle sorting within eukaryotic cells. The second type of function for lipid segregation deals with the bending which is imposed on a membrane by a transfer of phospholipids. We discuss the two aspects separately .

Phosphatidylserine interacts specifically with various membrane-bound enzymes. For example, protein kinase C activity depends strongly upon PS concentration (Newton and Koshland, 1990; Rando, 1988). Similarly, G proteins and annexins that correspond to a whole family of enzymes involved in many different cell functions are PS-dependent enzymes (Geisow and Walker, 1986; Klee, 1988). Thus, within a cell, these enzymes can select their target membranes if PS is asymmetrically distributed on organelle's surface. However, the best example of a physiological modulation of enzyme activity controlled by "flippases" concerns a plasma membrane and is the conversion from prothrombin to thrombin induced by the sudden reorientation of PS at the outer surface of platelets. Platelet stimulation gives rise to a partial lipid scrambling with the appearance of PS on the outer monolayer of the plasma membrane as well as the release of shedded vesicles with a scrambled lipid distribution (see above for references). This phenomenon creates a PS interface on the external monolayer of the platelet membrane that stimulates the conversion of prothrombin to thrombin. This conversion is the first stage of the cascade of events involved in blood coagulation.

The progressive appearance of PS on the outer surface of aged erythrocytes may signal cell elimination. This is based on experiments that showed that ghosts with an artificially scrambled lipid distribution are recognized preferentially by macrophages and endothelial cells (Schlegel et al., 1985; Schroit et al., 1985; Tanaka and Schroit, 1983). Recently Fadok and collaborators provided evidence for macrophage recognition of PS on apoptotic lymphocytes and demonstrated the existence of PS receptors on macrophages (Fadok et al., 1991). Progressive oxidation of the aminophospholipid translocase and the consecutive unpairing of the mechanism of maintenance of lipid asymmetry would be responsible for the lipid scrambling in aged erythrocytes. Up until now, however, it has not been possible to measure a significant variation of PS distribution in "aged" red cells, isolated on the grounds of their sedimentation characteristics (Herrmann and Devaux, 1990). So the above mentioned theory is only valid if a very small fraction of PS on the outer layer is sufficient to tag "old" cells as such.

Aging may not be the only cell dysfunctioning accompanied by lipid reorientation. Sickled red blood cells have a measurable fraction of PS on their outer leaflet, simultaneously, these cells have a shorter life time in blood circulation than normal red cells (Lubin et al., 1981). Also, tumorigenic, undifferentiated murine erythroleukemic cells express seven to eightfold more PS in their outer leaflet than do their undifferentiated non tumorigenic counterparts. Increased expression of PS in the tumorigenic cells directly correlated with their ability to be recognized and bound by macrophages (Connor et al., 1989). Similarly, elevated expression of PS in the outer membrane leaflet of human tumor cells as well as recognition by activated human blood monocytes was reported recently (Utsugi et al., 1991).

Yet another hypothesis to explain the role of transmembrane lipid asymmetry is the different fusogenic properties of various lipid interfaces. PS and SM resist membrane fusion because of the stability of the corresponding bilayer (or monolayer) structure and also because of the long-range hydration shell of these lipids. On the contrary, PS (and PE) interfaces form smaller layers of bound water, thereby allowing closer contact betwen different membranes (Rand and Parsegian, 1989); secondly they are more fusogenic, particularly in the presence of Ca^{2+}, because of their propensity to undergo phase transitions at physiological temperature from bilayer to non-bilayer phase (Cullis and de Kruijff, 1979).

Thus, the normal lipid polarization of plasma membranes prevents cell-cell fusion because their outer layer contains PC and SM. Interestingly, myoblasts are one of the few exceptions. These cells contain a large proportion of aminophospholipids on the outer monolayer of their plasma membrane. Myoblasts are, precisely, cells that fuse together (Sessions and Horwitz, 1983). However, the actual physical principle that explains the capacity of cells to fuse has been questioned in recent work from Blumenthal's laboratory. Indeed, Herrmann et al. (1990) attempted to elucidate the role of phospholipid head-group specificity in modulating virus fusion to erythrocyte ghosts by using lipid scrambling or chemical modification of the outer leaflet. The results showed that fusion of influenza virus as well as vesicular stomatitis virus (VSV) was affected by the presence of unsaturated fatty acyl chains in the outer leaflet, but not by specific phospholipids. Moreover, in the case of myoblasts, the fluidity of the plasma membrane increases rapidly before the onset of fusion (Prives and Shinitzky, 1977). Thus, lipid packing properties of the interacting membranes may be the important factor.

Within a cell, fusion of Golgi-derived vesicles with the plasma membrane is facilitated by the fluid PS-PE interface. Real fusion competence also requires that exocytic vesicles expose similar aminophospholipids on their outer cytosolic face. This seems to be the case for chromaffin granules and would involve the activity of an aminophospholipid translocase located in a subcellular organelle (Zachowski et al., 1989).

Cullis'group carried out very elegant experiments on model systems in which the fusion between large unilamellar vesicles was modulated in the presence of Ca^{2+} by the transmembrane reorientation of a small fraction of phosphatidic acid (PA) under the influence of a change in pH (Eastman, 1991). Of course in real cells, specific proteins are certainly required for the targetting of the fusion events, but the lipid interface provides one form of recognition between membranes of the various organelles.

A different phenomenon may be the most important biological function associated with transmembrane phospholipid redistribution. This topic deals with shape changes caused by the transfer of lipids from one monolayer to the other of a membrane. In 1974 Sheetz and Singer, on the basis of experiments with human erythrocytes, proposed the bilayer couple hypothesis. This theory accounts for red cell shape changes by the selective dilation of the outer or inner monolayer resulting from the intercalation of drugs on either side of the cell. The actual shape of erythrocytes results from the simultaneous action of the bilayer and the cytoskeleton. Platelets, which have different cytoskeleton proteins, undergo different shape changes when the same amphiphilic drugs are intercalated. But, for both systems, the addition of a small percentage of lipids on one leaflet (~ 1%) triggers the formation of membrane protrusions. We have already mentioned that shape changes have been triggered in large unilamellar vesicles devoided of proteins by the reorientation of a very small percentage of lipids. Thus, while in erythrocytes the activity of the aminophospholipid translocase seems limited to the maintenance of an overall discoid shape, in cells provided with a higher translocase activity that transports aminophospholipids (PS and PE) from the outer to the inner leaflet, the invaginations could lead to endocytic vesicles(for more details on such a model see Devaux (1991). The proteins considered to provide the driving force for membrane pitching off (clathrin) could in fact be present in order to localize the regions of the membrane where the invaginations start. Extrapolation of such an idea leads one to suggest that Golgi membranes should contain an aminophospholipid translocase, which could participate in the formation of Golgi vesicles, oriented towards the cytosol. As for vesicles formed from the ER, they could be generated by a missmatch between rates of lipid synthesis on one side of the membrane and rates of lipid flip-flop by a type IV phospholipid flippase.

SUMMARY AND CONCLUSION

Transmembrane asymmetry has been extensively studied in eukaryotic cells. It is as yet only clearly demonstrated in the plasma membrane of a few cells. Subcellular organelles have evidence of lipid asymmetry, but very little consistent quantitative data. Proteins involved in transmembrane passage of lipids comprise enzymes of lipid metabolism and also the so-called phospholipid flippases that are either passive or active putative lipid transporters. The aminophospholipid translocase that pumps aminophospholipids from the outer to the inner monolayer of the plasma membrane of eukaryotes is a Mg^{2+}-ATP dependent protein with a high lipid selectivity. Lipid asymmetry provides an asymmetrical environment to membrane enzymes. Thus, PS (and PE) reorientation could be a way of controlling or triggering specific enzymes. Also, the asymmetrical distribution of phospholipids most likely determines the fusion competent membranes and/or which sides of membranes should fuse. Finally, the lipid pump as well as all enzymes responsible for the net transmembrane flux of phospholipids may provide the driving force for membrane bending, notably during the formation of endocytic vesicles.

Clearly, real progress in this area will be made only if the proteins of the flippase family are purified and antibodies obtained that will permit the recognition and localization of these proteins in various cells. Also, specific inhibitors as well as mutants would allow one to infer more directly what are the real functions of these proteins. At a late stage, the protein purification will eventually permit speculation on the mechanism of action of a pump that must transport simultaneously hydrophilic and hydrophobic groups through a membrane.

Acknowledgments

Work supported by grants from the Centre National de la Recherche Scientifique (URA 526), the Institut National de la Santé et de la Recherche Médicale (N° 900104) and the Université Paris VII.

REFERENCES

Andrick, C., Bröring, K., Deuticke, B., Haest, C. W. M. 1991. *Biochim. Biophys. Acta* 1064:235-241.

Backer, J. M., Dawidowicz, E. A. 1987. *Nature* 327:341-343.

Barsukov, L. I., Bergelson, L. D., Spiers, M., Hauser, J., Semenza, G. 1986. *Biochim. Biophys. Acta* 882:87-99.

Bergmann, W. L., Dressler, V., Haest, C. W. M., Deuticke, B. 1984. *Biochim. Biophys. Acta* 772:328-336.

Bevers, E. M., Comfurius, P., Zwaal, R. F. A. 1983 *Biochim. Biophys. Acta* 736:57-66.

Bevers, E. M., Tilly, R. H. J., Senden, J. M. G., Comfurius, P., Zwaal, R. F. A. 1989 *Biochemistry* 28:2382-2387.

Bishop, W. R., Bell, R. M. 1985. *Cell* 42:51-60.

Bishop, W. R., Bell, R. M. 1988. *Annu. Rev. Cell Biol.* 4:579-610.

Boeghein, J. P. J., van Linde, M., Op den Kamp, J. A. F., Roelofsen, B. 1983. *Biochim. Biophys. Acta* 735:438-442.

Bretscher, M. S. 1972. *Nature New Biol.* 236:11-12.

Bütikofer, P., Lin, Z. W., Chiu, D. T.-Y., Lubin, B., Kuypers, F. A. 1990. *J. Biol. Chem.* 265:16035-16038.

Cerbon, J. Calderon, V. 1991. *Biochim. Biophys. Acta* 1067:139-144.

Chandra, R., Joshi, P. C., Bajpai, V. K., Gupta, C. M. 1987. *Biochim. Biophys. Acta* 902:253-262.

Chap, H. J., Zwaal, R. F. A., van Deenen, L. L. M. 1977. *Biochim. Biophys. Acta* 467:146-164.

Classen, J., Haest, C. M. W., Tournois, H., Deuticke, B. 1987. *Biochemistry* 26:6604-6612.

Comfurius, P., Senden, J. M. G., Tilly, R. H. J., Schroit, A. J., Bevers, E. M., Zwaal, R. F. A. 1990 *Biochim. Biophys. Acta* 1026:153-160.

Connor , J., Bucana, C., Fidler, I. J., Schroit, A. J. 1989. *Proc. Natl. Acad. Sci. USA* 86:3184-3188.

Connor, J., Gillum, K., Schroit, A. J. 1990. *Biochim. Biophys. Acta* 1025:82-86.

Connor, J., Schroit, A. J. 1987. *Biochemistry* 26:5099-5105.

Connor, J., Schroit, A. J. 1988. *Biochemistry* 27:848-851.

Connor, J., Schroit, A. J. 1990. *Biochemistry* 29:37-43.

Cullis, P. R., de Kruijff, B. 1979. *Biochim. Biophys. Acta* 559:399-420.

Daleke, D. L., Cornely-Moss, K. A., Smith, C. M. 1991. *Biophys. J.* 59:381a.

Daleke, D. L., Huestis, W. H. 1985. *Biochemistry* 24:5406-5416.

Daleke, D. L., Huestis, W. H. 1989. *J. Cell. Biol.* 108:1375-1385.

De Kruijff, B., van Zoelen, E. J. J., van Deenen, L. L. M. 1978. *Biochim. Biophys. Acta* 509:537-542.

Devaux, P. F. 1991. *Biochemistry* 30:1163-1173.

Dressler, V., Haest, C. W. M., Plasa, G., Deuticke, B. and Erusalimsky, J. D. 1984. *Biochim. Biophys. Acta* 775:189-196.

Eastman, S. J. 1991. *Studies generation and function of phospholipid asymmetry.* Ph.D. Thesis, University of Vancouver, BC, Canada.

Etemadi, A. -H. 1980 *Biochim. Biophys. Acta* 604:423-475.

Fadok, V. A., Voelker, D. R., Campbell, P. A., Cohen, J. J., Bratton, D. L., Henson, P. M. 1991. *J. Immunol.* 148:2207-2210.

Farge, E., Devaux, P. F. 1992. *Biophys. J.*, 61:347-357.

Fontaine, R. N., Harris, R. A., Schroeder, F. 1980. *Neurochem.* 34:269-277.

Fontaine, R. N., Schroeder, F. 1979. *Biochim. Biophys. Acta* 558:1-12.

Ganong, B. R. Bell, R. M. 1984. *Biochemistry* 23:4977-4983.

Gascard, P. Tran, D., Sauvage, M., Sulpice, J. -C., Fukami, K., Takenawa, T., Claret, M., Giraud, F. 1991. *Biochim. Biophys. Acta* 1069:27-36.

Geisow, M. J., Walker, J. H. 1986. TIBS 11:420-423.

Gordesky, S. E., Marinetti, G. V., Love, R.. 1975. *J. Membr. Biol.* 20:111-132.

Henseleit, U., Plasa, G., Haest, C. 1990. *Biochim. Biophys. Acta* 1029:1217-1227.

Herrmann, A., Clague, M. J., Puri, A., Morris, S. J., Blumenthal, R., Grimaldi, S. 1990. *Biochemistry* 29:4054-4058.

Herrmann, A., Devaux, P. F. 1990. *Biochim. Biophys. Acta* 1027:41-46.

Herrmann, A., Zachowski, A., Devaux, P. F. 1990. *Biochemistry* 29:2023-2027.

Higgins, C.F., Gottesman, M.M. 1992. *TIBS* 17:18-21.

Higgins, J.A., Evans, W. H. 1978. *Biochem. J.* 174:563-567.

Hirata, F., Axelrod, J. 1978. *Proc. Natl. Acad. Sci. USA* 75:2348-2352.

Hjelmstad, R. H., Bell, R. M. 1991. *Biochemistry* 30:1731-1740.

Hubbell, W. L. 1990. *Biophys. J.* 57:99-108.

Hullin, F., Bosant, M.-J., Salem Jr., N. 1991. *Biochim. Biophys. Acta* 1061:15-25.

Kawashima, Y., Bell, R. M. 1987. *J. Biol. Chem.* 262:16495-16502.

Kitagawa, S., Matsubayashi, M., Kotani, K., Usui, K., Kametani, F. 1991. *J. Memb. Biol.* 119:221-227.

Klee, C. 1988. *Biochemistry* 27:6645-6653.

Kramer, R. M., Hasselbach, H. -J., Semenza, G. 1981. *Biochim. Biophys. Acta* 643:233-242.

Lubin, B., Chiu, D., Bastacky, J., Roelofsen, B., van Deenen, L. L. M. 1981. *J. Clin. Invest.* 67:1643-1649.

Martin, O. C., Pagano, R. E. 1987. *J. Biol. Chem.* 262:5890-5898.

Middelkoop, E., Lubin, B. H., Op den Kamp, J. A. F., Roelofsen, B. 1986. *Biochim. Biophys. Acta* 855:421-424.

Morrot, G., Hervé, P., Zachowski, A., Fellmann, P., Devaux, P. F. 1989 *Biochemistry* 28:3456-3462.

Morrot, G., Zachowski, A., Devaux, P. F. 1990. *FEBS lett.* 266:29-32.

Myher, J. J., Kuksis, A., Pind, S. 1989. *Lipids* 24:396-407.

Newton, A.. C., Koshland, D. E. 1990. *Biochemistry* 29:6656-6661.

Op den Kamp, J. A. F. 1979. *Ann. Rev. Biochem.* 48:47-71.

Pelletier, X., Mersel, M., Freysz, L., Leray, C. 1987. *Biochim. Biophys. Acta* 902:223-228.

Perret, B., Chap, H. J., Douste-Blazy, L. 1979. *Biochim. Biophys. Acta* 556:434-446.

Post, J. A., Langer, G. A., Op den Kamp, J. A. F., Verkleij, A. J. 1988. *Biochim. Biophys. Acta* 943:256-266.

Prives, J., Shinitzky, M. 1977. *Nature* 268:761-763.
Rand, R. P., Parsegian, V. A. 1989. *Biochim. Biophys. Acta* 988:351-376.
Rando, R. R. 1988. *FASEB J.* 2:2348-2355.
Record, M., El Tamer, A., Chap, H., Douste-Blazy, L. 1984. *Biochim. Biophys. Acta* 778:449-456.
Sandra, A., Pagano, R.E. 1978. *Biochemistry* 17:332-338.
Schlegel, R. A.., Prendergast, T. W., Williamson, P. 1985 *J. Cell Physiol.* 123:215-218.
Schrier, S.L., Zachowski, A., Devaux, P.F. 1992. *Blood* 79:782-786.
Schroit , A. J., Madsen, J. W., Tanaka, Y. 1985. *J. Biol. Chem.* 260:5131-5138.
Seigneuret, M., Devaux, P. F. 1984 *Proc. Natl. Acad. Sci. USA.* 81:3751-3755.
Seigneuret, M., Zachowski, A., Herrmann, A., Devaux, P. F. 1984. *Biochemistry* 23:4271-4275.
Sessions, A., Horwitz, A. F. 1983. *Biochim. Biophys. Acta* 728:103-111.
Sheetz, M. P., Singer, S. J. 1974. *Proc. Natl. Acad. Sci. USA* 71:4457-4461.
Siegel, D. P., Banschbach, J., Yeagle, P. L. 1989. *Biochemistry* 28:5010-5019.
Sune, A., Bette-Bobillo, P., Bienvenüe, A., Fellmann, P., Devaux, P. F. 1987. *Biochemistry* 26:2972-2978.
Tanaka, K. I., Ohnishi, S. -I. 1976. *Biochim. Biophys. Acta* 426:218-231.
Tanaka, Y., Schroit, A. J. 1983. *J. Biol. Chem.* 258:11335-11343.
Tilley, L., Cribier, S., Roelofsen, B., Op den Kamp, J. A. F., van Deenen, L. L. M. 1986. *FEBS lett.* 194:21-27.
Tournois, H., Leunissen-Bijvelt, J., Haest, C. W. M.., de Gier, J., de Kruijff, B. 1987. *Biochemistry* 26:6613-6621.
Utsugi, T., Schroit, A. J., Connor, J., Bucana, C. D., Fidler, I. J. 1991. *Cancer Res.* 51:3062-3066.
Van Meer, G., Gahmberg, C. G., Op den Kamp, J. A. F., van Deenen, L. L. M. 1981. *FEBS lett.* 135:53-55.
Venien, C., Le Grimellec, C. 1988. *Biochim. Biophys. Acta* 942:159-168.
Verkleij, A. J., Zwaal, R.. F. A., Roelofsen, B., Comfurius, P., Kastelijn, D., van Deenen, L. L. M. 1973. *Biochim. Biophys. Acta* 323:178-193.
Wang, C. T., Shia, Y. I., Chen, J. C., Tsai, W. J., Yang, C. C. 1986. *Biochim. Biophys. Acta* 856:244-258.
Williamson, P., Algarin, L., Bateman, J., Choe, H. -R., Schlegel, R. A. 1985. *J. Cell. Physiol.* 123:209-214.
Williamson, P., Bateman, J., Kozarsky, K., Mattocks, K., Hermanowicz, N., Choe, H. R., Schlegel, R. 1982. *Cell* 30:725-735.
Williamson, P., Kulick, A., Zachowski, A., Schlegel, R. A., Devaux, P. F. 1992., *Biochemistry* 31:6355-6360.
Zachowski, A., Devaux, P. F. 1990. *Experientia* 46:644-656.
Zachowski, A., Henry, J. -P., Devaux, P.. F. 1989. *Nature* 340:75-76.
Zachowski, A., Herrmann, A., Paraf, A., Devaux, P. F. 1987. *Biochim. Biophys. Acta* 897:197-200.
Zachowski, A., Morot Gaudry-Talarmain, Y. 1990. *J. Neurochem.* 55:1352-1356.

PHOSPHATIDYLINOSITOL TRANSFER PROTEIN AND MEMBRANE VESICLE FLOW

Karel W.A. Wirtz and Gerry T. Snoek

Center for Biomembranes and Lipid Enzymology
State University of Utrecht
NL-3584 CH Utrecht, The Netherlands

INTRODUCTION

In mammalian cells, as in yeast, the endoplasmic reticulum is the main site of phosphatidylinositol (PI) biosynthesis (Bishop and Bell, 1988; Rana and Hokin, 1990; Paltauf and Daum, 1990). Concomitantly with the receptor-mediated degradation of phosphatidylinositol 4,5-bisphosphate (PIP_2) in the plasma membrane, PI is continuously converted into phosphatidylinositol 4-phosphate (PIP) and PIP_2 by both plasma membrane-bound and cytosolic kinases (Berridge, 1987; Rana and Hokin, 1990). If we accept that the formation of PIP_2 is restricted to this membrane (see also Helms et al., 1991) then the question comes up by what mechanism the steady-state level of PI in the plasma membrane is maintained. Similar to what has been proposed for the transport of both cholesterol in mammalian cells (Urbani and Simoni, 1990; Lange, 1991) and certain phospholipids in *Dictyostelium* (Da Silva and Siu, 1981) it is conceivable that PI is transported from the endoplasmic reticulum to the plasma membrane by intracellular vesicle flow. In another proposed mechanism, transfer of PI to the plasma membrane occurs directly by a monomolecular insertion reaction involving the phosphatidylinositol transfer protein (PI-TP) (Wirtz et al., 1978; Van Paridon et al., 1987; Helmkamp, 1990). Recently, evidence was provided that in a PI-TP-deficient strain of *Saccharomyces cerevisiae* transport of PI to the plasma membrane proceeded normally (Gnamusch et al., 1992). In the same study it was demonstrated that, if transport of PI occurs by vesicle flow, these vesicles are most definitely different from the vesicles involved in protein secretion. This shows that in yeast and, for this matter, also in mammals there is no evidence for the direct involvement of PI-TP in the transport of PI from intercellular stores to the plasma membrane. On the other hand, recent studies have convincingly demonstrated that PI-TP has an important role in the cell, more specifically in the functioning of the Golgi complex.

Here we will discuss some of the properties of PI-TP from mammals and yeast, its role in yeast Golgi functioning and its localization and regulation in mammalian cells.

PROPERTIES

All mammalian tissues thus far examined contain PI-TP (Helmkamp, 1990; Wirtz, 1991). This protein forms a monomolecular complex with PI and *in vitro* transfers it between membranes. The first PI-TP to be purified was that from bovine brain (Helmkamp et al., 1974; Demel et al., 1977). This protein has a molecular weight of 33,000 and consists of two isoforms with isoelectric points of 5.3 and 5.6, closely resembling rat and human PI-TP (George and Helmkamp, 1985; Venuti and Helmkamp, 1988). This resemblance is reflected in the cross-reactivity of the anti-rat PI-TP antibody with PI-TP in bovine and human brain cytosol [Venuti and Helmkamp, 1988). In fact, PI-TP appears strongly conserved as anti-bovine PI-TP antibody reacted with a 35-36 kDa protein in the membrane-free cytosol from mammals, birds, reptiles, amphibians and insects (Dickeson et al., 1989). In further support of a strongly conserved primary structure, rat PI-TP cDNA hybridizes with genomic DNA of *Drosophila melanogaster* and of a host of higher eukaryotes (Dickeson et al., 1989).

An intriguing aspect of PI-TP is that it has a dual specificity transferring PI between membranes, but also phosphatidylcholine (PC) although at a significantly lower rate (Kasper and Helmkamp, 1981; Somerharju et al., 1983). From competition binding experiments using fluorescently labeled PI and PC, it was inferred that PI-TP binds PI 16 times more efficiently than PC (Van Paridon et al., 1987). We have argued that this difference in relative affinity makes PI-TP ideally suited to maintain the PI/PC ratio in intracellular membranes. In this concept, a depletion of PI in the plasma membrane as a result of agonist-induced PIP_2 degradation (see above) will be instantaneouly corrected by PI-TP inserting its bound PI in exchange for PC in that very membrane. This would equally apply to other intracellular menbranes upon which PI-TP can act. By this action PI-TP would behave as a sensor of relative levels of PI and PC in membranes. Recently, some observations have indicated that this concept may possibly held true for the yeast Golgi membranes (Cleves et al., 1991).

In *in vitro* systems, both yeast and mammalian PI-TP express a similar capacity to transfer PI and PC between membranes (Szolderits et al., 1989). In addition, both proteins have the same molecular weight and a similar isoelectric point. On the other hand, mammalian and yeast PI-TP lack amino acid sequence homology (Salama et al., 1990; Dickeson et al., 1989) as well as immunological cross-reactivity (Szolderits et al., 1989). It is of further note that computer searches have failed to identify any other proteins with which rat PI-TP shares significant sequence homologies.

PI-TP IN YEAST

Recently, the interest in the physiological role of PI-TP has strongly increased with the observation that PI-TP in yeast is identical to the SEC14 protein (Bankaitis et al., 1990). SEC14p facilitates secretory protein transport through the Golgi complex, apparently by acting at the level of the late Golgi compartment (Bankaitis et al., 1989; Salama et al., 1990). So it was observed that the temperature-sensitive $sec14^{ts}$ mutants exhibit a marked extension of Golgi-derived structures under nonpermissive conditions (Novick et al., 1980). Immunolocalization studies using the antibodies against SEC14p and KEX2p, an integral membrane protein of the yeast Golgi, showed that PI-TP colocalizes with KEX2p (Cleves et al., 1991a). Hence, PI-TP in yeast is associated with Golgi structures. The importance of this protein for the cell follows from the observation that deletion of the gene is lethal for the organism (Aitken et al., 1990). The fundamental question whether the function of PI-TP in yeast is to actually transfer PI and PC between membranes or whether it has another, as yet, undefined cellular activity, cannot be answered. Fact is that in parellel with the dysfunction of the Golgi secretory protein pathway in sec14^{ts} yeast mutant under restricive conditions, the PI/PC transfer activities in the cell-free extracts prepared from this mutant show a marked temperature-sensitive lability (Bankaitis et al., 1990). Specifically, PI transfer activity in wild-type and mutant strain was comparable when assayed at the permissive temperature of 25 °C. In contrast to the wild-type strain, PI transfer activity in the extract from the sec14^{ts} strain was almost completely inhibited when assayed at the restrictive temperature of 37 °C. Very interestingly, a mutation in the CDP-choline pathway for PC biosynthesis can bypass the requirement for PI-TP in the sec14^{ts} mutant (Cleves et al., 1991a,b). Suppression was not observed with mutations in the methylation pathway for PC biosynthesis. This has been interpreted to indicate that PC biosynthesis via the CDP-choline pathway is restricted to the Golgi and that PI-TP is there to remove PC in exchange for PI so that an optimal PI/PC ratio in the Golgi is maintained. In the suppression mutant it then appears that the Golgi expresses the correct PI/PC ratio obviating any need for PI-TP. It has also been suggested that as a consequence of the inactivation of the CDP-choline pathway in the endoplasmic reticulum, the PI/PC ratio in this membrane would be elevated and subsequently in the Golgi which depends for its supply of phospholipids on bulk lipid flow from the endoplasmic reticulum. This again would obviate the need for PI-TP in the suppression mutant. Be it as it may, these observations clearly indicate that the proper functioning of the Golgi complex depends on PI-TP activity connected to PC biosynthesis in some poorly understood way.This presupposes that the relative PI/PC composition of the yeast Golgi is critical to the secretory competence of these membranes and that its maintenance requires PI-TP. On the other hand, one cannot exclude that PI-TP itself is a factor in the budding of vesicles from the trans-Golgi and that its function in this process is regulated by it containing either bound PI or PC.

PI-TP IN MAMMALIAN CELLS

To study the function of PI-TP in mammalian cells we have selected Swiss mouse 3T3 fibroblast cells because they are very active in phosphoinositide metabolism, they are

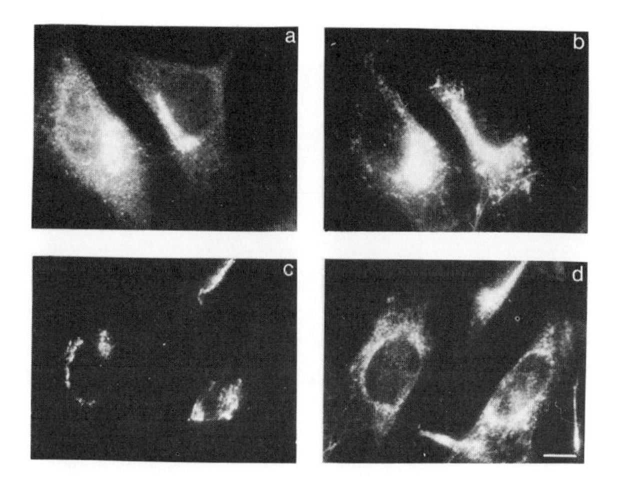

Figure 1. Immunofluorescence localization of PI-TP (panel a,c) and the Golgi marker (panel b,d) in exponentially growing (a,b) and in semi-quiescent (c,d) Swiss mouse 3T3 cells. Bar: 5 µm.

responsive to a wide variety of agonists and they can be easily switched from a quiescent to an exponentially growing state resulting in increased PI-synthesis and transport. Due to the conserved sequence of PI-TP, the antibody against bovine brain PI-TP is cross-reactive with PI-TP from the 3T3 cells. In our studies, however, we have used protein A-purified antibodies elicited against predicted epitope in rat brain PI-TP (Dickeson et al, 1989). By indirect immunofluorescence it was demonstrated that in exponentially growing 3T3 cells, PI-TP is localized in the cytoplasm associated with structures around the nucleus and in the nucleus itself (Fig. 1a). In semi-quiescent cells (cells cultured in medium containing 0.5% foetal calf serum) the most prominent labeling is found in the structures around the nucleus (Fig. 1b). The pattern of labeling around the nucleus indicates an association of PI-TP with the Golgi complex (Snoek et al., 1992). This was established by double-labeling experiments in which the antibody against PI-TP (visualized with the goat-ant-rabit IgG coupled to FITC) labeled the same region in the 3T3 cell as Ricin-TRITC, a marker for proteins which are specifically processed in the Golgi system. Further evidence for the colocalization of PI-TP with the Golgi complex was obtained by showing that treatment of the cells with brefeldin A resulted in a rapid dispersal of PI-TP labeling throughout the cell which could be reversed by the subsequent removal of this drug from these cells (Lippincott-Schwartz et al., 1989).

In quiescent cells (confluent and serum-starved) PI-TP labeling was barely detectable (Fig. 2a). When these cells were stimulated by phorbol 12-myristate, 13-acetate (PMA) or bombesin, a remarkable increase of Golgi-associated PI-TP labeling was observed within 10 min of incubation (Figs. 2b,c). This rapid response strongly suggests that PI-TP already present in the cell, is directed towards the Golgi possibly as a result of some chemical modification. Since PMA can activate protein kinase C directly (Nishizuka et al., 1986) and bombesin indirectly through activation of phospholipase C-dependent breakdown of PIP_2 (Cook et al., 1990), it was investigated whether PI-TP is a substrate for protein kinase C both *in vitro* and *in vivo*. Incubation of PI-TP from bovine brain with purified protein kinase C from rat brain gave a Ca^{++}- and phosphatidylserine-dependent phosphorylation (G.T. Snoek, to be published). This is in agreement with the five potential phosphorylation sites (Thr-59, Ser-166, Thr-169, Thr-198, Thr-251) for protein kinase C which were predicted from the amino acid sequence of PI-TP from rat brain. Phosphorylation of PI-TP *in vivo*

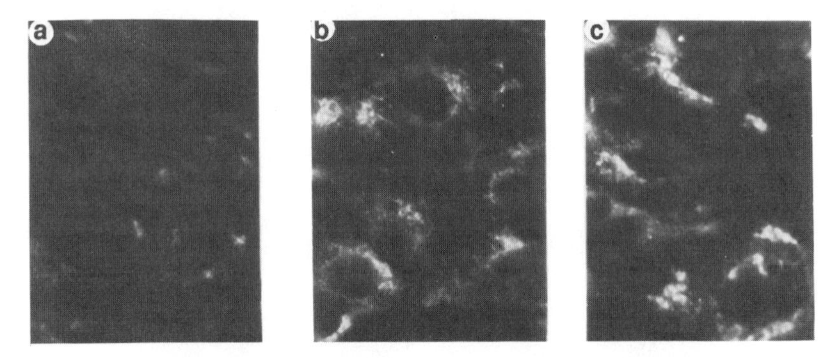

Figure 2. Immunofluorescence localization of PI-TP in quiescent Swiss mouse 3T3 cells (panel a) and in quiescent cells incubated for 10 min with PMA (50 ng/ml, panel b) or bombesin (10 nM, panel c).

was investigated in semi-quiescent 3T3 cells prelabeled with ^{32}P-phosphate. PI-TP was immunoprecipitated from control cells and from cells stimulated with PMA for 10 min. In control cells some phosphorylation of PI-TP can be observed (Fig. 3, lane 5). After stimulation with PMA for 10 min, the phosphorylation of PI-TP is increased (Fig. 3, lane 6). Phosphorylation of total soluble proteins in control and PMA-stimulated cells is shown in lanes 1 and 2. Proteins that bind non-specifically to IgG, were removed by incubating the cytoplasmic fractions with pre-immune IgG (Fig 3, lanes 3,4). Although purified PI-TP is readily phosphorylated by protein kinase C *in vitro*, it is not yet known whether protein kinase C is responsible for the phosphorylation of PI-TP *in vivo*, or whether other protein kinases are (also) involved (Thomas 1992). These data, however, do suggest that the intracellular redistribution of PI-TP observed after cell stimulation, is in some way related to the phosphorylation of PI-TP.

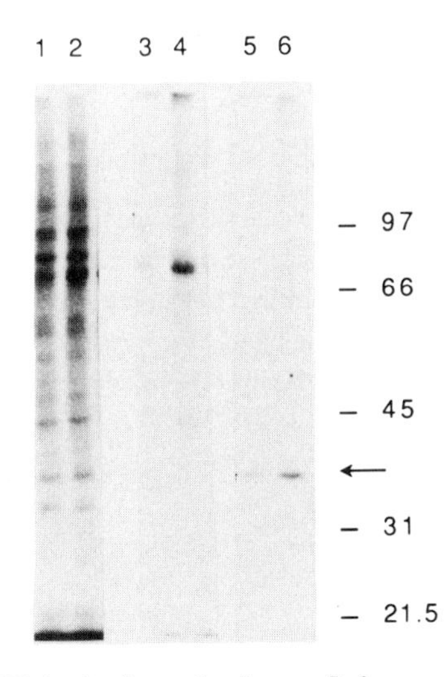

Figure 3. Phosphorylation of PI-TP *in vivo* in semi-quiescent Swiss mouse 3T3 cells. Samples were analyzed by 15% SDS-PAGE folowed by autoradiography. Lanes 1,2: total ^{32}P-labeled proteins in the cytoplasmic fraction from 3T3 cells; lanes 3,4: ^{32}P-labeled proteins cleared from the cytoplasmic fractions by pre-immune IgG; lanes 5,6: ^{32}P-labeled PI-TP immunoprecipitated from the precleared cytoplasmic fractions by anti-PI-TP antibody. Lanes 1,3,5: control cells; lanes 2,4,6: cells incubated for 10 min with PMA, 50 ng/ml.

In the scheme depicted in Fig. 4, transfer of PI to the plasma membrane may occur by the direct action of PI-TP and/or by vesicle flow. In this transfer process, phosphorylation of PI-TP could be a mechanism to regulate its function by affecting either its association with the Golgi complex (thereby, controlling vesicle flow in conjunction with the PI/PC composition of these vesicles) or directly its phospholipid transfer activity. Our current investigations are aimed to disclose whether the phosphorylation of PI-TP has any effect on these proposed functions.

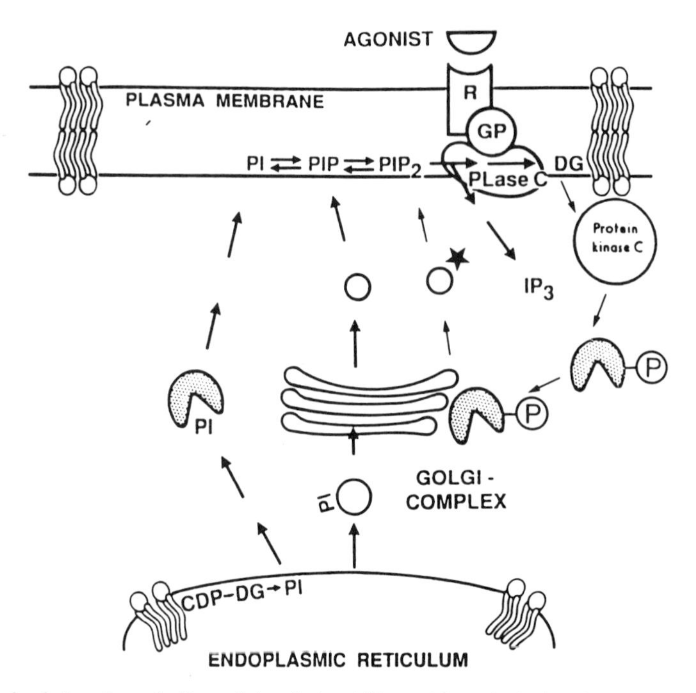

Figure 4. Stimulus-induced metabolism of phosphoinositides and hypothetical pathways to maintain PI-levels in the plasma membrane. CDP-DG, cytidine-diphosphate-diacylglycerol; DG, diacylglycerol; IP_3, inositol 1,4,5-trisphosphate; R, receptor; GP, GTP-binding protein; PLase, phosphoinositide-specific phospholipase C; the packman is PI-TP; vesicle with asterix represents a modified vesicle due to the phosphorylation of PI-TP. Amended version of a figure published by J.B. Helms (Thesis, University of Utrecht, 1991).

REFERENCES

Aitken, J.F. and Van Heusden, G.P.H., Temkin, M. & Dowhan, W., 1990, The gene encoding the phosphatidylinositol transfer protein is essential for cell growth, J. Biol. Chem. 266:4711.

Bankaitis, V.A., Malehorn, D.E., Emr, S.D. and Greene, R., 1989, The *Saccharomyces cerevisiae* SEC14 gene encodes a cytosolic factor that is required for transport of secretory proteins from the yeast Golgi complex, J. Cell Biol. 108:1271.

Bankaitis, V., Aitken, J.R., Cleves, A.E. and Dowhan, W., 1990, An essential role for a phospholipid transfer protein in yeast Golgi function, Nature 347, 561.

Berridge, M.J., 1987, Inositol trisphosphate and diacylglycerol: two interacting second messengers, Annu. Rev. Biochem. 56:159.

Bishop, W.R. and Bell, R.M., 1988, Assembly of phospholipids into intracellular membranes: biosynthesis, transmembrane movement and intracellular translocation, Ann. Rev. Cell Biol. 4:579.

Cleves, A.E., McGee, T.P., Whitters, E.A., Champion, K.M., Aitken, J.R., Dowhan, W., Goebl, M. and Bankaitis, V.A., 1991a, Mutations in the CDP-choline pathway for phospholipid biosynthesis bypass the requirement for an essential phospholipid transfer protein, Cell 64:789.

Cleves, A.E., McGee, T. and Bankaitis, V.A., 1991b, Phospholipid transfer proteins: a biological debut, Trends Cell Biol. 1:30

Cook, S.J., Palmer, S., Plevin, R. and Wakelam, M.J.O., 1990, Mass measurement of inositol 1,4,5-trisphosphate and *sn*-1,2-diacylglycerol in bombesin-stimulated Swiss 3T3 mouse fibroblasts, Biochem. J. 265:617.

Da Silva, N.S. and Siu, C.-H.. 1981, Vesicle-mediated transfer of phospholipids to plasma membrane during cell aggregation of *Dictyostelium discoideum*, J. Biol. Chem. 256:5845.

Demel, R.A., Kalsbeek, R., Wirtz, K.W.A. and Van Deenen, L.L.M., 1977, The protein-mediated net transfer of phosphatidylinositol in model systems, Biochim. Biophys. Acta. 466:10.

Dickeson, S.K., Lim, C.N., Schuyler, G.T., Dalton, T.P., Helmkamp, G.M. and Yarbrough, L.R., 1989, Isolation and sequence of cDNA clones encoding rat phosphatidylinositol transfer protein, J. Biol. Chem. 264:16557.

George, P.Y. and Helmkamp, G.M., 1985, Purification and characterization of a phopsphatidylinositol transfer protein from human platelets, Biochim. Biophys. Acta. 836:176.

Gnamusch, E., Kalaus, C., Hrastnik, C., Paltauf, F. and Daum, G., 1992, Transport of phospholipids between subcellular membranes of wild-type yeast cells and of the phosphatidylinositol transfer protein-deficient strain *Saccharomyces cerevisiae sec 14*, Biochim. Biophys. Acta 1111:120.

Helmkamp, G.M., 1990, Transport and metabolism of phosphatidylinositol in eukaryotic cells, in: Subcellular Biochemistry, Hilderson, H.J., ed., Vol. 16, pp. 129-174, Plenum Press, New York

Helmkamp, G.M., Harvey, M.S., Wirtz, K.W.A. and Van Deenen, L.L.M., 1974, Phospholipid exchange between membranes. Purification of bovine brain proteins that preferentially catalyze the transfer of phosphatidylinositol, J. Biol. Chem. 249:6382.

Helms, J.B., De Vries, K.J. and Wirtz, K.W.A., 1991, Synthesis of phosphatidylinositol 4,5-bisphosphate in the endoplasmic reticulum of Chinese hamster ovary cells, J. Biol. Chem. 266:21368.

Kasper, A.M. and Helmkamp, G.M., 1981, Intermembrane phospholipid fluxes catalyzed by bovine brain phospholipid exchange protein, Biochim. Biophys. Acta 664:22.

Lange, Y., 1991, Disposition of intracellular cholesterol in human fibroblasts, J. Lipid Res. 32:329.

Lippincott-Schwartz, J., Yuan, L.C., Bonifacio, J.S., Klausner, R.D., 1989, Rapid redistribution of Golgi proteins into the ER in cells treated with brefeldin A: evidence for membrane cycli9ng from Golgi to ER, Cell 56:801.

Nishizuka, Y., 1986, Studies and perspectives of protein kinase C, Science 233:305.

Novick, P., Field, C. and Schekman, R., 1980, Identification of 23 complementation groups required for post-translational events in the yeast secretory pathway, Cell. 21:205.

Paltauf, F. and Daum, G., 1990, Phospholipid transfer in microorganisms, in: Subcellular Biochemistry, H.J. Hilderson, ed., Vol. 16, pp 279-299, Plenum Press, New York.

Rana, R.S. and Hokin, L.E., 1990, Role of phosphoinositides in transmembrane signaling, Physiol. Rev. 70:115.

Salama, S.R., Cleves, A.E., Malehorn, D.E., Whitters, E.A. & Bankaitis, V.A., 1990, Cloning and characterization of the *Kluyveromyces lactis* SEC14: a gene whose product stimulates Golgi secretory function in *S. cerevisiae*, J. Bacteriol. 172:4510.

Snoek, G.T., De Wit, I.S.C., Van Mourik, J.H.G. and Wirtz, K.W.A., 1992, The phosphatidylinositol transfer protein in 3T3 mouse fibroblast cells is associated with the Golgi system, J. Cell Biochem. 49:339.

Somerharju, P., Van Paridon, P.A. and Wirtz, K.W.A., 1983, Phosphatidylinositol transfer protein from bovine brain. Substrate specificity and membrane binding properties. Biochim. Biophys. Acta, 731:186.

Szolderits, G., Hemetter, A., Paltauf, F. and Daum, G., 1989, Membrane properties modulate the activity of a phosphatidylinositol transfer protein from the yeast, *Saccharomyces cerevisiae*, Biochim. Biophys. Acta 986:301.

Thomas, G., 1992, MAP kinase by any other name smells just as sweet, Cell 68:3.

Urbani, L. and Simoni, R.D., 1990, Cholesterol and vesicular stomatitis virus G protein take separate routes from the endoplasmic reticulum to the plasma membrane, J. Biol. Chem. 265:1919.

Van Paridon, P.A., Gadella, T.W.J., Somerharju, P.J. and Wirtz, K.W.A., 1987, On the relationship between the dual specificity of the bovine brain phosphatidylinositol transfer protein and membrane phosphatidylinositol levels, Biochim. Biophys. Acta, 903:68.

Venuti, S.E. and Helmkamp, G.M., 1988, Tissue distribution, purification, and characterization of rat phosphatidylinositol transfer protein, Biochim. Biophys. Acta. 946:119.

Wirtz, K.W.A., Helmkamp, G.M. and Demel, R.A., 11978, in Protides of the biological fluids, H.Peeters, ed., Vol. 25, pp 25-32. Pergamon Press, London.

Wirtz, K.W.A., 1991, Phospholipid transfer proteins, Annu. Rev. Biochem. 60:73.

MODIFICATION OF CELLULAR PHOSPHOLIPID COMPOSITION AND CONSEQUENCES FOR MEMBRANE STRUCTURE AND FUNCTION

Joseph Donald Smith
Department of Chemistry
University of Massachusetts Dartmouth
North Dartmouth, MA 02747 (USA)

Over the past few years, my laboratory has been engaged in studies on phospholipid metabolism using the ciliate protozoan *Tetrahymena thermophila* as a model system for eukaryotic phospholipid metabolism (Smith, 1983; Smith, 1984; Smith, 1986b; Smith, et al., 1992b; Smith and O'Malley, 1978). *Tetrahymena* is characterized by having high concentrations of phosphonolipids — 2-aminoethylphosphonoglyceride and 2-aminoethylphosphonoceramide — as well as phosphatidylcholine and phosphatidylethanolamine as the major phospholipids (Smith, 1985; Smith and Giegel, 1981; Smith and Giegel, 1982; Smith, et al., 1992; Smith and O'Malley, 1978; Thompson, 1972; Thompson, et al., 1971).

Among our studies are experiments aimed at altering the phospholipid composition of the organism to determine the effect of such modifications on phospholipid metabolism and to try to elucidate potential functions for each of the phospholipids in the cell membrane (Smith, 1986a; Smith and Barrows, 1988; Smith and Giegel, 1981; Smith and Giegel, 1982; Smith, et al., 1992b; Smith and Ledoux, 1990; Smith, et al., 1992; Smith and O'Malley, 1978). When different phospholipid headgroups or analogues are added to the culture medium, *Tetrahymena* incorporates these compounds into its phospholipids. Some of the compounds used are shown in Figure 1.

Of particular interest are the results obtained with the various phosphonic acids. Growth with 2-aminoethylphosphonate, the natural compound, brings about an increase in the 2-aminoethylphosphonoglyceride and a corresponding decrease in phosphatidylethanolamine in a dose-dependent manner (Smith, 1984; Smith, 1986b; Smith and O'Malley, 1978). There is no apparent effect on cell viability or morphology. However, there is a dramatic alteration in the utilization of the two different pathways for formation of phos-

New Developments in Lipid-Protein Interactions and Receptor Function
Edited by K.W.A. Wirtz *et al.*, Plenum Press, 1993

2--Aminoethylphosphonate (AEP)

Ethanolamine Phosphate

3--Aminopropylphosphonate (APP)

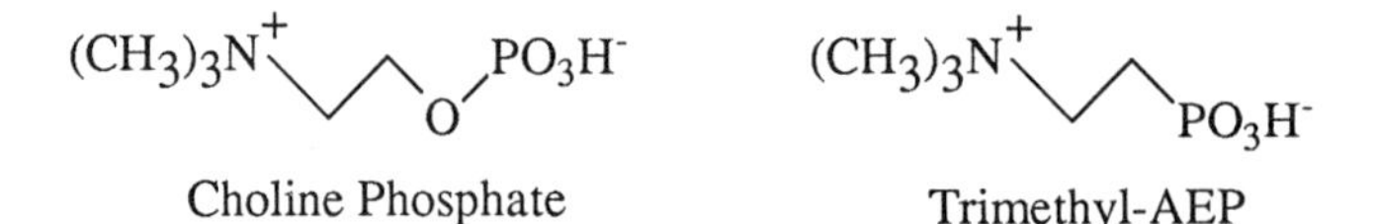

Figure 1. Phosphonates used to modify cellular phospholipid composition compared to the normal phospholipid headgroups: choline phosphate and ethanolamine phosphate.

phatidylcholine. Phosphatidylethanolamine methylation, which normally supplies about 60% of the phosphatidylcholine, is reduced to 25% with a corresponding increase in the cholinephosphotransferase pathway (Smith, 1986b).

3-Aminopropylphosphonate behaves as an analogue of phosphoethanolamine, its isostere, rather than of 2-aminoethylphosphonate (Smith, 1986b; Smith and Giegel, 1981). 3-Aminopropylphosphonoglyceride is formed at the expense of phosphatidylethanolamine, not 2-aminoethylphosphonoglyceride, and the formation of the new phosphonolipid leads to growth inhibition and ultimately to cell death although cell morphology is normal. Neither of the phosphonolipids is a substrate for the phosphatidylethanolamine N-methyltransferase (Smith, 1983). In neither case is there alteration in phosphatidylcholine levels.

N,N,N-Trimethyl-2-aminoethylphosphonate, an analogue of phosphocholine, is incorporated into the phospholipids of *Tetrahymena* at the expense of phosphatidylcholine (Smith and Giegel, 1982). While much higher medium concentrations are required to achieve phospholipid alterations (25 - 50 mM compared to 5 mM 3-aminopropylphosphonate), probably because of transport difficulties, the results are even more dramatic. The cells develop a grossly altered morphology - loss of the pyriform shape and major membrane convolutions which are clearly seen under optical microscopy. The phosphonolipid formed from this compound replaced phosphatidylcholine with minimal alterations in the other phospholipids (Smith and Giegel, 1982).

The phospholipid modification studies with phosphonates in *Tetrahymena* suggested that these compounds could be useful for comparable work in animal cells, particularly since these compounds arc natural products (Hilderbrand and Henderson, 1983), are found in the human diet (Hilderbrand and Henderson, 1983) and have even detected in human tissue (Tan and Tan, 1989). There has been one report on the effect of 2-aminoethylphosphonate on rats in vivo (Nair, et al., 1978). Large doses of the compound resulted in elevated serum cholesterol and altered activity of several liver enzymes but changes in phospholipid levels were not reported (Nair, et al., 1978).

Previous lipid modification studies with animal cells in culture demonstrated that various ethanolamine or choline analogs, including N-methylethanolamine, N,N-dimethylethanolamine, 3-amino-1-propanol, ℓ-2-amino-1-butanol and 2-amino-2-methyl-1-propanol, were incorporated into cellular phospholipids when the alcohols were included in the culture medium of fibroblasts (Glaser, et al., 1974; Maziere, et al., 1990; Schroeder, 1980; Schroeder, et al., 1976). For the most part, the new phospholipids were formed at the expense of both phosphatidylethanolamine and phosphatidylcholine, with relatively minor changes in phosphatidylserine, phosphatidylinositol, sphingomyelin, lysophosphatidylcholine and cardiolipin.

Most of the membrane-associated enzymes assayed in lipid-altered cells, Na^{+}-K^{+}-ATPase, 5'-nucleotidase, NADPH-cytochrome c reductase, glucose-6-phosphatase, and succinate-cytochrome c reductase, showed minimal differences in activity from control cells (Schroeder, et al., 1976). However the activities of acyl-CoA: cholesterol acyltransferase and diacylglycerol acyltransferase decreased about 50% in monomethylethanolamine and dimethylethanolamine-grown cells (Maziere, et al., 1990). It should be noted that since fibroblasts lack phosphatidylethanolamine N-methyltransferase activity (Coleman, et al., 1978), there may be long-term viability problems in cells cultured with inadequate medium levels of choline.

For our own studies examining the effect of phosphonates on animal cells, we selected mouse 3T3-L1 fibroblasts which have well-defined phospholipid metabolism (Coleman, et al., 1978) and which can be induced to differentiate into adipocytes by a number of factors including insulin (Chiang, et al., 1987), tumor necrosis factor (Calvo, et al., 1992), progesterone (Rondinone, et al., 1992) and 3-deazaadenosine (Chiang, 1981; Chiang, et al., 1987).

Cells were grown in standard Dulbecco's modified Eagle medium supplemented by heat-inactivated calf serum (for growth) or by heat-inactivated fetal bovine serum (for differentiation studies) to which appropriate concentrations of 2-aminoethylphosphonate were added (Smith, et al., 1992a). The phosphonate was added at the time of inoculation and maintained through the entire culturing period unless otherwise mentioned. When cell were grown to confluence in the presence of 10 mM 2-aminoethylphosphonate, the 2-aminoethylphosphonoglyceride was found to comprise about 10% of the total phospholipids and was formed at the expense of both phosphatidylethanolamine and phosphatidylcholine; none of the minor phospholipids was affected.

For differentiation studies, 10 μg/ml insulin was added to experimental flasks, with and without 2-aminoethylphosphonate, while controls without insulin were run in parallel. Medium containing insulin and/or 2-aminoethylphosphonate was changed every 48 hr. After the indicated time, cells were harvested and triglyceride (as a measure of differentiation) and DNA contents of each tissue-culture flask were determined. Triglyceride was determined using a triglyceride kit (Sigma Chemical Co.) and DNA by the method of Burton (Burton, 1956).

As shown in Figure 2, triglyceride formation in 3T3-L1 cells was inhibited by the presence of 2-aminoethylphosphonate, whether the differentiation was induced by insulin (Fig. 2B) or was spontaneous (Fig. 2A). At 1 mM 2-aminoethylphosphonate the differentiation response was about 50% of the controls while the differentiation was totally inhibited at 5 mM or 10 mM 2-aminoethylphosphonate. This is more clearly seen as the dose-response curves plotted in Figure 3.

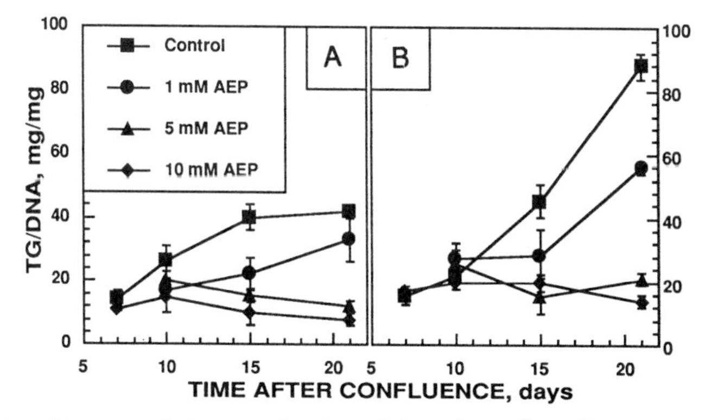

Figure 2. Triglyceride accumulation as a function of time after cell confluence measured as TG/DNA (mg/mg) per tissue culture flask. Cells were cultured and triglyceride and DNA were determined as described in Methods. In AEP-grown cells, AEP was present throughout the experiment. Values represent means ± S.D. for triplicate determinations. A. cells without insulin; B. cells with insulin. Reproduced from (Smith, et al., 1992a), by permission.

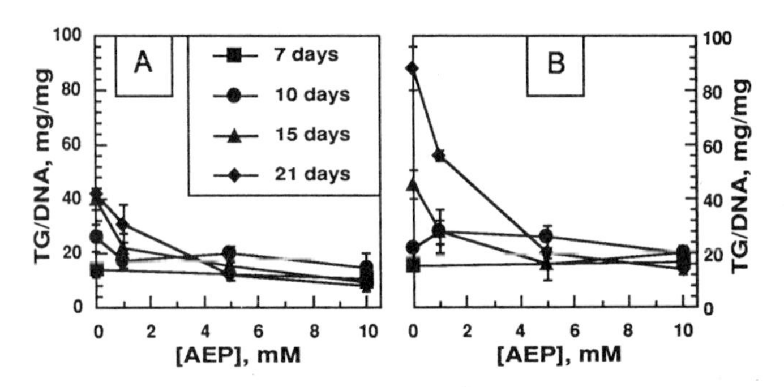

Figure 3. Triglyceride accumulation as a function of AEP concentration. Cells were cultured and triglyceride and DNA were determined as described in Methods. In AEP-grown cells, AEP was present throughout the experiment. The time represents days after cell confluence had been attained. Values represent means ± S.D. for triplicate determinations. A. cells without insulin; B. cells with insulin. Reproduced from (Smith, et al., 1992a), by permission.

To determine whether the 2-aminoethylphosphonate had to be present in the medium throughout the entire growth and differentiation periods, experiments were carried out where 2-aminoethylphosphonate was present only during a specific time period. When the phosphonate was present only during the 48 hr from inoculation to cell confluence but not during the subsequent differentiation period, the insulin-induced differentiation was still inhibited indicating that once the phosphonolipid was formed, there was no wash-out of the phosphonate (Smith, et al., 1992a).

Although we have no evidence as to where in the signal transduction pathway the phosphonolipid might affect the process, it is worth noting that phosphonolipids are resistant to phospholipase D activity (lacking a hydrolyzable bond between the phosphorus and the base) (Proll and Clark, 1991) and are known to inhibit both phospholipase A_2 (Marshall, et al., 1991) and phospholipase C activities (Shashidhar, et al., 1990) and the insulin receptor itself is sensitive to its phospholipid environment (Arnold and Newton, 1991; Arnold and Newton, 1992). Further work will be required to determine whether phosphonolipids affect differentiation induced by other factors as they do the insulin-induced differentiation and to define the step in the process which is actually affected.

ACKNOWLEDGMENTS

My research on *Tetrahymena* is partially supported by a grant from the National Science Foundation (DCB89-04979). I should also like to thank Dr. P. K. Chiang and his colleagues at Walter Reed Army Institute of Research, Washington, DC, for their hospitality and support for the studies in 3T3 cells.

REFERENCES

Arnold, R.S. and Newton, A.C. (1991). Inhibition of the insulin receptor tyrosine kinase by sphingosine, *Biochemistry* **30**: 7747-7754.

Arnold, R.S. and Newton, A.C. (1992). Regulation of insulin receptor function by lipids, *FASEB J.* **6**: A89.

Burton, K. (1956). A Study for the Conditions and Mechanism of the Diphenylamine Reaction for the Colorimetric Estimation of Deoxyribonucleic Acid, *Biochem. J.* **62**: 315-323.

Calvo, J.C., Chernick, S. and Rodbard, D. (1992). Mouse mammary epithelium produces a soluble heat-sensitive macromolecule that inhibits differentiation of 3T3-L1 preadipocytes, *Proc. Soc. Exp. Biol. Med.* **201**: 174-179.

Chiang, P.K. (1981). Conversion of 3T3-L1 fibroblasts to fat cells by an inhibitor of methylation: effect of 3-deazaadenosine, *Science* **211**: 1164-1166.

Chiang, P.K., Brown, N.D., Padilla, F.N. and Gordon, R.K. (1987). "Induction of differentiation of 3T3-L1 fibroblasts to adipocytes by 3-deazaadenosine and insulin," *in:* "Tumor Cell Differentiation." J. Aarbakke, P.K. Chiang and H.P. Koeffler, eds., The Humana Press, Clifton, N.J. pp. 231-240.

Coleman, R.A., Reed, B.C., Mackall, J.C., Student, A.K., Lane, M.D. and Bell, R.M. (1978). Selective changes in microsomal enzymes of triacylglycerol, phosphatidylcholine and phosphatidylethanolamine biosynthesis during differentiation of 3T3-L1 preadipocytes, *J. Biol. Chem.* **253**: 7256-7261.

Glaser, M., Ferguson, K.A. and Vagelos, P.R. (1974). Manipulation of the phospholipid composition of tissue culture cells, *Proc. Nat. Acad. Sci. USA* **71**: 4072-4076.

Hilderbrand, R.L. and Henderson, T.O. (1983). "Phosphonic acids in nature," *in:* "The Role of Phosphonates in Living Systems." R.L. Hilderbrand, ed., CRC Press, Inc., Boca Raton, FL pp. 5-30.

Marshall, L.A., Bolognese, B., Yuan, W. and Gelb, M. (1991). Phosphonate-phospholipid analogues inhibit human phospholipase A_2, *Agents Actions* **34**: 106-109.

Maziere, C., Auclair, M., Mora, L. and Maziere, J.C. (1990). Modification of phospholipid polar head group with monomethylethanolamine and dimethylethanolamine decreases cholesteryl ester and triacylglycerol synthesis in cultured human fibroblasts, *Lipids* **25**: 311-315.

Nair, C.R., Stith, I.E., Nair, R.R. and Das, S.K. (1978). The effects of aminoethyl phosphonic acid on hepatic microsomal drug metabolism and ultrastructure of normal and cholesterol fed rats, *J. Nutr.* **108**: 1234-1243.

Proll, M.A. and Clark, R.B. (1991). Potent Gi-mediated inhibition of adenylyl cyclase by a phosphonate analog of monooleylphosphatidate, *Mol .Pharmacol.* **39**: 740-744.

Rondinone, C.M., Baker, M.E. and Rodbard, D. (1992). Progestins stimulate the differentiation of 3T3-L1 preadipocytes, *J. Steroid Biochem. Mol .Biol.* **42**: 795-802.

Schroeder, F. (1980). Regulation of aminophospholipid asymmetry in murine fibroblast plasma membranes by choline and ethanolamine analogues, *Biochim. Biophys. Acta* **599**: 254-270.

Schroeder, F., Perlmutter, J.F., Glaser, M. and Vagelos, P.R. (1976). Isolation and characterization of subcellular membranes with altered phospholipid composition from cultured fibroblasts, *J. Biol. Chem.* **251**: 5015-5026.

Shashidhar, M.S., Volwerk, J.J., Keana, J.F.W. and Griffith, O.H. (1990). Inhibition of phosphatidylinositol-specific phospholipase-C by phosphonate substrate analogues, *Biochim. Biophys. Acta* **1042**: 410-412.

Smith, J.D. (1983). Effect of modification of membrane phospholipid composition on the activity of phosphatidylethanolamine n-methyltransferase of *Tetrahymena, Arch. Biochem. Biophys.* **223**: 193-201.

Smith, J.D. (1984). Incorporation of serine into the phospholipids of phosphatidylethanolamine-depleted *Tetrahymena, Arch. Biochem. Biophys.* **230**: 525-532.

Smith, J.D. (1985). Differential selectivity of cholinephosphotransferase and ethanolaminephosphotransferase of *Tetrahymena* for diacylglycerol and alkylacylglycerol, *J. Biol. Chem.* **260**: 2064-2068.

Smith, J.D. (1986a). Effect of dimethylaminoethylphosphonate on phospholipid metabolism in *Tetrahymena, Biochim. Biophys. Acta* **878**: 450-453.

Smith, J.D. (1986b). Phosphatidylcholine homeostasis in phosphatidylethanolamine-depleted *Tetrahymena, Arch. Biochem. Biophys.* **246**: 347-354.

Smith, J.D. and Barrows, L.J. (1988). Use of 3-aminopropanol as an ethanolamine analogue in the study of phospholipid metabolism in *Tetrahymena, Biochem. J.* **254**: 301-302.

Smith, J.D. and Giegel, D.A. (1981). Replacement of ethanolamine phosphate by 3-aminopropylphosphonate in the phospholipids of *Tetrahymena, Arch. Biochem. Biophys.* **206**: 420-423.

Smith, J.D. and Giegel, D.A. (1982). Effect of a phosphonic acid analog of choline phosphate on phospholipid metabolism in *Tetrahymena, Arch. Biochem. Biophys.* **213**: 595-601.

Smith, J.D., Gordon, R.K., Brugh, S.A. and Chiang, P.K. (1992a). Inhibition of the insulin-induced differentiation of 3T3-L1 fibroblasts into adipocytes by phosphonolipids, *Biochem. Arch.* **8**: 339-344.

Smith, J.D., Hardin, J.H. and Patterson, C.H. (1992b). Effect of high exogenous levels of ethanolamine and choline on phospholipid metabolism in *Tetrahymena, Biochem. Arch.* **8**: 121-125.

Smith, J.D. and Ledoux, D.N. (1990). Effect of the methylation inhibitors 3-deazaadenosine and 3-deazaaristeromycin on phosphatidylcholine formation in *Tetrahymena, Biochim. Biophys. Acta* **1047**: 290-293.

Smith, J.D., Ledoux, D.N. and Steinmeyer, N.A. (1992). Combined effects of 2-aminoethylphosphonate and 3-deazaadenosine on phosphatidylcholine biosynthesis in *Tetrahymena, Biochem. Arch.* **8**: 23-27.

Smith, J.D. and O'Malley, M.A. (1978). Control of phosphonic acid and phosphonolipid synthesis in *Tetrahymena, Biochim. Biophys. Acta* **528**: 394-398.

Tan, S.A. and Tan, L.G. (1989). Distribution of ciliatine (2-aminoethylphosphonic acid) and phosphonoalanine (2-amino-3-phosphonopropionic acid) in human tissues, *Clin. Physiol. Biochem.* **7**: 303-309.

Thompson, G.A., Jr. (1972). *Tetrahymena pyriformis* as a model system for membrane studies, *J. Protozool.* **19**: 231-236.

Thompson, G.A., Jr., Bambery, R.J. and Nozawa, Y. (1971). Further studies on the lipid composition and biochemical properties of *Tetrahymena pyriformis* membrane systems, *Biochemistry* **10**: 4441-4447.

PHOSPHOLIPID HEADGROUPS AS SENSORS OF ELECTRIC CHARGE

Joachim Seelig

Biocenter of the University of Basel
Department of Biophysical Chemistry
CH-4056 Basel, Switzerland

MEMBRANE PROPERTIES

In 1839 the anatomist T. Schwann published his "Microscopic Investigations on the Similiarity of Structure and Growth of Animals and Plants" in which he provided the first evidence that animals and plants are composed of the same elements, the cells. The new cell theory immediately led to the question of how the cells could manage to move matter from one cell to the other. Around 1900 the biologist E. Overton investigated the transport rate of more than 300 different organic compounds in animal and plant cells. He observed that all compounds which were easily soluble in oil or similar solvents could move through the living protoplast with high speed whereas other compounds which were easily soluble in water but not in ether, alcohol, or oil migrated only slowly. Based on this selective solubility of plant and animal cells he concluded that the outer surface of the cell was impregnated by a substance which had solubility properties similar to those of a fatty oil. In particular, he suggested that the outer cell layer was composed of a mixture of lecithin and cholesterol [1].

Overton was unable to make any predictions on the molecular organization of the lipid membranes. We now know that the biological membrane is only 10 nm thick, i.e. about 5 times smaller than the wavelength of the visible light. However, even without the availability of the electron microscope two Dutch physicians, E. Gorter and F. Grendel, could derive the molecular picture of the biological membrane from indirect experiments already in 1925, long before the advent of electron microscopy. Gorter and Grendel extracted the lipid component of red blood cells with aceton and layered the lipids on a water surface. Using a movable barrier the monomolecular layer was compressed until a distinctive increase in pressure could be observed. The corresponding monolayer area was measured and compared with the surface area of the erythrocytes. For erythrocytes of different origin Gorter and Grendel always found the same result: the monolayer area was twice as large as the erythrocyte area. They concluded that the red blood cells were covered by a layer of fatty substances that was two molecules thick[2]. They postulated that the polar groups of the lipids were in contact with water whereas the hydrocarbon chains were pointing towards the inner part of the double layer. In subsequent years, many other methods such as X-ray diffraction, electron microscopy, and magnetic resonance spectroscopy have provided a more direct

New Developments in Lipid-Protein Interactions and Receptor Function
Edited by K.W.A. Wirtz *et al.*, Plenum Press, 1993

picture of the membrane organization and have confirmed the bilayer model of Gorter and Grendel.

Organizing the phospholipids of the biological membrane into a phospholipid double layer provides a highly efficient permeability barrier for charged particles. Metal ions and highly charged peptides and proteins cannot penetrate into the hydrophobic interior of the membrane since the activation energy for this process is extremely large. On the other hand, water molecules are sufficiently nonpolar to easily diffuse through the phospholipid bilayer and the water permeability coefficient has been determined by different physical chemical techniques to be $P_{H_2O} = 3 \times 10^{-3}$ cm/sec [3].

For small unilamellar vesicles, prepared by sonification of lipids in buffer, this means that the total water content in the interior of the vesicle is exchanged in less than 0.2 msec with the bulk water of the surrounding. A lipid bilayer also contains a considerable number of water molecules dissolved in the membrane phase and the total water concentration can be estimated to be about 5 µM. This is approximately also the water concentration in pure organic solvents such as hexane when brought into equilibrium with water.

The biological membrane is, however, not only composed of lipids. At the membrane surface peptides and proteins can be attached and there is an equally large variety of integral membrane proteins which span the membrane from one surface to the other. These proteins are, in part, channels regulating the transport of matter and the production of energy. Others act as receptors regulating the flow of information between the neighboring cells.

Most lipids and proteins in biological membranes are in constant movement. The fluidity as measured by spectroscopic probe methods corresponds to a macroscopic viscosity found, for example, in olive oil (1-10 poise). The molecular state of the lipid membrane can thus be characterized as that of a liquid crystalline phase. Indeed, compared to other biological substances only lipids form liquid crystalline phases. Moreover, the lipid bilayer is only one possibility among many different types of liquid crystalline arrangements. The polymorphism of lipids can easily be detected with diffraction methods or with nuclear magnetic resonance. In addition to the lipid bilayer, cylindrical and cubic structures as well as micellar arrangements have been identified [4,5].

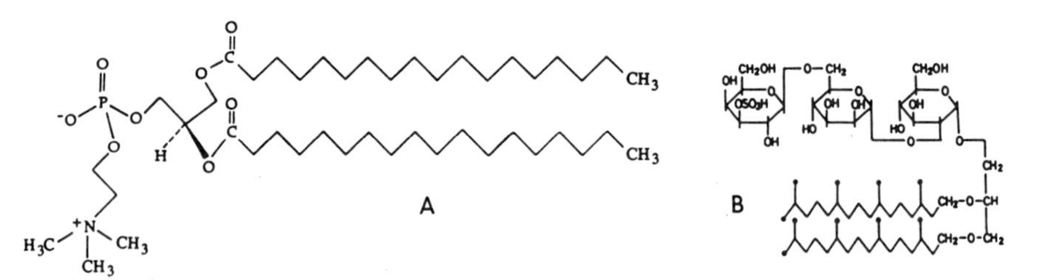

Figure 1. The chemical structures of (A) a common phospholipid, phosphatidylcholine (lecithin), and (B) a bacterial glycosulfolipid.

A large variety of lipids has been isolated and identified in biological membranes. For example, in the human erythrocyte membrane about 400 chemically different species of lipid have been found [6]. Figure 1 shows the chemical structure of one of the most common lipids, lecithin, and that of a rather unusual bacterial lipid, a phosphoglycosulfolipid. Today more than thousand different lipids, i.e. lipids with chemically different structures have been identified [7]. The variability of lipids is caused by their different fatty acyl chains (saturated, unsaturated, branched fatty acids, variation of the chain length), by their differences in headgroups, and by the different types of linkages between polar groups and fatty acids. Chemical syntheses are known only for a very limited number of these lipids.

In order to appreciate the complexity of lipid structures it should be recalled that nucleic acids have only four building stones and that all proteins are composed of not more than 24 amino acids. We do not know why nature has provided such a large variety of lipids. If the role of lipids were only to provide a fluid environment for the proteins, i.e. to act as a grease to keep the protein machinery in motion, or if their only function were to create a permeability barrier, this could be achieved with a very much reduced number of lipids. An alternative explanation for lipid complexity would be to postulate a specific interaction between membrane proteins and a given type of lipids, in analogy to the well-known substrate-enzyme interactions. However, this simple model is contradicted by experimental results which indicate that the function of many membrane proteins can be reconstituted by quite different lipid mixtures. Only for a very limited number of membrane proteins has it been possible to demonstrate a specific and unique interaction between lipids and proteins.

On the other hand, one may speculate that the physical-chemical properties of the membrane such as the surface dipole potential or the internal membrane pressure, can be influenced by the chemical composition of the lipid phase and, in turn, might regulate the functional state of membrane-bound proteins. It was observed already by Overton that chemically quite different compounds such as alcohols, ethers, chloroform, etc. could induce dizziness and anesthesia in tadpoles, albeit, at quite different concentrations in the aqueous phase. Overton then took into account the different lipid/water distribution coefficients and obtained the surprising result that anesthesia required a minimum concentration of the anesthetic substance in the membrane which was of the order of 30 mM, independent of the chemical constitution (Meyer-Overton rule). Around 1950 an interesting reverse experiment. was reported. If anesthesized tadpoles which were resting motionless at the bottom of a water tank, were put under nitrogen pressure of about 90 atmospheres they were found to move again [8]. The change in the membrane properties as induced by the anesthetic molecules could obviously be reversed by increasing the pressure. This pressure reversal of anesthesia does not require a specific one-to-one-interaction between lipids and proteins but the physical-chemical properties of the membrane are collectively changed which, in turn, is felt by the membrane proteins.

Here we are particularly interested in the role of the lipid polar groups in membrane-protein interactions. We shall discuss this problem in three steps:

- We shall describe the conformation of phospholipids in model membranes and in biological membranes,
- we then demonstrate how the headgroup orientation can be changed by electric charges, and
- we shall discuss biological consequences and regulatory mechanisms which follow from the reorientation of phospholipid headgroups.

STRUCTURE AND DYNAMICS OF LIPID HEADGROUPS

Phospholipids have a molecular weight of ~ 800 Dalton and are thus rather small biomolecules. Nevertheless, phospholipids have eluded most efforts of crystallisation and only three crystal structures of naturally occurring lipids have been solved so far. The first crystal structure, published in 1974, was that of phosphatidylethanolamine [9]. Since then only two further structures, namely those of phosphatidylcholine and phosphatidylglycerol have been added [10,11]. All three lipids were found to crystallize in bilayer structures, and the

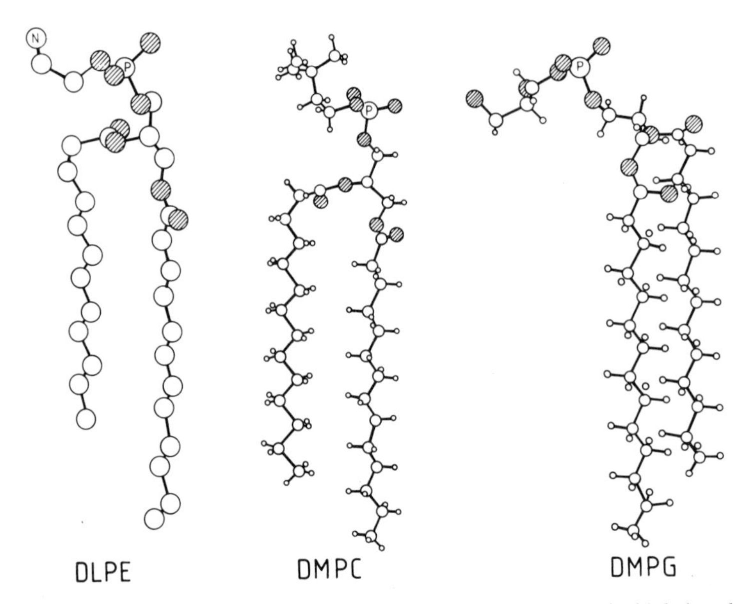

Figure 2. The crystal structures of some naturally occurring lipids. PE = phosphatidylethanolamine, PC = phosphatidylcholine, PG = phosphatidylglycerol.

molecular conformations as derived from the X-ray analysis are shown in figure 2. The three structures have a number of features in common. We note (i) the more or less parallel orientation of the phospholipid headgroup with respect to the bilayer surface and (ii) the different initial orientations of the sn-1 and sn-2 chain in the same molecule. While the sn-1 chain is in the all-trans-configuration and is in line with the glycerol backbone, the sn-2 chain starts out parallel to the surface of the membrane and has a bend at the C-2-segment of the hydrocarbon chain.

The crystallization of the three lipids mentioned above was achieved in rather unusual solvents such as glacial acetic acid or chloroform-water mixtures. It could be argued, therefore, that the crystal structures are not relevant for a lipid membrane in contact with an aqueous environment. Here two experimental methods have proven to be especially fruitful, namely neutron scattering methods and deuterium magnetic resonance. A prerequisite for the application of these methods is a specific deuteration of the lipid segment of interest which can be achieved either by chemical synthesis or by biochemical incorporation. Due to the different neutron scattering factors of protons and deuterons it is then possible to determine the average position of the deuterated segments with an accuracy of about 1 Å [12]. On the other hand, deuterium magnetic resonance provides information on the average lipid conformation, on the fluctuations of the individual segments, and on the packing of the lipids in the membrane [13].

The application of these methods to liquid crystalline model membranes, bacterial membranes, and cell cultures has led to the following results:

- The choline, ethanolamine and glycerol headgroups are oriented approximately parallel to the membrane surface.
- There are no significant differences between the headgroup structure in model membranes and those of intact biological systems.
- The conformational difference between the sn-1 and sn-2 hydrocarbon chain remains unaltered in the liquid crystalline state.

- The rate of rotational diffusion of the phospholipid headgroups at the membrane surface is one to two orders of magnitude slower than in aqueous solution. This can be explained by an extended hydrogen bonding network at the membrane surface.

LIPID HEADGROUPS AND ELECTRIC CHARGE

Lipid headgroups are rather sensitive to the electric charge at the membrane surface[14]. Figure 3 shows deuterium NMR spectra of phospholipid membranes with various surface

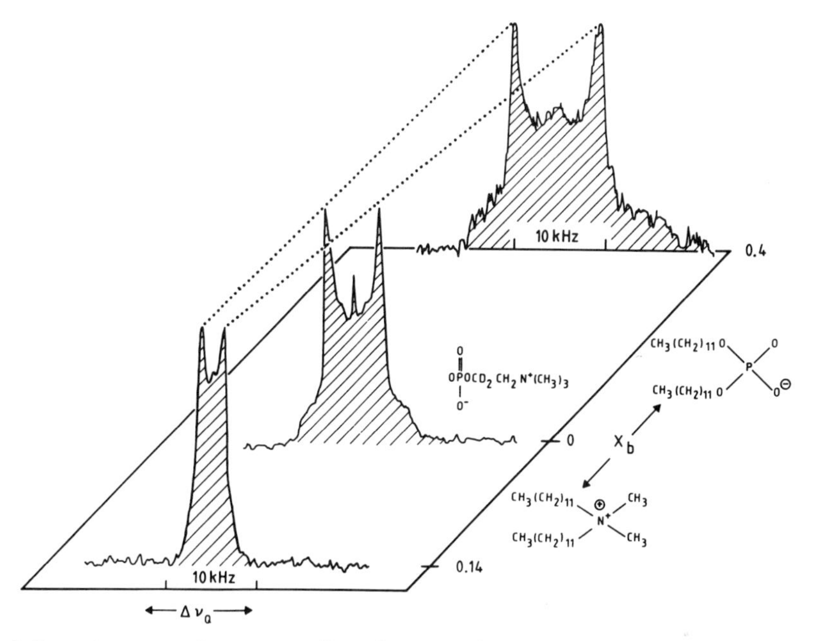

Figure 3. Deuterium magnetic resonance (D-nmr) spectra of a lipid membrane composed of zwitter-ionic lecithin mixed with either negatively or positively charged amphiphiles. The lipid was deuterated at the α–segment of the choline headgroup. Addition of positive electric charge *decreases* the quadrupole splitting, negative charge *increases* it.

charges. In all three spectra the phosphatidylcholine (lecithin) molecule was deuterated as the choline headgroup and dispersed in water to form coarse lipid bilayers. The unusual form of the deuterium NMR spectra is a peculiarity of the deuterium nucleus. The separation of the two most intense peaks in the spectrum (marked by dashed lines in figure 3) is the so-called quadrupole splitting. In the present context, the quadrupole splitting is simply used as an empirical parameter to indicate changes in the headgroup conformation. The quadrupole splitting is hence comparable, for example, with the change in the UV absorption upon the denaturation of a protein. For the pure lipid membrane without surface charge, the deuterated α–methylene group of the choline moiety is characterized by a quadrupole splitting of 6.0 kHz (middle spectrum in figure 3). If a positive electric charge is brought to the membrane surface the quadrupole splitting *decreases* by up to 7 kHz (lower spectrum). On the other hand, a negative surface charge has the opposite effect and *increases* the quadrupole splitting of the headgroup segment under investigation. In figure 3 the change of the electric surface charge was achieved by mixing the lipid with either positively charged dialkylammoniumbromide or negatively charged dialkylphosphate. Both compounds are amphiphilic and even in pure form aggregate spontaneously into lipid doublelayers when dispersed in water. They can also be mixed with lecithin in any ratio but without inducing changes in the hydrophobic part of the membrane.

In order to explain the variation of the D-nmr parameters in terms of a molecular model, we recall that the choline headgroup ($^{-}OPOCH_2CH_2N^{+}(CH_3)_3$) as well as the ethanolamine headgroup($^{-}OPOCH_2CH_2NH_3^{+}$) (for which similar experimental results were obtained) are electrically neutral but are nevertheless characterized by a large electrical dipole moment directed from the negative phosphate to the positive ammonium charge. The size of this dipole moment is approximately 20 Debye. In the absence of charges the phospholipid headgroup dipoles are extended almost parallel to the membrane surface. The quantitative analysis of the quadrupole splittings then demonstrates, that a positive electric surface charge moves the N^{+}-end of the dipole into the water phase, a negative surface charge moves the dipole closer to the hydrocarbon interior. The angular change can amount to as much as ±20°.This is demonstrated schematically by figure 4.

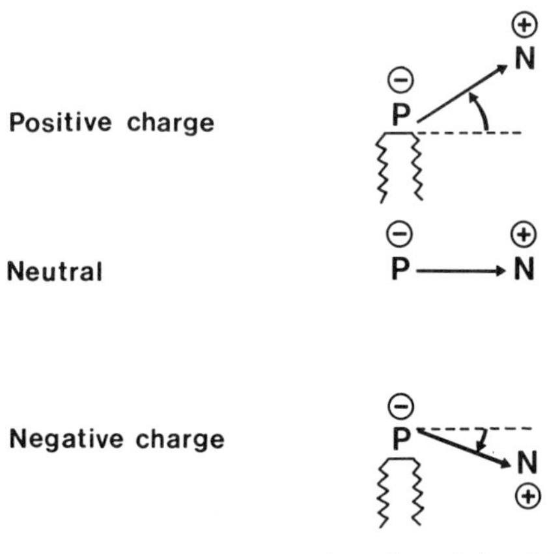

Figure 4. Influence of electric surface charge on the orientation of the $^{-}P-N^{+}$ dipole of the choline and ethanolamine headgroup. Positive charge moves the dipole into the aqueous phase, negative charge into the water phase.

This reorientation of the lipid headgroup would be of limited interest, where it is restricted only to rather unusual amphiphilic compounds. However, further experiments have proven this change to be a rather general phenomenon. All positively charged substances which are absorbed at the membrane surface or which can intercalate between lipids, i.e. metal ions (Ca^{2+}, Mg^{2+}, La^{3+}) [15-18], hydrophobic ions (tetraphenylphosphonium[19], local anesthetics)[20,21] or positively charged peptides (melittin, somatostatin) induce changes in the phospholipid headgroups comparable to those observed for dimethyldialkylammonium. In contrast, addition of anions, i.e. chaotropic anions (I^-, SCN^-)[28], hydrophobic anions (tetraphenylborate)[14], and negatively charged phospholipids[29] lead to a change in the headgroup orientation which corresponds to that induced by dialkylphosphates. Since the absorbed electrical charges are not distributed evenly on the membrane surface the observed changes in the headgroup orientation vary distinctly, depending on the position and distance of the charges. In addition, molecules which are electrically neutral but are characterized by a large dipole moment can also cause a change in the headgroup orientation of the lipids[30]. Taken together, a large set of experimental data has been accumulated by now which demonstrates unambiguously that some lipid headgroups, in particular the choline and the ethanolamine headgroup, are rather sensitive to changes in their local electrical environment. These headgroups may therefore be considered as "molecular voltmeters". Different results have, however, been obtained for the phosphoglycerol headgroup. This headgroup appears to be much less sensitive to electric charges and conformational changes are small under most experimental conditions.

BIOLOGICAL REGULATIONS VIA PHOSPHOLIPIDS?

Phosphatidylcholines and phosphatidylethanolamines are among the most abundant natural lipids and the observed changes in the lipid headgroups allow speculations about possible collective mechanisms of action. The most prominent mechanism is presumably a conformational change of the membrane proteins via the electric dipole field. Electric measurements on proteins in solution have demonstrated that the equilibrium of the α-helix-coil transition can be shifted by electric fields of the order of 10^5 V/cm. Electric fields of this size can easily be generated at the membrane surface. If the lipid headgroups are oriented exactly parallel to the membrane, their corresponding dipole field cannot penetrate deeply into the membrane interior. However, a rotation of the dipole away from the membrane by about 20° generates a dipole component perpendicular to the membrane surface of about 6 D and the corresponding dipole field is about 90 mV/5 nm = $1.8 \cdot 10^6$ V/cm. This dipole field will enhance or attenuate the existing electric fields and is sufficiently large to initiate the postulated changes in protein conformation.

Equally important for regulation mechanisms could be the fact that chemically quite differente substances can bind to the surface of the membrane. The binding affinities vary over a very large range and different binding mechanisms are possible. In biological membranes this binding could be accompanied by a change in the membrane surface potential due to relatively high contribution of negatively charged lipids in most biological membranes. The negatively excess charge leads to a surface potential of about -50 mV in the absence of binding ligands. However, in the presence of, for example, 1 mM Ca^{2+} the membrane potential may increase to -20 mV due to Ca^{2+}-binding.[31] At the same time the orientation of the lipid headgroups can change. Alternatively, a change in the membrane potential can regulate the local concentration of metal ions. Finally, the observed physical mechanism could lead also to a change in the internal pressure of the membrane. It will be the matter of future research to demonstrate if and how nature takes advantage of this possible mechanisms of regulation.

REFERENCES

1. E. Overton, *Vierteljahresschr. Naturforsch. Ges. Zürich* XLIV, 88 (1899).
2. E. Gorter and F. Grendel, *J. Exp. Med.* 41, 439 (1925).
3. A. Finkelstein. "Water Movement through Lipid Bilayers, Pores, and Plasma Membranes. Theory and Reality", Wiley, New York (1987).
4. V. Luzzati, *in:* "Biological Membranes", Vol. 1, pp. 71-123 (Chapman, D., ed.) Academic Press, New York (1968).
5. P.R. Cullis and B.de Kruijff, *Biochim. Biophys. Acta* 559, 399 (1979).
6. R.F.A. Zwaal, R.A. Demel, B. Roelofsen, and L.L.M. van Deenen, *Trends Biochem. Sci.* 1, 112 (1976).
7. F.D. Gunstone, J.L. Harwood, and F.B. Padley: The Lipid Handbook. Chapman & Hall, London 1986.
8. F.H. Johnson und E.A. Flagler, *Science* 112, 91-92 (1950); cf. also A.D. Bangham, *in:* G. Weissmann und

R. Clairborne (Hrsg.): Cell Membranes; Biochemistry, Cell Biology and Pathology. HP Publishing, New York 1975, S. 24-34.
9. B.P. Hitchock, R. Mason, K.M. Thomas und G.G. Shipley, *Proc. Natl. Acad. Sci.* USA 71, 3036 (1974).
10. R.H. Pearson and I. Pascher, *Nature* (London)281, 499 (1979).
11. I. Pascher, S. Sundell, K. Harlos und H. Eibl. *Biochim. Biophys. Acta* 896, 77 (1987).
12.G. Büldt, H.U. Gally, A. Seelig, J. Seelig, und G. Zaccai, *Nature* (London) 271, 182 (1978).
13. J. Seelig and P.M. Macdonald, *Acc. Chem. Res.* 10, 221 (1987).
14. J. Seelig, P.M. Macdonald, and P.G. Scherer, *Biochemistry* 26, 7535 (1987).
15. Ch. Altenbach and J. Seelig, *Biochemistry* 23, 3913 (1984).
16. Ch. Altenbach, Thesis, University of Basel, (1984).
17. P.M. Macdonald and J. Seelig, *Biochemistry* 26, 6292 (1987).
18. P.M. Macdonald and J. Seelig; *Biochemistry* 26, 1231 (1987)
19. C. Altenbach and J. Seelig, *Biochim. Biophys. Acta* 818, 410 (1985).
20. Y. Boulanger, S.Schreier,, and I.C.P. Smith, *Biochemistry* 20, 6824 (1981).
21. A. Seelig, P.R. Allegrini, and J. Seelig, *Biochim. Biophys. Acta* 939, 267 (1988).
22. E. Kuchinka and J. Seelig (1989) *Biochemistry* 28, 4216 (1989).
23. G. Beschiaschvili and J. Seelig, *Biochemistry* 29, 52 (1990).
24. M. Roux, J.M. Neumann, R.S. Hodges, P. Deveaux, and M. Bloom, *Biochemistry* 28, 2313 (1989).
25. C.E. Dempsey and A. Watts, *Biochemistry* 26, 5803 (1987).
26. G. Beschiaschvili and J. Seelig, *Biochemistry* 29, 10995 (1990).
27. G. Beschiaschvili and J. Seelig, *Biochim. Biophys. Acta* 1061, 78 (1991).
28. P.M. Macdonald and J. Seelig, *Biochemistry* 27, 6769 (1988).
29. P.G. Scherer and J. Seelig, *EMBO J.* 6, 2915 (1987).
30. Bechinger, B. and J. Seelig, *Biochemistry* 30, 3923 (1991).
31. J. Seelig, *Cell. Biol. Int. Rep.* 14, 353 (1990).

MEMBRANE PHOSPHOLIPIDS ACT AS DNA/RNA RECEPTORS DURING FORMATION OF SPECIFIC DNA-NUCLEAR MEMBRANE CONTACTS AND GENE EXPRESSION: A HYPOTHESIS BASED ON THE STUDY OF INTERACTION BETWEEN PHOSPHOLIPID VESICLES AND DNA OR POLYNUCLEOTIDES

Renat Zhdanov[1,2] and Vasily Kuvichkin[1]

[1]Institute for Biotechnology, Nauchnyi Proezd, 8
Moscow 117246 Russian Federation and
[2]University of Ancona, Ancona 60131 Italy

INTRODUCTION

Interaction of DNA with biomembranes was discovered for all types of organisms: viruses, phages and bacteria, plant and animal cells (e.g.,Moyer,1979), participating, by our opinion, in signal transduction. In 70-s, along with lipid-protein interactions (e.g., Watts and De Pont, 1985) DNA-membrane interactions have also been studied. DNA-nuclear membrane contacts (DNA-MC) have been considered important for matrix synthesis of nucleic acids, DNA packing in bacteria and in eucaryotic chromosomes, translocation of nucleic acids through biomembranes during virus infection and transfection, as well as for DNA replication and transcription (e.g., Bush, 1974, Moeyer, 1979). These works covered mainly DNA-membrane proteins interactions, and small attention has been paid to that of DNA and biomembrane phospholipids. Only a limited number of works was dedicated to the part of nuclear lipids for stabilization or destabilization of DNA (e.g., Manzoli et al., 1972; Manzoli et al., 1974; Alesenko and Burlakova, 1976; Alesenko and Pantaz, 1983). Interaction of nucleic acids with lipids has been also studied by a number of techniques: spectrophotometry (Brosius and Reisner, 1986), IR-spectroscopy (Shabarchina et al., 1979), ^{31}P- (Victorov et al.,1984) and ^{13}C-NMR (Budker et al., 1986), microcalorimetry (Bichenkov et al., 1988a) and spin labeling (Krasilnikov et al., 1989), though without relation to the problem of DNA-MC. Contemporary data on interaction of polynucleotides and nucleic acids with lipids (Sukhorukov et al., 1980; Bonora et al., 1981; Grepachevsky et al., 1986) and phospholipid liposomes (Hoffman et al., 1978; Budker et al., 1978, 1980; Gruzdev et al., 1982; Kuvichkin, 1983; Kuvichkin and Sukhomudrenko, 1987; Kuvichkin et al., 1989; Kuvichkin, 1990; Zhdanov and Kuvichkin, 1991; Zhdanov, 1992; Zhdanov et al. 1993 a, b, c) reveal the existence of considerable interaction between them in the presence of bivalent metal ions, characterized by the possible coordination of phosphoryl groups of polynucleotides and phospholipids through metal ion bridges.

At the same time new data has been appearing, testifying to importance of lipid-nucleic acid interactions in realization of cell functions. In particular, some Z-DNA binding proteins were found to be at the same time lipid-binding ones (Krishna et al., 1990). Though the translocation of single-stranded oligonucleotide $(dT)_{16}$ into cells was recently proposed to involve receptor proteins and adsorption endocytosis mechanism (Vlasov et al, 1989), polymorphic behavior of gram-negative bacteria membrane was discovered to be important for translocation of DNA through it (Borovjagin et al., 1987; Sabelnikov et al., 1988). Series of papers was dedicated to the study of DNA translocation through membrane bilayer with the use of large liposomes as model phospholipid membrane (e.g., Budker et al., 1987; Bichenkov et al., 1988b).

New Developments in Lipid-Protein Interactions and Receptor Function
Edited by K.W.A. Wirtz *et al.*, Plenum Press, 1993

Present paper formulates mechanism of formation of DNA-nuclear membrane contacts, suggesting the leading part of membrane phospholipid - nucleic acid interactions in this process, as well as participation of small nuclear (s.n.) RNA/RNP in endowing DNA-MC with specific features. RNA, complementary to intervening sequences (introns) of functional genes and released during RNA splacing, could act as such RNA. This hypothesis is based on the study of interaction between DNA or polynucleotides and organized phospholipid surfaces in the presence of metal ions by various biochemical and biophysical techniques.

EXPERIMENTAL

Materials

Dimiristoylphosphatidylcholine, dipalmitoylphosphatidylcholine, polyadenylic acid (polyA) (S^{20} 8.0-8.5), polyuridilyc acid (polyU) (6-7 S, S^{20} 6) and calf thymus DNA (highly polymerized), Triton X-100 and HEPES were purchased from Serva (Heidelberg, F.R.G.). Salmon sperm DNA (20,000 kDa) was obtained from Sigma (St.Louis, MO, U.S.A.). L-α-Phosphatidylcholine (PC) (90%) was obtained from Reakhim (Moscow, Russia) and Khar'kov (Ukraine).

Commercial preparations of polyA and polyU (Reanal, Hungary) have been purified from low molecular mass impurities by precipitation of high molecular mass fraction from aqueous solution by cold ethanol, solubilization of precipitate in 1 mM EDTA with following dialysis for 4 days and lyophilization of the final solution. Average molecular mass of the preparations, determined by sedimentation technique in 10 mM tris.HCl buffer solution, pH 7.5 was 50 kDa. Sedimentation constants (S^{20}) were 5.1 ($\pm$0.2) for polyA and 5.3 ($\pm$0.2) for polyU. PolyA:polyU duplex was prepared by keeping equimolar polyA/polyU solution for 30 min at 40°C and than overnight at 4°C.

All the solvents used were of the purest quality. All metal salts were of reagent grade and obtained from Souzreaktiv (Moscow, Russia).

Phospholipid Liposome Preparation

An ethanol or chloroform solution of phosphatidylcholine was evaporated under reduced pressure at 40°C and dried in vacuum for 2 hours. Then lipids were rehydrated with buffer solution under nitrogen , submitted to intensive vortexing for 5 min and sonicated for 10 min at 0°C using ultrasound desintegrator (UZDN, Russia) at 22 kHz under nitrogen flow. The sonicated mixture was centrifugated at 12,000 rpm, and the precipitate was removed. The study of properties of the complexes of nucleic acids, phospholipid vesicles and metal ions ("triple complexes") by variety of techniques was conducted not earlier than 30 min after preparation of corresponding mixtures.

BASIC PRINCIPLES OF INTERACTIONS OF NUCLEIC ACIDS WITH ORGANIZED PHOSPHOLIPID SURFACES

Formation of Two Types of "Triple Complexes" in the Presence of Me Ions.

Addition of double-stranded polynucleotides and nucleic acids to suspension containing liposomes and bivalent metal ions results in liposomes' aggregation/ adhesion expressed by the increased turbidity of mixture, caused by the formation of the "triple complex" of the 1st type. It is convenient to determine the liposome aggregation degree in such conditions by measuring suspension turbidity at 600 nm (Budker et al., 1980; Kuvichkin and Sukhomudrenko, 1987) (Figure 1). Parameter α may be used for description of aggregation process as well as for study of the influence of different factors on it. Aggregation degree α increses with the increase of the time of sonication of multilamellar liposomes (Figure 1) (i.e. with the decrease in vesicles' size) and, for the same DNA concentration, has almost triple values for smaller vesicles. The following row of parameter α values (in parenthess) was obtained for different metal ions (20 mcg/ml of salmon sperm DNA and 200 mcg/ml of PC liposomes): Mn (3.9), Mg (3.7), Ca (3.7), Co (3.6), Ni (3.4), Sr (3.0), Zn (2.7), Fe (2.6), Ba (1.2) (Kuvichkin and Sukhomudrenko, 1987). This series of α values correlates well with the row of binding constants of these cations for PC liposomes (McLauglin et al, 1978). The maximum values of parameter α were obtained for liposomes prepared from amphiphilic phospholipids: phosphatidylcholine, phosphatidylethanolamine and sphingomyelin.

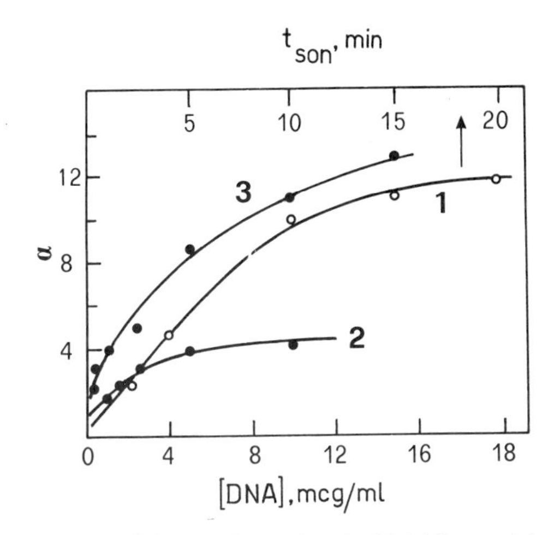

Figure 1. Dependence of degree of egg phosphatidylchline vesicles aggregation (parameter α) in the presence of Mg ions from liposome sonication time and DNA concentration. **1** - dependence of parameter α value from liposome sonication time (20 mcg/ml of salmon sperm DNA); **2** - dependence of parameter α value from DNA concentration at 5 min of sonication; **3** - the same, but sonication time is 20 min (10 mM HEPES, pH 7.2; Beckman-26 spectrophotometer) (rebuilt basing on data of Kuvichkin and Sukhomudrenko, 1987). $MgCl_2$ solution was added to the PC vesicles' (Reakhim) suspension, while stirring, till the final concentration of 5 mM. Then DNA solution of the necessary concentration was added, and after 1-2 sec of stirring the kinetics of the turbidity increasing at 600 nm was observed. Parameter $\alpha = D_c/D_o$ characterizes liposomes' aggregation degree and degree of their interaction with DNA (D_o - initial optical density at 600 nm, D_c - final optical density at 600 nm after mixing with DNA).

According to the previous data complexes of double-stranded DNA with liposomes and metal ions are being destroyed by DNase I, phospholipase C, spermidine, EDTA, 0.3 M NaCl and pH value drop to 3.5 as well as by compounds, able to intercalate between DNA bases.

"Triple complexes" of the 2nd type are formed during interaction of liposomes with single-stranded polynucleotides and metal ions. Liposomes' coordination with such polynucleotides occurs without their aggregation and noticeable increase of turbidity (α 1.2). There is an adhesion of these polynucleotides to liposomes' surface.

Properties of the complexes of the 1st and the 2nd types are well illustrated by isotherms of adsorption of Mn ions by the PC liposomes - polynucleotide system in Scatchard coordinates (Figure 2). EPR spectrum of Mn ions (6 components) during their binding to polynucleotide-liposome system widens and doesn't influence Mn ion EPR spectrum. It makes possible to determine concentration of unbound Mn ions from the amplitude of EPR spectrum of this mixture.

Equation for these Scatchard plots is as follows (Reuben and Gabbay, 1975):

$$r/m = k\ (1/n - r),$$

where $r = /Mn/_b\ /\ /P/_t$; $/P/_t$ is the total phosphate concentration; n - number of phosphate moieties, participating in the binding of Mn ions; k - binding constant; $m = /Mn/_f$. Subscripts b, f and t mean bound, free and total, correspondingly.

According to this equation, in the absence of competition the curve, demonstrating dependence of r/m from r has to be linear. If it does not, it means that biomacromolecule has a few types of binding sites. Results, represented on Figure 2, testify to considerable differences in binding of Mn ions with polynucleotides or with polynucleotides and

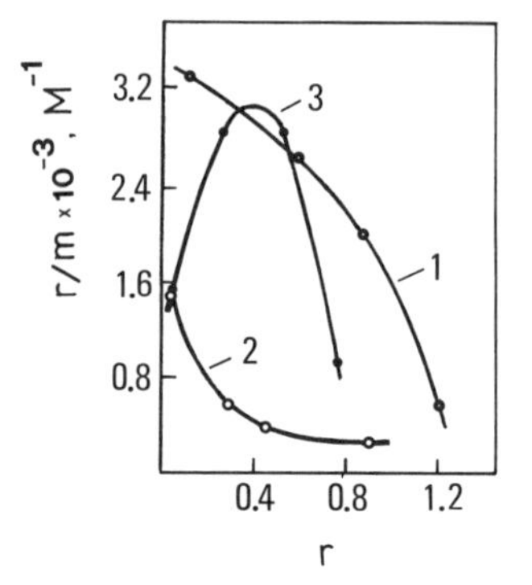

Figure 2. Adsorption isotherms for titration of nucleic acids with Mn ions revealed by EPR technique. **1** - polyA (Serva) and egg PC liposomes; **2** - calf thymus DNA; **3** - DNA and egg PC liposomes. Concentrations were: 200 mcg/ml (0.31 mM of b.p.) of DNA; 200 mcg/ml (0.61 mM of b.p.) of polyA and 2.5 mg/ml (3.25 mM) of egg PC (5 mM of $MgCl_2$, 10 mM tris.HCl/10 mM NaCl, pH 7.2; Bruker ER-200 radiospectrometer) (rebuilt basing on data of Kuvichkin et al, 1989). Titration was done by the following technique: 0.1 M solution of $MnCl_2$ (1-30 mcL) was added to the mixture of nucleic acid and PC liposomes (Khar'kov), total volume of mixture reaching 300 mcL with addition of buffer solution. The mixture was kept for 30 min, after that a capillar (inner diameter 1 mm) was filled with the mixture and the measurements of amplitude of the third spectral component were conducted.

multilamellar liposomes. Non-linearity of the curves 1 and 2, showing DNA and polyA titration by Mn ions, indicates that these macromolecules have more than one type of binding center. Taking into account small degree of saturation of phospholipid surface by Mn ions, the convex shape of the curve 3 may testify to the cooperative interaction between Mn ion binding sites. Dissociation constant of complex formed by Mn ions with liposomes and DNA (per 1 b.p.) can be estimated as equal to 2.82 mM (n = 0.9) (at large r values) presuming that only 10% of phospholipids participate in binding (Budker et al., 1980; Zhdanov et al., 1993c), while ions form bridges between phosphoryl groups of phospholipids (average m.m. is 770) and DNA (average m.m. of base pair is 643; m.m. of c.t. DNA is 1,800 kDa). This value of dissociation constant, corresponding to weak binding, is still almost 40-fold less than the value of the same constant (111 mM), characterizing interaction between phospholipids and DNA , intrapped in the inner volume of liposomes in the absence of bivalent metal cations (Brosius and Riesner, 1986; Bichenkov et al., 1988b). As it follows from analysis of the curve 3 (Figure 2), during "complex" formation binding of Mn ions with native DNA approaches that of Mn ions and s.s. polyA, which could testify to the local unwinding of DNA helix. Thus, taking into account differences in binding of Mn ions with liposomes and DNA or polyA and basing on turbidimetry data, it becomes possible to demostrate existence of two types of "triple complexes".

Double-Stranded Polynucleotides Can Interact with Phospholipid Vesicles without Me Ions

Above mentioned phenomenon has been established as result of microcalorimetric study of the interaction between polyA:polyU duplex, dimiristoylphosphatidylcholine (DMPC) vesicles and Mg ions at various ratio of the components (Figure 3) (Zhdanov et al., 1993c). In the absence of Mg ions meltings of polyA:polyU duplex and DMPC vesicles in their mixture happen almost independently from each other (unshown). However, small increase happens in T_m value of polynucleotide duplex (from 44.6(±0.05)°C till 45.3°C) at the increasing of the DMPC/polyA:polyU ratio (from 0.5 : 1 to 2 : 1) (Figure 3, curve 3), which may testify to the existence of interaction between phospholipid vesicles and double-stranded polynucleotides in the absence of bivalent metal ions. At the same time the value of calorimetric enthalpy of gel - liquid crystal phase transition, ΔH, in multilamellar DMPC liposomesdecreases from 5.0 (±0.1) $J.K^{-1}.M^{-1}$ (DMPC) to 4.3 $J.K^{-1}.M^{-1}$ for polyA:polyU/DMPC complex (1:2). This finding indicates that polyA:polyU duplex and DMPC liposomes interact even without metal ions, with stabilization of polynucleotide duplex.

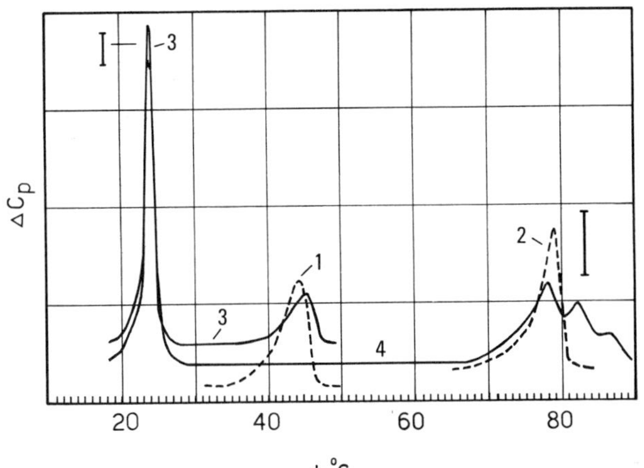

Figure 3. Temperature dependence of apparent specific heat for various poly-nucleotides - phospholipid vesicles systems. **1** - polyA:polyU; **2** - polyA:polyU and $MgCl_2$ (2.5 mM); **3** - polyA:polyU and DMPC liposomes; **4** - polyA:polyU, DMPC liposomes and $MgCl_2$ (2.5 mM) (ΔCp bar is 0.5 $J.K^{-1}.M^{-1}$, short label is only for curve **3**), concentrations being as follows: 1 mM of (A):(U) base pair of polyA:polyU duplex (Reanal) and 1 mM of DMPC (for curve **3** - 2 mM DMPC), 10 mM tris.HCl/10 mM NaCl, pH 7.5; cell volume was 1.5 ml). DMPC liposomes were prepared as described earlier (Permyakov et al., 1989). Differential scanning calorimetry was performed with a Privalov DASM-4M adiabatic differential microcalorimeter (Pushchino, Russia) using scan rate 1°C/min and overpressure $1.5x10^5$ Pa.

Local Unwinding of Nucleic Acid Duplex in the Presence of Liposomes and Me Ions

Preliminary biochemical data indicated the increased sensitivity of DNA, included in "triple complex" with liposomes and Mg ions, to the treatment by reagents which are specific to single-stranded DNA sites (dimethylsulphate, S1 endonuclease) (Budker et al., 1978 and 1980). This findings may be interpreted as an appearence of single-stranded sites in DNA helix. To confirm this suggestion it was convenient to use circular dichroism technique, which is very informative for the study of DNA secondary structure. In the case of "triple complexes" it is useful also due to the fact that CD-spectra in relatively wide range do not depend from turbidity of liposomes' suspension.

As it is seen from the Figure 4, CD spectra of DNA and DNA in the presence of liposomes correspond completely (curve 1). Addition of Mg ions (5 mM) to DNA decreases for 12% the intensity (amplitude) of spectrum positive band with maximum at 276 nm (curve 2). In the case of "triple complex" DNA - liposomes - Mg ions (curve 3) both positive and negative bands of CD spectrum decrease, aproximately for 25%. These changes in DNA CD spectra may be explained by the superposition of two phenomena : 1) local unwinding of some areas of DNA duplex, which is being indicated by the increased intensity of CD positive band for the "triple complex" comparing to the initial DNA and 2) DNA adhesion on liposomes' surface, accompanied by decrease in concentration of free DNA in solution. CD DNA spectrum does not change its shape during triple complexation (curve 3), but decrease in DNA helix sites does take place. Character of conformational changes does not

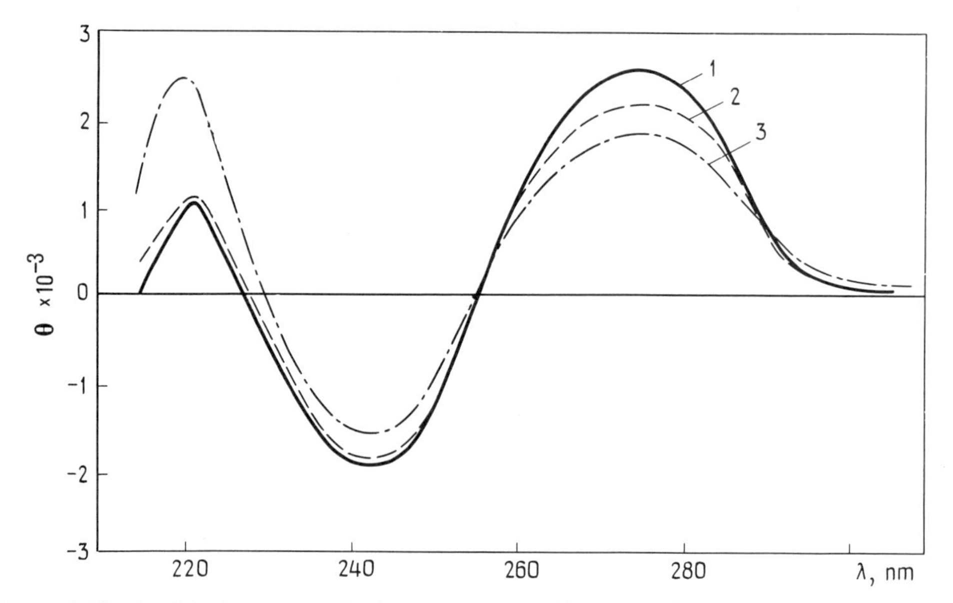

Figure 4. Circular dichroism spectra of: salmon sperm DNA (20 mcg/ml) (**1**), DNA and egg PC (Reakhim) liposomes (200 mcg/ml) (**1**), DNA and 5 mM $MgCl_2$ (**2**) and DNA, PC liposomes (100 mcg/ml) and 5 mM $MgCl_2$ (**3**) (10 mM HEPES, pH 7.2; JASCO-500A spectropolarimeter, 1 cm width and 2 ml volume quartz cell) (Kuvichkin and Sukhomudrenko, 1987).

resemble helix-roll transition which would be characterized by rise of CD positive band with maximum at 276 nm.

In the case of interaction of ribopolynucleotide duplex with PC vesicles and Mg ions (according to microcalorimetric data) the decrease of polyA:polyU duplex T_m value for 1°C has been observed (Zhdanov et al., 1993a) testifying to existence of local disturbances in polynucleotide secondary structure. So, though there is a complex character of polyA:polyU helix melting, the presence of main maximum in melting curve of polyA:polyU ("triple complex") at 78°C (Figure 4, curve 4) (comparing with T_m 79.1°C for polyA:polyU-Mg ions on curve 2) may indicate to the local unwinding of polynucleotide helix.

Interaction with DNA Results in Apperance of New Phase in Membrane Phospholipid Bilayers. There is at Least Partial Fusion of Phospholipid Vesicles' Bilayers

Microcalorimetry (Kuvichkin and Sukhomudrenko, 1987; Bichenkov et al., 1988a; Vojcikova et al, 1989; Zhdanov et al., 1993a) and magnetic resonance (Zhdanov et al., 1993c) demonstrate changing of properties of phospholipid bilayers in "triple complex" comparing with initial phospholipid vesicles. Thus, T_m value of multilamellar DMPC vesicles increases with formation of "triple complex" between vesicles, DNA and Mg ions from 23.0°C to 23.5°C (Kuvichkin and Sukhomudrenko, 1987). At the same time results of the above cited papers show that shape of the phospholipid (PC, DMPC, DPPC) melting curve in "triple complex" deforms, the single peak spliting with appearance of the second one, slightly moved to the area of higher temperatures. This finding implicates appearance of a new phase in phospholipid bilayers. Using this technique Vojcikova et al. (1989) succeded even in registering formation of a new phase during interaction between vesicles and DNA without Mg ions.

Appearance of a new phase during formation of "triple complex" between polyA:polyU helix, Mg ions and multilamellar egg vesicles spin labeled with 5- and 12-doxylstearic acid was registered by electron paramagnetic resonance technique (Zhdanov et al., 1993c). For this "triple complex" Arrhenius plot of temperatrure dependence of 5-doxylstearic acid order parameter S is nonlinear. This nonlinearity is caused by the presence of sharp bend in the plot at 32°C, associated with considerable change in the apparent activation parameters of spin label structural flexibillity. This bend is absent for spin-labeled vesicles and spin-labeled vesicles - polyA:polyU system. However, analysis of EPR spectra of 16-doxylstearic acid labeled vesicles as well as of corresponding "double or triple complexs" with polyA:polyU helix demonstrated that parameters of these spectra do not register formation of complexes of the 1st and the 2nd types. Bend's temperature for all the three cases are the same, indicating involvement in the interaction between polynucleotide helix and vesicles of only those groups which are situated near phospholipid bilayer surface. In general it corresponds with microcalorimetric data. Figure 3 shows relative stability of phospholipid's melting curve in the "triple complex". At the same time fast changes, such as liposomes' fusion or partial fusion of bilayers with final storation of bilayer's structure, are not excluded. These changes happen so fast that they could not be registered by techniques we used, taking into account time scale. Though for a longer period of complex conservation (3 days) small changes in phospholipid bilayer may accumulate and be finally registered (Vojcikova et al., 1989).

Optical or fluorescent microscopy was initially applied for studying "triple complexes" using DNA or phospholipid specific dyes. Pictures we observed permitted to suggest that along with liposome's aggregation during formation of "triple complexes" fusion of liposomes does happen, their size increasing by 10 and more times. Formation of very big flat structures (bands and spirals) has been observed. However, the most informative method for this purpose happens to be electron microscopy freeze-etching technique, because it doesn't influence the "triple complex" structure (Kuvichkin, 1990). As it follows from the figure 5b, addition of Mg ions to liposomes doesn't cause considerable changes in liposomes' sizes and shapes. Calf thymus DNA addition to liposomes causes formation of liposomes' aggregates, but doesn't change their size and form. During formation of "triple complexes" (Figure 5d,e) structure and shape of liposomes changed considerably, the streached elipsoides of rotation with one or several rims (10-20 nm thick) being formed. After addition of EDTA, binding Mg ions, a dissociation of "triple complex" is being observed.

MODELS OF DNA - MEMBRANE CONTACTS

Mechanism of DNA - Mg Ions - Vesicles "Triple Complexation"

The proposed model takes into account all the above mentioned data and can explain many of the structures and processes in cell (Figure 6). By our opinion interactions in the "triple complex" may be as follows. Liposomes prepared from neutral phospholipids due to the forces of repulsion are known to be distanced from each other by 20-30 A. With d.s. DNA being 20 Å thick, contact is possible between liposomes and DNA situated between them. Even few contacts between phosphates of phospholipids and both DNA strands through Me ion bridges may weaken sufficiently interaction between DNA strands (hydrogen bonds, stacking interactions), causing their local (a few b.p.) unwinding (Figure 6.1). This, in its turn, anables other phosphate moieties of both DNA strands to come into contact with phospholipids phosphates, resulting in additional unwinding of helix. Interaction spreads through DNA helix, causing approach of liposomes' bilayers and, finally, their contact, which is considered to be sufficient for liposomes' fusion (Figure 6.2). The following course of events is shown on Figure 6. Fusion of two vesicles with formation of a phospholipid bilayer between them and a rim of single-stranded DNA girding the fusion area is possible.

Model of DNA - Nuclear Membrane Contacts, Based on Phenomenon of Membrane Phospholipid - DNA Interactions

What is the correlation between the "triple" complex" structures we obtained *in vitro* and the cell structures *in vivo* ? At first we will examine the main components of DNA-MC up to our knowledge , basing on analysis of up-to-date results on membrane-associated DNA. They are at least the following: 1) DNA (moderate repeatances, 200-400 b.p., in eucaryotic cells and promoter/operator sites in procaryotic cells; 2) small nuclear RNA (4-8 S), differing from ribosomal, tRNA and mRNA; 3) phospholipids of two bilayer lipoprotein membranes (inner and outer nuclear membrane, Golgi, etc); 4) bivalent metal ions (Mg, Ca).

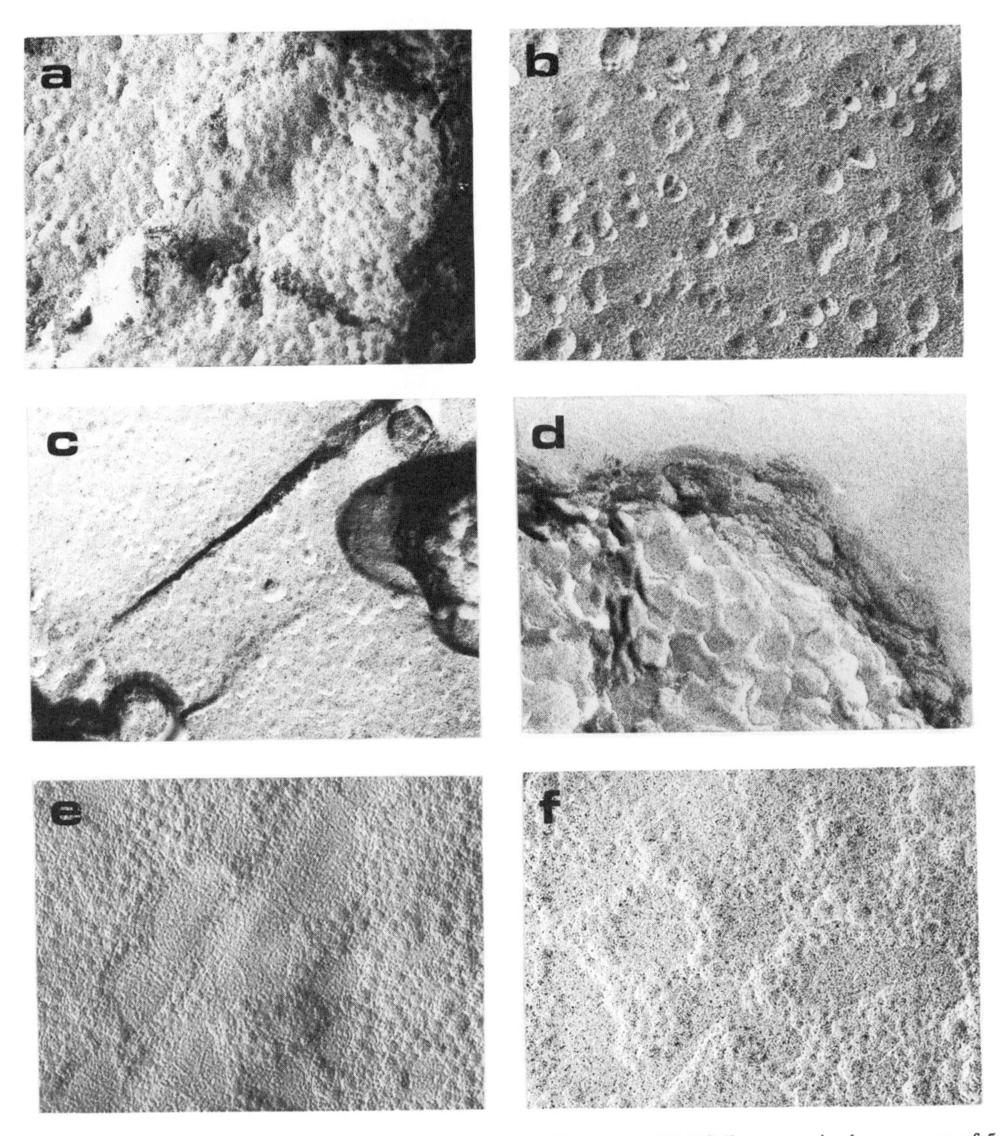

Figure 5. Freeze-etching electron micrographs of: (**a**) PC liposomes, (**b**) PC liposomes in the presence of 5 mM $MgCl_2$ (50, 000x), (**c**) PC liposomes and DNA, (**d**) PC liposomes and DNA in the presence of 5mM $MgCl_2$, (**e**) PC liposomes and DNA in the presence of 10 mM $MgCl_2$ and (**f**) PC liposomes and DNA in the presence of 5 mM $MgCl_2$ and 5 mM EDTA. PC liposomes (100 mg/ml, 20 min of sonication) and calf thymus DNA (0.5 mg/ml) preparations were prepared in 10 mM tris.HCl, pH 7.5 one hour prior experiment. EDTA was added 20 min after "triple complex" formation. The samples were fractured (-100°C and 3.10^{-6} Torr) and Pt-carbon (angle 30°, 2-3 sec) and carbon (90°, 5-6 sec) shadowed. Magnification for all samples excepting (**b**) was 100, 000x.

Nuclear pores, "Bayer's junctions" in gram-negative bacteria and mesosomes in gram-positive bacteria are supposed to be the sites for DNA-membrane interactions. Besides the process of "triple complexation", differences between the model situation and the one *in vivo* consist in supercoil DNA organization (nucleosomes) in cell as well as in the presence in cell of proteins, other lipids and low-molecular weight ligands, etc. According to the latest data it is not flat membranes which are initially responsible for formation of DNA-MC *in vivo*, as it was supposed earlier, but membrane vesicles forming nuclear envelop, mesosomes and "Bayer's junctions".

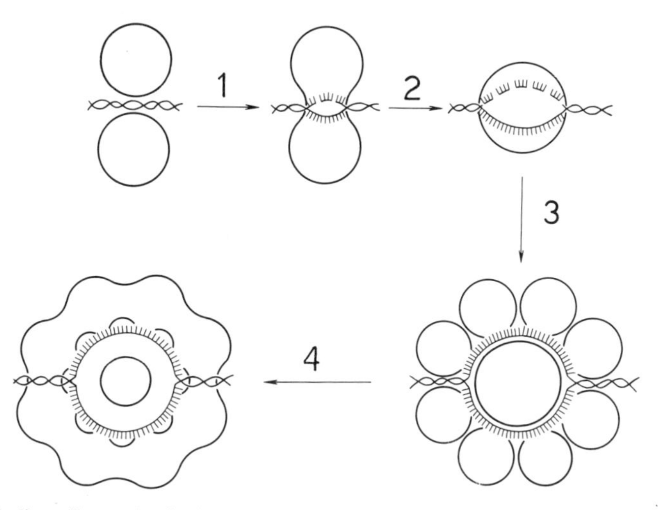

Figure 6. Unwinding ("preactivation") of double-stranded polynucleotides and formation of pre-pore complex as a result of the interaction with membrane vesicles. **1** - partial fusion of two membrane vesicles with DNA unwinding at the fusion site; **2** - complete fusion of membrane vesicles with the formation of single-stranded DNA region along vesicle's equator; **3** - aggregation of eight vesicles with DNA; pre-pore formation; **4** - fusion of peripheral vesicles and the central one; model pore complex.

Taking into account the above described data, our ideas concerning the process of nuclear pores formation are as follows. The stage of formation of "triple complex" (Figure 6.2) will be considered as the initial event of forming nuclear pore. Taking into account the possibility of vesicles' spontaneous fusion, it is possible to expect fusion between the vesicles surrounding "triple complex", as well as between them and the central one (Figure 6.3). Finally, the structure is formed, resembling nuclear pore with central granule and diaphragm covering the pore's hole (Figure 6.4). Such structures, fusing with the similar ones and other vesicles, may form a closed nuclear envelop. Chromosomal DNA, decompacting during interphase, act as DNA during DNA-MC. Octagonal symmetry of nuclear pores may be explained by participation of histone octamer, connected with the pore's periphery, or by fusion of eight vesicles (the structures shown on Figure 6.4), as well as by participation of nuclear lamina proteins. Thus nuclear pores provide not only for transport of bioactive compounds through the nuclear membrane, but also for mechanism of effective functioning of genome.

"DNA/RNA Reception" by Nuclear Membrane Phospholipids

DNA-preactivation by its interaction with membrane vesicles, resulting in local unwinding of DNA helix, is the main point of the above described model of DNA-MC. These s.s. DNA regions serve as main sites of reduplication and transcription initiation, and, therefore, of genome expression. Nuclear membrane phospholipids can destabilize d.s. DNA, providing its unwinding in any MC-zone, but it is small nuclear RNA/RNP (Mattaj, 1988), which is necessary for faciliating of DNA unwinding and providing for specificity of DNA-MC (R-loop). Figure 7 shows the scheme of formation of nuclear pores and "Bayer's junctions". During chromatin decondensation nucleosome DNA enters into contact with inner nuclear membrane in the presence of small nuclear RNA (U1 and U2 types) (Figure 7, A), which hybridizes with one strand of DNA, leaving the other one opened for the action of RNA polymerase and transcription factors. The inner nuclear membrane comes then into contact with the outer one, causing their fusion (Figure 7, B) and the following formation of eucaryotic pore complex (Figure 7, C and D; Figure 8). As it is small nuclear RNA/RNP that can provide for specificity of attachment to inner nuclear membrane of DNA regulatoric sites, it is possible to consider s.n. RNA-phospholipid membrane as DNA receptors. RNA, complementary to intervening sequences (introns) of functional genes and released during RNA splacing, could act as such RNA. This is how "Bayer's junctions" and eucaryotic pore complexes are being formed. Quantitative consideration of formation of nuclear pores,

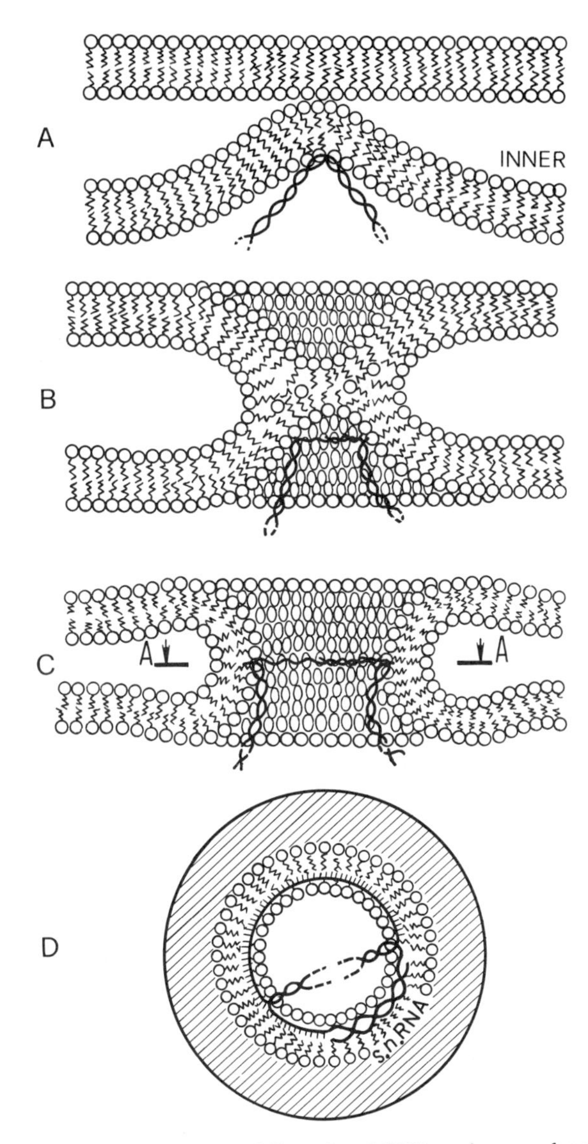

Figure 7. Scheme of formation of "interphase" DNA-nuclear membrane contact (nuclear pore). **A** - initial stage: contact of DNA (nucleosomal, chromosomal) with the inner nuclear membrane or procaryotic cytoplasmatic membrane (e.g., during chromatin decondensation after formation of closed nuclear envelope in the DNA-small nuclear RNA/RNP triplex region); **B** - intermediate stage of formation of eucaryotic pore complex (closed pore), possible formation of procaryotic "Bayer's junctions", partial fusion of two membranes with simultaneous unwinding of DNA in the triplex region, formation of R-loop; **C** - final stage of formation of eucaryotic pore complex (breaking of membrane diaphragm and DNA unwinding under influence of forces of surface tension till the size of unwinded sites exceeds that of small nuclear RNA); **D** - the same stage as **C**, but represented as a slice at plane A.

basing on the proposed model of DNA - phospholipid membrane contacts (formation of R-loop in lipid surrounding of membrane), gives value of radii of nuclear pore openings (308 Å for C-s.n.RNA and 361 Å for D-s.n.RNA) which correlates well with experimental data. These our ideas correspond well with "replicon theory", proposed earlier by Jacob et al

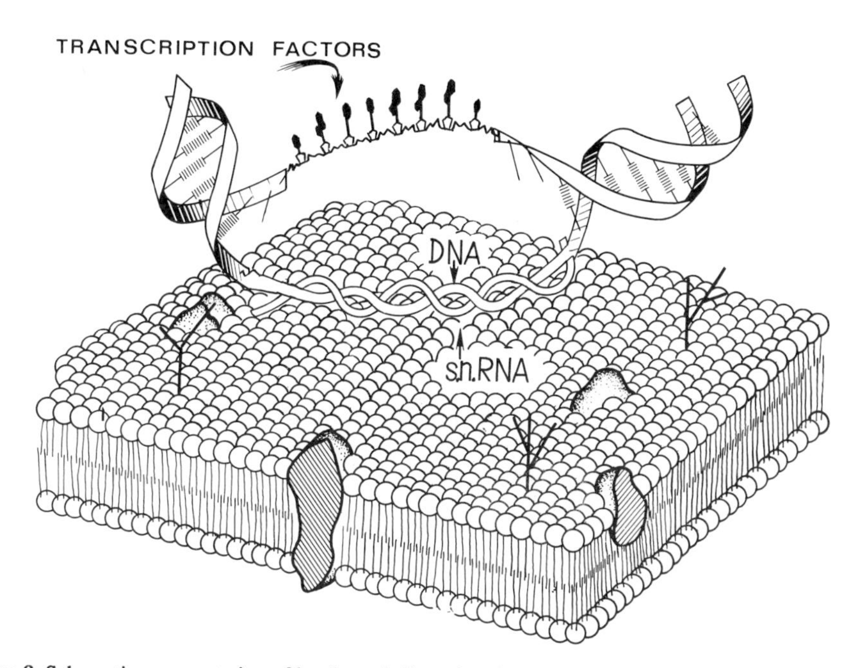

Figure 8. Schematic representation of local unwinding of nucleosome DNA during the interaction with small nuclear RNA and nuclear membrane phospholipids in the presence of calcium or/and magnesium ions.

(1963) and do not contradict contemporary data on nucleus structure and functioning (e.g., Gasser and Laemmli, 1987; Boy de la Tour and Laemmli, 1988; Jack and Eggert, 1992).

Scheme for Expression of Functional Genes During Direct Transfer

The fact of direct non-vectorial transfer of functional genes into procaryotic and eucaryotic cells has been recently established (Felgner and Rhodes, 1991). Though up to now all consequences of the influence of this expression into living beings (immune response against functional genes and products of their expression) have not been yet studied, it is reasonable to suggest application of this direct gene transfer to therapy as the basis for creation of novel types of medicines and vaccines (e.g., Felgner and Rhodes, 1991; Zhdanov et al., 1992). What is exactly the expression mechanism of that genes? Present concepts of modes of gene expression (protein repressors, transcription factors, recognizing regulatoric gene sequences) are not sufficient to explain the mechanism of expression as the result of direct gene transfer. In our opinion phospholipid membrane - nucleic acid interactions can be essential for expression of functional genes. Membrane activation of functional genes (unwinding of nucleic acid duplex during interaction with phospholipid membrane), as well as that of genome, can be important in the mechanism of their expression (local DNA unwinding during its interaction with membrane vesicles, existing in prophase of mitosis in the presence of bivalent metal cations). These single-stranded DNA sites serve as main places for reduplication and transcription initiation (Manzoli et al., 1979).

Thereby the possibility of phospholipid modulation of activity of many membrane - bound enzymes of template synthesis should be also taken into account. Getting into a cell during mitosis (prophase), exogenic DNA (functional gene) interacts with membrane vesicles, formed from nuclear membrane. Then it may even form some kind of a nuclear pre-pore (Figure 6.3), incorporated into nuclear membrane, and exist as an "extrachromosome" with its own transcription (reduplication) initiation sites. Figure 6 shows formation of the model pore complex (6.4) as result of DNA - membrane phospholipid interactions. Taking into account this pre-activation, it is possible to explain the 10-fold increase in transfection of S. collicolos with phage ΨC31, caused by the addition of liposomes of the certain size (200 nm) to DNA solution (Rodicio et al., 1982). Thus, for the effective transfection we consider sufficient for plasmid DNA to get bound with small vesicles at their surface in the presence of metal cations, when the conditions of DNA-phospholipid complex formation are being formed. In this case one should, however, try to avoid formation of DNA coat on the liposomes' surface, which can prevent their fusion with cells (Hug and Sleight, 1991).

Acknowledgements: The authors are very grateful to Prof. Alexander A. Baev, Prof. Lev D. Bergelson and Prof. Georgyi G. Georgiev (Moscow) for fruitful discussions.

REFERENCES

Alesenko, A.V., and Burlakova, E.B., 1976, Role of phospholipids in DNA synthesis in cells of mammals, *Doklady Akad. Nauk SSSR* 229: 199

Alesenko, A.V., and Pantaz, E.A., 1983, Differences in the phospholipid composition in the nucleus and chromatin during the course of rat liver cell proliferation after hepatectomy, *Biokhimiya* 48: 263

Bichenkov, E.E., Budker, V.G., Korobeinikova, I.K., Savchenko, E.V., and Filimonov, V.V., 1988a, Interaction of DNA with phosphatidylcholine liposomes. DNA melting and phase transition of lipid membrane in complex, *Biol. Membrany* 5: 843

Bichenkov,E.E., Budker, V.G., Vainer, L.M., and Kruppa, A.I., 1988b, Interaction of DNA with phosphatidylcholine liposomes, *Biol. Membrany* 5: 501

Bonora, B., Palka, G., Jovine, R., Caramelli, E., and Manzoli, F.A., 1981, Interactions among DNA, metallic ions and lipids, *Physiol. Chem. & Physics* 13: 19

Borovjagin, V.L., Sabelnikov, A.G., Tarakhovsky, Y.S., and Vasilenko, I.A., 1987, Polymorphic behavior of gram-negative bacteria membranes, *J. Memb. Biol.* 100: 229

Boy de la Tour, E., and Laemmli, U.K., 1988, The metaphase scaffold is helically folded: sister chromatides have predominantly opposite helical handedness, *Cell* 55: 937

Brosius, B., and Riesner,D., 1986, Influence of lipid membranes on the conformational transitions of nucleic acids, *J. Biomol. Structure & Dynamics* 4: 271

Budker, V.G., Kazachkov, Yu. A., and Naumova, L.A., 1978, Polynucleotides adsoption on mitochondrial and model lipid membrane in the presence of bivalent cations, *FEBS Lett.* 95: 143

Budker, V.G., Godovikov, A.A., Naumova, L.P., and Slepneva, I.A., 1980, Interaction of polynucleotides with natural and model membranes, *Nucleic Acid Res.* 8: 2499

Budker, V.G., Bichenkov, E.E., Voldman, Ya,Yu., and Vainer, L.M., 1986, 13C-NMR study of polynucleotide-phosphatidylcholine complexes, *Biol. Membrany* 3: 299

Budker, V.G., Sokolov, A.B., Vainer, L.M., and Krainev, A.G., 1987, DNA translocation through model phospholipid membranes. 1. Mechanism of translocation, *Biol. Membrany* 4: 55

Busch, H. (ed.), 1974, " The Cell Nucleus", vol. 1, Academic, New York

Felgner, P.L., and Rhodes, G., 1991, Gene therapeutics, *Nature* 349: 351

Gasser, S.M., and Laemmli, U.K., 1987, A "glimpse" at chromosomal order, *Trends Genet.* 3: 16

Grepachevsky, A.A., Manevich, E.M., and Bergelson, L.D., 1986, DNA-phospholipid interaction. A study with the aid of lipid-specific fluorescent and photo-activable probes, *Bioorg. Khim.* 12: 947

Gruzdev, A.D., Khramtsov, V.V., Weiner L.M., and Budker, V.G., 1982, Fluorescent polarization study of the interaction of biopolymers with liposomes, *FEBS Lett.* 137: 227

Hoffman, R.M., Margolis, L.B., and Bergelson, L.D., 1978, Binding and entrapment of high molecular weight DNA by lecithin liposomes, *FEBS Lett.* 93: 365

Hug, P., and Sleight, R.G., 1991, Liposomes for the transformation of eucaryotic cells, *Biochim. Biophys. Acta* 1097: 1

Jack, R.S., and Eggert,H., 1992, The elusive nuclear matrix, *Eur. J. Biochem.* 209: 503

Jacob, F., Brenner, A., and Cuzin, F., 1963, Regulation of desoxyribonucleic acid replication in bacteria, Cold Spring Harbor Symp. Quant. Biol. 28: 329

Krishna, P., Kennedy, B.P., Waisman, D.M., van de Sande, J.H., and McGhee, J.D., 1990, Are many Z-DNA binding proteins also lipid binding proteins?, *Proc. Natl. Acad. Sci. USA* 87: 1292

Krasilnikov, V.A., Mil', E.M., Alesenko, A.V., and Binykov, V.M., 1989, Comparison of DNA interactions with sphingomyelin, spermin and magnesium ions by spin probe method, *Biofizika* 34: 953

Kuvichkin, V.V., 1983, Theoretical model of DNA-membrane contacts, *Biofizika* 28: 771

Kuvichkin, V.V., 1990, Ultrastructural study of DNA-liposomes-Mg ions comlexes, *Biofizika* 35: 256

Kuvichkin, V.V., and Sukhomudrenko, A.G., 1987, Interaction of natural and synthetic polynucleotides with liposomes in the presence of bivalent cations, *Biofizika* 32: 628

Kuvichkin, V.V., Volkova, L.A., Naryshkina, E.P., and Isangalin, F.S., 1989, Study of bivalent cations by ESR and ^{1}H-NMR spectroscopy, *Biofizika* 34: 405

Manzoli, F.A., Muchmore, J.H., Bonora, B., Sabioni, A., and Stefoni, S., 1972,Interaction between sphingomyelin and DNA, *Biochim. Biophys. Acta* 277: 251

Manzoli, F.A., Muchmore, J.H., Bonora, B., Capitani, S., and Bartoli, S., 1974, Lipid- DNA interactions,II. Phospholipids, cholesterol, glycerophosphorylcholine, sphingosine and fatty acids, *Biochim. Biophys. Acta* 340: 1

Mattaj, J.W., 1988 , UsnRNP assembly and transport, in: "Strusture and Function of Major and Minor Small Nuclear Ribonucleoproteins Particles", M.L.Birnstiel, ed., Springer, Berlin Heidelberg, New York

McLauglin, A., Grathwohl, C., and McLauglin, S., 1978, The adsorption of divalent cations to phosphatidylcholine bilayer membranes, *Biochim. Biophys.Acta* 513: 338

Moyer, M.P., 1979, The association of DNA and RNA with membranes, *Int. Rev. Cytol.* 61: 1

Permyakov, E.A., Kreimer, D.I., Kalinichenko, L.P., Orlova, A.A. and Shnyrov, V.I., 1989, Interactions of parvalbumins with model phospholipid vesicles, *Cell Calcium* 10: 71

Reuben, J., and Gabbay, E., 1975, Binding of Mn (II) to DNA and the competitive effects of metal ions and organic cations. EPR study, *Biochemistry* 14: 1230

Rodicio, M.R., and Chater, K.F., 1982, Small DNA free liposomes stimulate transfection Streptomyces lividaus protoplasts, *J. Bacteriol.* 151: 1078

Sabelnikov, A.G., Vasilenko, I.A., Mishina, I.M., and Shvets, V.I., 1988, Interaction of posomes with E. coli cells: the effect of exogenous DNA uptake, *Biol. Membrany* 5: 407

Shabarchina, L.I., Sukhorukov, B.I., and Kuvichkin, V.V., 1979, IR-spectroscopic study of the DNA-lipid interaction, *Biofizika* 24: 990

Sukhorukov, B.I., Kuvichkin, V.V., and Shabarchina, L.I., 1980, On structure and function of DNA-membrane contacts in cell, *Biofizika* 25: 270

Viktorov, A.V., Grepachevsky, A.A., and Bergelson, L.D., 1984, DNA-phospholipid interaction. 31-P-NMR investigation, *Bioorg. Khim.* 10: 935

Vlasov, V.V., Deeva, E.A., Ivanova, E.M., and Yakubov, L.Ya, 1989, Possible participation of specific receptors in nucleic acid transport into cell, *Doklady Akad. Nauk SSSR* 308: 998

Vojcikova, L., Svajdlenka, E., and Balgavy, P., 1989, Spin label and microcalorimetric studies of the interaction of DNA with unilamellar phosphatidylcholine liposomes, *Gen. Physiol. Biophys.* 8: 399

Watts, A., and De Pont,J.J.H.H.M. (eds.), 1985, Progress in Protein-Lipid Interactions, Vol. 1, Elsevier, Amsterdam

Zbarsky, I.B., and Kuz'mina N.A., 1991, Skeleton Structure of Cell Nucleus, Nauka, Moscow (in Russian)

Zhdanov, R.I., and Kuvichkin, V.V., 1991, On mechnism of expression of functional genes during direct non-vectorial transfer, in: "Basic Directions of Modern Biotechnology", R.G.Vasilov, ed., NPO "Biotekhnologiya", Moscow, 31

Zhdanov, R.I., 1992, Spin-labeled medicines: enzymes, biomembranes and potential pharmaceuticals. An overview, in: "Bioactive Spin Labels", R. Zhdanov, ed., Springer, Berlin, Heidelberg, New-York

Zhdanov, R.I., Kovalev, I.E., Beburov, M. I., and Fedchenko, V.I., 1992, Potential medicine, based on human preproinsulin gene, In: "Man and Drug", Chuchalin, A.M. and Vasilov, R.G., eds, Moscow

Zhdanov, R.I., Akopyan, V.I., Kuvichkin, V.V., and Shnyrov, V.I., 1993a, Membrane phospholipid - nucleic acid interactions. Microcalorimetric study of interaction between multilamellar PC liposomes, polyA:polyU duplex and Mg ions. Energy of triple complexation, Manuscript in preparation

Zhdanov, R.I., Volkova, L.A., and Artemova, L.G., 1993b, Membrane phospholipid -

nucleic acid interactions. Spin labeling and turbidimetric study of the interaction of phosphatidylcholine vesicles with polyA:polyU duplex: Membrane bilayers, *Appl. Magn. Res.* 4 (in press)

Zhdanov, R.I., Volkova, L.A., Petrov, A.I., and Kuvichkin, V.V., 1993c, Membrane phospholipid - nucleic acid interactions. Spin labeling and spectro- photometric study of the interaction of PC vesicles with polyA:polyUduplex. Nucleic acid component, *Appl. Magn. Res.* 4 (in press)

MODE OF PHOSPHOLIPID BINDING TO THE MEMBRANE ACTIVE PLANT TOXIN PHORATOXIN-A

Ofer Markman, Usha Rao, Karen A. Lewis, Gregory J. Heffron, Boguslaw Stec and Martha M. Teeter

Department of Chemistry , Merkert Chemistry Center
Boston College,
Chestnut Hill MA. 02167 USA

INTRODUCTION

Phoratoxin-A, a mistletoe toxin, is a member of a family of small (5000 kD MW) plant toxins. One inactive member of this family, crambin, has had its crystal structure determined at atomic resolution (0.83Å, Teeter, et al., 1993). This chapter will focus mostly on the toxins, especially phoratoxin- A. These proteins are active both in membrane lysis (Bohlmann, H. and Apel, K., 1991) as well as on the intracellular level (Carrasco, L., *et al,*. 1991). They cause membrane leakage and cell lysis perhaps by interacting directly with phospholipids, although bacteriostatic activity was also shown. The bacteriostatic activity can be related to intracellular targets, such as calmodulin for which binding and a reasonable atomic model has been demonstrated (Rao, U., *et al.* , 1992).

We present conclusions about the toxin structures derived from X-ray crystallographic studies of the toxins and homologous proteins (Teeter, M., *et al.*, 1993, Teeter, M. *et al.*, 1990, Rao, U., unpublished). We also present results from extensive molecular modeling of phoratoxin-A. Binding of phospholipid to this protein and other members of the family was confirmed and monitored by NMR. These experiments allowed us to model specifically the phospholipid binding, which may be crucial to the toxic activity.

New Developments in Lipid-Protein Interactions and Receptor Function
Edited by K.W.A. Wirtz *et al.*, Plenum Press, 1993

THE STRUCTURE OF PHORATOXIN-A

The Sequence

Phoratoxin-A is a member of a family of plant toxin purified from seeds, stems and leaves of a variety of plants. The toxins related to it are 45-47 amino acids long, share 30% identity and have 50 -90% homology with the family. Crambin is a non-toxic protein homologous to the toxins, the role of which is not yet understood. The structure of crambin though is well known (Teeter, M.,*et al.* , 1993). Its X-ray structure was defined at 0.83Å resolution. The significant sequence differences (Table 1) nevertheless explain its reduced toxicity, and thus shed light on the toxins' activity.

The Three Dimensional Structure

The folding motif of phoratoxin-A consist of two anti parallel amphiphatic helices perpendicular to a two-stranded beta sheet and the C-terminal coil region. The overall shape of the protein is reminiscent of the Greek capital letter Γ (gamma) with a groove created between the helices and the β-sheet (Hendrickson and Teeter, 1981). It contains three disulfide bridges, unlike thionins (purothionins and hordothionins) which have an additional disulfide bridge (positions 12-31 in Table 1). Phoratoxin-A and other members of the family exhibit typical properties of membrane-active proteins, *i.e.* they are compact, highly basic, with one weakly polar flat face. Crambin is neutral and highly hydrophobic. Despite the differences, the folding and overall topology of these toxins is identical to that of crambin (Whitlow, M. and Teeter, M., 1985; Clore, M., et al., 1987).

Molecular packing in the crystal of α_1–purothionin provided clues to understand the organization of phoratoxin-A in solution. The aggregates consist of two types of strong intermolecular contacts to form hydrophobic and hydrophilic dimers. As suggested from the structure of the thionins (Stec, B., *et al.*, 1993), the hydrophilic dimer is bridged by bound inorganic phosphate ion (Figure 1). In phoratoxin-A we also modeled the phosphate ion close to the guanidinium group of Arg 10 which is salt bridged to the COO^- terminus (His 46). The bound phosphate ion in α_1–purothionin is hydrogen bonded to the NZ of Lys 1 and backbone carbonyl 44 (because of an insertion this residue is equivalent to 45 in phoratoxin-A). Its presence seems to be necessary for the lattice formation. Accessible surface area calculations suggest the possibility of a stable dimer in solution (which buries 673Å^2 of surface area in the α_1–purothionin (Janin, J., *et al.*, 1988; Stec, B., *et al.*, 1993)) organized around a phosphate. Further a stable tetramer could be formed by by two hydrophilic dimers which associate to bury hydrophobic surfaces (hydrophobic dimer surface area covers 621Å^2). Phosphate stabilizes the structure of these basic toxins

Table 1. Comparison of the amino acid sequences of seed–specific thionins of cereals (purothionin, hordothioinin, avenothionin), leaf thionins of barley (DB4, DC4, DG3), *Pyrularia* toxin, viscotoxin and related thionins and crambin[☆]

Protein	Position (1 5 10 15 20 25 30 35 40 45)	Ref[‡]
α_1–Purothionin	K S C C R S T L G R N C Y N L C R A R G - - A Q K L C A G V C R C K I S S G L S C P K G F P K	1,2,3
α_2–Purothionin	K S C C R T T L G R N C Y N L C R S R G - - A Q K L C S T V C R C K L T S G L S C P K G F P K	1
β–Purothionin	K S C C K S T L G R N C Y N L C R A R G - - A Q K L C A N V C R C K L T S G L S C P K D F P K	2,3,4
α–Avenothionin	K S C C R N T L G R N C Y N L C R S R G - - A P K L C A T V C R C K I S S G L S C P K D F P K	5
β–Avenothionin	K S C C K N T L G R N C Y N L C R A R G - - A P K L C S T V C R C K L T S G L S C P K D F P K	5
α–Hordothionin	K S C C R S T L G R N C Y N L C R V R G - - A Q K L C A G V C R C K L T S S G K C P T G F P K	6,7,7a
β–Hordothionin	K S C C R S T L G R N C Y N L C R V R G - - A Q K L C A N A C R C K L T S G L K C P S S F P K	8
DB4	K S C C K D T L A R N C Y N T C H F A G G - S R P V C A G A C R C K I I S G P K C P S D Y P K	9
DC4	K S C C K D T L A R N C Y N T C R F A G G - S R P V C A G A C R C K I I S G P K C P S D Y P K	9
DG3	K S C C K N T T G R N C Y N A C R F A G G - S R P V C A T A C G C K I I S G P T C P R D Y P K	9
Pyrularia toxin	K S C C R N T W A R N C Y N V C R L P G T I S R E I C A K K C D C K I I S G T T C P S D Y P K	10
Viscotoxin A2	K S C C P N T T G R N I Y N T C R F G G G - S R E V C A S L S G C K I I S A S T C P S D Y P K	11
Viscotoxin A3	K S C C P N T T G R N I Y N A C R L T G A - P R P T C A K L S G C K I I S G S T C P S D Y P K	12
Viscotoxin B	K S C C P N T T G R N I Y N T C R L G G G - S R E R C A S L S G C K I I S A S T C P S D Y P K	13
Viscotoxin 1–PS	K S C C P N T T G R N I Y N T C R F G G G - S R E V C A R I S G C K I I S A S T C P S D Y P K	14
Phoratoxin A	K S C C P T T T A R N I Y N T C R F G G G - S R P V C A K L S G C K I I S G T K C D S G W N H	15,16
Phoratoxin B	K S C C P T T T A R N I Y N T C R F G G G - S R P I C A K L S G C K I I S G T K C D S G W N H	15,16
Ligatoxin	K S C C P S T T A R N I Y N T C R L T G T - S R P T C A S L S G C K I I S G S T C D S G W N H	17
Denclatoxin	K S C C P T T A A R N G Y N I C R L P G T - P R P V C A A L S G C K I I S G T G C P P G Y R H	18
Crambin 1	T T C C P S I V A R S N F N V C R L P G T - P E A L C A T Y T G C I I I P G A T C P G D Y A N	19,20
Crambin 2	T T C C P S I V A R S N F N V C R L P G T - S E A I C A T Y T G C I I I P G A T C P G D Y A N	19,20

[☆]Residues conserved in the family and hence responsible for similar folding are blocked. Those residues which are absolutely conserved only in the toxins but different in crambin are shaded.

[‡]References: 1. Jones & Mak, 1977; 2. Ohtani *et al.*, 1975; 3. Ohtani *et al.*, 1977; 4. Mak & Jones, 1976; 5. Bekes & Lasztity, 1981; 6. Reimann–Philipp *et al.*, 1989; 7. Ponz *et al.*, 1986; 7a. Ozaki, *et al.*, 1980; 8. Hernandez–Lucas *et al.*, 1986; 9. Bohlmann & Apel, 1987; 10. Vernon *et al.*, 1985; 11. Olson & Samuelson, 1974; 12. Samuelsson *et al.*, 1968; 13. Samuelsson & Pettersson, 1977; 14. Samuelsson & Jayawardene, 1974; 15. Mellstrand & Samuelsson, 1974; 16. Thunberg, 1983; 17. Thunberg & Samuelsson, 1982; 18. Samuelsson & Pettersson, 1977; 19. Teeter *et al.*, 1981; 20. Vermueulen *et al.*, 1987.

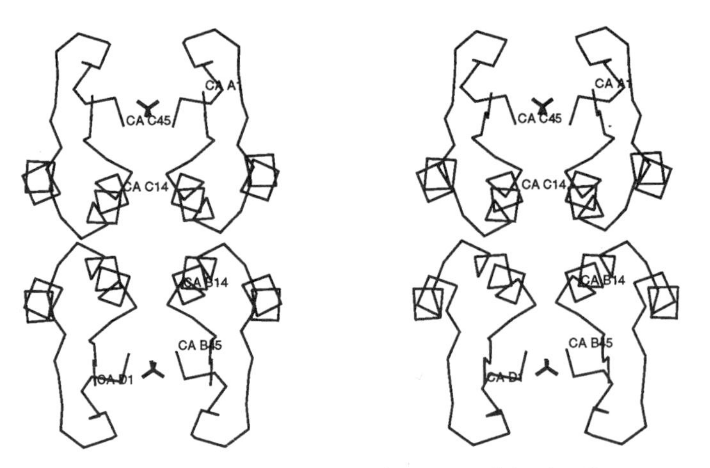

Figure 1. Stereodiagram of the tetramer of α1-purothionin. Cα atoms only. the phosphate ion is also represented in bold.

by neutralizing the positive charge of the monomers and making numerous Van der Waals contacts possible(Figure 1).

SOLUTION STUDIES OF PHORATOXIN-A AND OTHER TOXINS.

The Binding of Phospholipid to the Toxin in Solution

To test the importance of the phosphate for the toxins in solution, we have performed several NMR experiments. ^{31}P NMR binding experiments of the above toxins to phospholipids provided us with qualitative information about their general specificity for phospholipid head groups (Table 2).

Table 2. Interaction of phospholipids with crambin and toxins as derived from ^{31}P NMR experiments.

CRAMBIN:	PA > PE, PC
PHORATOXIN :	PA > G3P > PC > PE
PUROTHIONIN	PC > PA > G3P > PE

PC- phosphatidylcholine, PA- phosphatidic acid, PE- phosphatidyl ethanolamine, G3P- glycerol-3-phosphate. (Stec, B., *et al.*, 1993)

A series of 2D NMR experiments (TOCSY) performed on α_1- and β-purothionin with G3P and dic7-PA (diheptanoyl-PA) provided us with information about sites which went through changes of conformation upon binding of the model phospholipid to the toxins. This data allowed us to dock the phospholipid molecule into the

Table 3. Important proton shift differences seen in α_1- purothionin with and without G3P.

Residue	Cα protons	Cβ protons	Cγ protons	Cδ protons	Cε protons
Lys 1	4.357	1.296(a)	1.617(a)	1.129(a)	**3.087 (3.009)**
Arg 10	4.320	**2.13 (N.D.)**	(a)	(a)	
Asn 11	4.536	2.911			
Tyr 13	3.58	3.12 3.07		(b)	(b)
Asn 14	4.220	**3.09, (N.D); 2.89 (N.D)**			
Arg 17	4.130	1.900 1.780	**1.501 (1.658), 1.360 (1.482)**	**2.930 (2.950)**	
Gln 22	3.792	2.069	**2.578 (2.598) 2.284 (2.265)**		

In parentheses, chemical shift after G3P binding (if change observed).
(a) See also Figure 2
(b) Chemical shifts of protons on Cδ Cε carbons which were degenerate in purothionin were degenerate in the presence of G3P. (Stec B. et al., 1993)
N.D. Not detected probably due to peak broadening

Assignment and differences in proton shifts of purothionin. Shifts of residues where important changes were observed are shown.

Functional groups:

Phosphate binding group: **Lys-1, Arg-10**
Dimer contact groups: **Asn-11, Asn-14, Arg-10**
MPD binding: **Arg-17, Gln-22, Tyr-13**

binding site, and in some cases even to determine side-chain conformation constraints (this work is still in progress).

We also used these experiments as a test for our hypothesis about the aggregation state of the toxins in the PL-bound monomer/phosphate-bound dimer complexes. Using the NMR results and the structural homology, we could determine the structural elements involved in the phospholipid binding in phoratoxin-A. We could also use them as constraints for the model building. The constraints are therefore the following: protein folding and hydrogen patterns taken form X-ray structures of homologous proteins (Stec, B. *et al.* , 1993), phospholipid binding mode as implied from the NMR data of the homologous protein purothionin, and differences in phospholipid specificity of the different toxins.

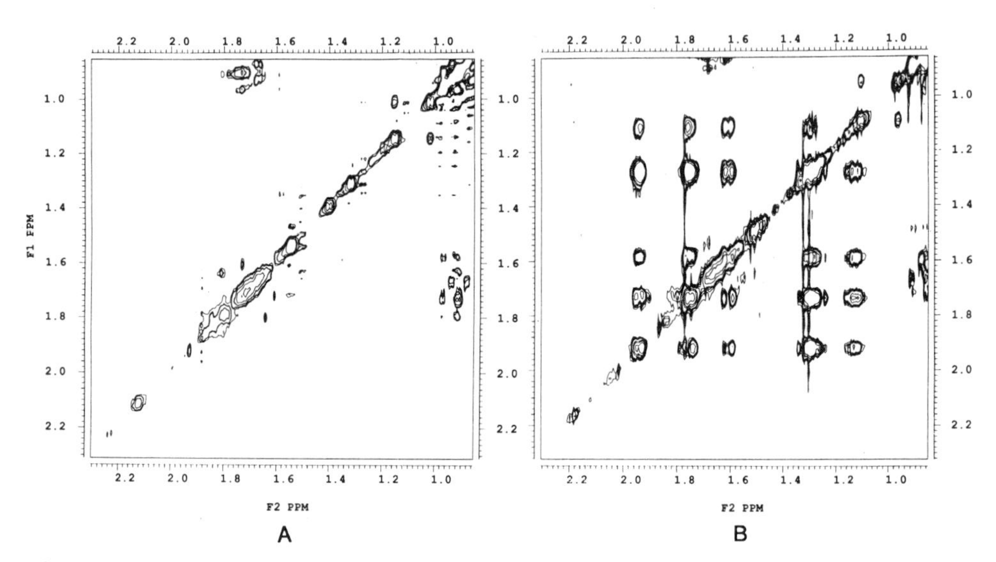

Figure 2. TOCSY spectra of α_1-purothionin in solution and with G3P in the solution. **A**. without G3P. **B**.with G3P. The spectra corresponding to side chain protons of Lys 1 and Arg 10 are shown.

MODELING PHORATOXIN-A

Building the Model of Phoratoxin-A

The crambin model was used to build a starting model of phoratoxin-A. Nonhomologous amino-acids were replaced and stereochemically refined using the Refine option in FRODO on an Evans & Sutherland interactive graphics display (Jones, T.A. & Liljas, L., 1984). Conserved side chains in crambin and purothionin were kept intact, including all conserved hydrogen bonds. Torsion angles of the most important non-conserved residues were constrained if possible to the same rotomer (Ponder, J. & Richards, F., 1987). These include Lysines 1, 28, 39 and Ile-12. Residues 41-46 were placed to fit the purothionin model at R=16% : His-46 was set to hydrogen bond with OG of Thr-7, Asp-41 reversed to g^- conformation (-40°), and

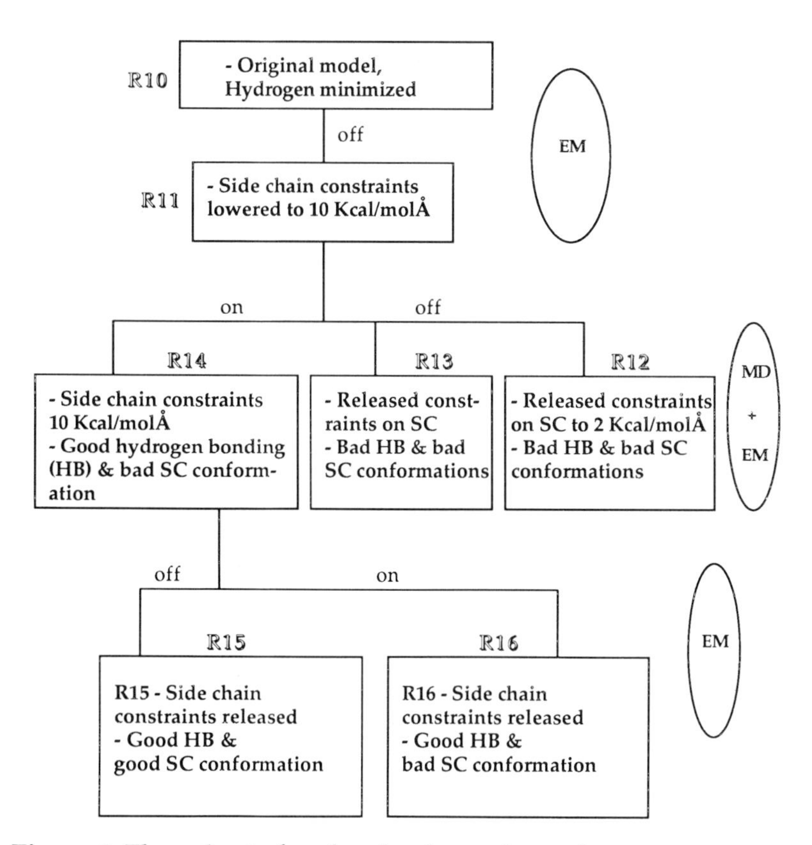

Figure 3. Flow chart of molecular dynamics and energy minimization of phoratoxin-A.
EM- Energy minimization;
MD - Molecular dynamics;
HB - Hydrogen Bonds;
SC - Side chains;
on - Free charge was turned "on".
Off - free charge was turned "off".

MD followed the following short protocol: First the structure was heated up to 300°K in two 25 fs intervals, it was then equilibrated for only 50 fs and was then run for 1 ps molecular dynamics.
EM used Powell option in X-Plor and was run until gradient lower then 0.1 Kcal/cycle or 1000 cycles was reached.

Trp-44 was fixed to a position as close as posible to Phe-43 of purothionin and Tyr-44 of crambin. Hydrogens were automatically added using the HBUILD procedure in X-PLOR (Brunger, A.T, 1992).

Energy minimization and molecular dynamics calculations were done using the program X-PLOR with the the forcefield CHARMM (Elber and Karplus, 1987; this program is using a united atom forcefield.).

The structure was subjected to constrained energy minimization in a stepwise manner. First, only hydrogen were released and the rest of the structure was constrained to 100 Kcal/molÅ (Note: If hydrogens were not minimized first, the structure did not reach energy minimum, because the Van der Waals repulsions were too high. R10, Figure 3). The amount of free charge (+9) in this structure created a challenging problem for the minimization and molecular dynamics. Formal charge "*in vacuo*" is badly represented in these methods as a result of lack of screening from the omission of solvent. On the other hand, charges are important for hydrogen bonds and electrostatic interaction. At first all charges on the basic residues Arg and Lys were turned "off" while the charges on the amino- and carboxy- terminus were left "on". The resulting structure was then subjected to harmonic constrained molecular dynamics at 300° K for 1 ps, followed by subsequent energy minimization. The models obtained from the following molecular dynamics and energy minimization were designated the names R10-R16 (Figure 3).

Molecular Dynamic and Energy Minimization

We have found that the geometrical quality of the model highly depends on the path by which the energy minimization (EM) and molecular dynamic (MD) are applied as was found before (Whitlow and Teeter , 1985). Therefore, we utilized a careful combination of MD and EM. While keeping Cα constrained, harmonic restraints for side chains and N and carbonyl backbone atoms were slowly released from 100 Kcal/molÅ in structure R10 to 10 Kcal/molÅ (R11- Figure 3). Then restraints were removed on side chains and molecular dynamics and energy minimization with the same Cα constraints were applied on the structure (Structures R12 and R13 with 2 and 0 Kcal/molÅ respectively).

To overcome the problem of charge representation, we tried several approaches. Structure R11 was subjected to molecular dynamics and energy minimization with charges on Arg and Lys residues turned "on". Energy minimization had side chains restrained to 10 Kcal/molÅ (R14). Using this setup, we achieved the correct hydrogen bonding network without great deformations resulting from the formal charge. To release the effects from nonscreened formal charge and harmonic restraints, structure R14 was subjected to molecular dynamics and then energy minimization with charges on Arg and Lys residues turned "off" (R15) or "on" (R16) and with restraints removed from side chains. The resulting minimum of energy in each of our models was biased by the setup of our experiment. The three dimensional model is shown in Figure 4. The energy results are shown in Figure 5.

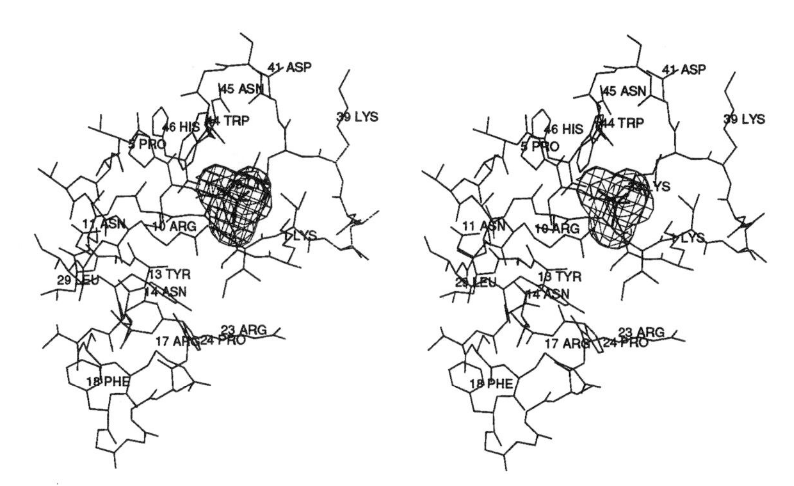

Figure 4. The structure of phoratoxin-A. Phosphate is covered by a space filling surface.

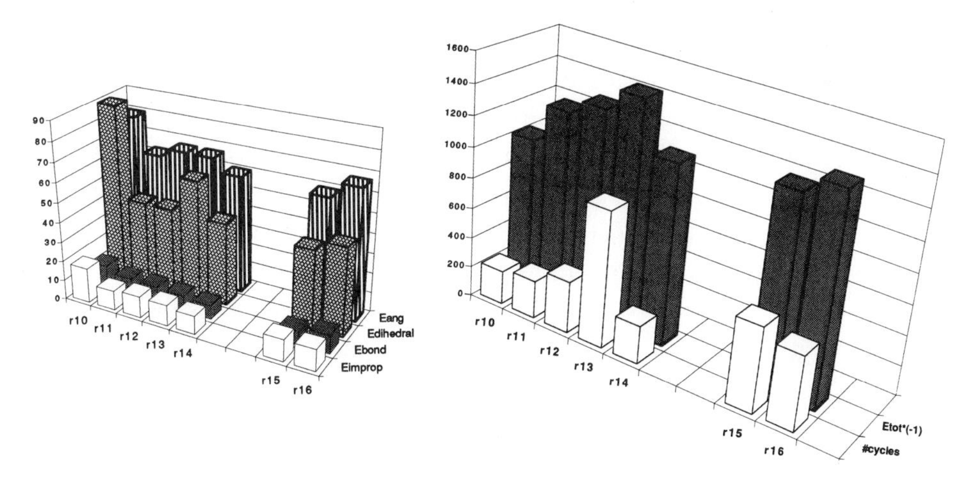

Figure 5. Comparison of phoratoxin models from molecular dynamics and minimization . Eang - angular energy; Ebond - bond energy; Edihedral - dihedral energy; Eimprop - improper energy. Also shown are the number of cycles in the EM run (#cycles) and the total (negative) energy (Etot(-1)).

Since the most biased terms were the harmonic and the electrostatic energy, we decided to examine the resulting models for their geometric energies which are not biased, *i.e.* bond, angular, and dihedral energy as well as improper energy (which is an energy used to restrain chirality - Figure 5).

From Figures 3 and 5, we can see the superiority of model R14 over models R10-13. This model was also better in conserving the hydrogen bond networks and side-chain conformation which are similar to the known structures of proteins in the family. Further modification of the energy terms did not result in a major improvement although model R15 (minimized with charges "off" after finishing minimization with charges "on") showed some reduction in the problems of the formal charge, and the already formed hydrogen bonds did not break. Comparing models R15 and R16 shows that the deformation caused by the formal charges is most pronounced in angles and dihedral energies.

Phospholipid molecule was then docked into the model in a similar conformation as it was docked in the other toxins (Stec, B. *et al.* , 1993).

LESSONS FROM PHORATOXIN STRUCTURE

We can thus suggest that the following structural features of phoratoxin-A are making it a good membrane attacking toxin. It has a good phospholipid binding site that would bind negatively charged phospholipids. This was supported both by the structural model and by preliminary NMR analysis. The site of phospholipid binding is in the groove between the alpha-helical part of the molecule and the beta-sheet part of it. The phosphate is binding to a positively-charged binding pocket in the vicinity of Lys 1 and the two termini of the molecule and the glycerol part of the phospholipid binding toward the random coil region (residue 36-44). Comparison of the models of phoratoxin-A and the models of known thionins also suggests that the differences are lying mostly in this random-coil region. Vernon has recently shown that pyrularia thionin (a homologous protein from pyrularia nut) is specifically directed against phosphatidyl-serine, a negatively charged phospholipid (personal communication). Phoratoxin-A is also an amphiphatic molecule, a feature frequently found in membrane binding proteins (Whitlow, M. *et al.*, 1986; Yashimura, T. *et al.*, 1992; de Kroon, A. I. P. M., et al., 1991). We also suggest that in common with other members of the family, the thionins, phoratoxin-A is likely to be in several aggregation states. It can exist as a monomer or be aggregated in dimers or tetramers depending on the presence of phosphate, phospholipids or other molecules yet to be found. The dimer is likely to be a hydrophilic dimer mediated by divalent anion as phosphate in the case of the thionins and the tetramer would be more of a "dimer of dimers" (Figure 1).

Thus the model we are suggesting not only explains the phospholipid binding but also suggest the organization of aggregates of phoratoxin-A.

Homologous Proteins and Analogous Systems.

The method presented here can be utilized for the modeling of other phospholipid binding proteins based on limited NMR data and homology to other well determined proteins from X-ray structures. It can be easily extended to other ligand binding proteins. This work also suggests solutions to some problems of energy minimization and molecular dynamic techniques and suggests the use of geometric parameters as a guide for the path to a better model. The use of previous structural knowledge from homologous protein and the use of low resolution NMR data to discriminate between models was shown to be of high value. Finally, the model presented here still remains to be tested on a high resolution X-ray structure of a complex of the toxin with phospholipids.

Future

The results from the modeling and NMR experiments encourage us to pursue crystallization of protein complexes with synthetic phospholipids (compounds with the relevant head group and two saturated acyl chains). So far, we have crystallized crambin complexed with synthetic PC and PA. Crystals obtained were thin needles with 0.1 X 0.2 mm width and diffracted to higher than 2Å resolution on Hamlin area detector. Complexes of phoratoxin and of α_1-purothionin with dipalmatoyl-phosphatidyl ethanolamine produced small needles suitable for further seeding. Determination of these structures may open up a much broader understanding of phospholipid binding to these toxins.

Furthermore, we have collected a TOCSY spectra of β-purothionin alone and with the synthetic phospholipid diheptanoyl-phosphatidic acid to elucidate the structural details of this complex in solution. This project requires the complete assignments of β-purothionin protons which is much easier since the crystal structure is known (Rao, U. *et al.* in preparation) and the assignment of the very homologous α_1-purothionin is also known (Clore, M., *et al* ., 1987).

ACKNOWLEDGMENTS

We would like to thank Luke A. Esposito for the purification of phoratoxin-A. Dr. M. F. Roberts' helpful discussions are gratefully acknowleged. We also wish to thank the NIH for support of this research (GM38114). Dr. Berne Jones' gift of α_1-purothionin is much appreciated.

REFERENCES

Bohlmann, H. and Apel, K., 1991, Thionins, *Ann. Rev. Plant Mol. Biol.* **42**, 227-40

Brunger A. T., 1992, X-Plor Version 3.0: a system for crystallography and NMR, Yale University.

Carrasco, L., Vazquez, D., Hernandez-Lucas, C., Carbonero, P., Garcia-Olmedo F., 1981, Thionins: plant peptides that modify membrane permeability in cultured mammalian cells. *Europ.J.Biochem.* **116**, 185-189.

Clore, G. M., Sukumaran, D. K. Grœnenborn, A. M., Teeter, M. M. Whitlow, M. Jones, B.L., ,1987, Nuclear magnetic resonance study of the solution structure of α1-purothionin: sequential resonance assignments, secondary structure, and low resolution tertiary structure. *J. Mol. Biol.*, **193**, 571-588.

Clore, G. M., Sukumaran, D. K. Nigles, M., Grœnenborn, A. M.,1987, Three dimentional structure of phoratoxin in solution: combined use of nuclear magnetic resonance, distance geometry, and restrained molecular dynamics. *Biochemistry*, **26**, 1732-45.

Elber, R and Karplus, M. 1987, Multiple conformational states of proteins: a molecular dynamics analysis of myoglobin. *Science* **235**, 318-321.

Hendrickson W.A. and Teeter M.M. , 1981, Structure of the hydrophobic protein crambindetermined directly from the anomalous scattering of sulfur. *Nature* **290**, 107-113.

Janin J. Miler S. and Chothia C., 1988, Surface, subunit interfaces and interior of oligomeric proteins. *J.Mol.Biol.*, **204**, 155-164.

Jones T.A. and Liljas L., 1984, Crystallographic refinment of macromolecules having non-crystallographic symmetry. *Acta Cryst.* , **A40**, 50-59.

de Kroon, A. I. P. M., Killian, J. A., de Gier, J., 1991. The membrane interaction of amphiphilic model peptides affects phosphotadylserine headgroup and acyl chain order and dynamics. *Biochemistry*, **30**, 1155-62.

Ponder J.W. and Richards F.M., 1987, Tertiary templates for proteins - use of packing criteria in the enumeration of allowed sequences for different structural classes. *J.Mol.Biol.* **193**, 775-791.

Rao U., Teeter M. M., Erickson-Viitanen S., DeGrado W., 1992, Calmodulin binding to a1-purothionin: Solution binding and modeling of the complex. *Proteins: Structure, Function and Genetics*, **14**, 127-138.

Stec, B., Rao, U., Markman, O., Heffron, G., Lewis, K.A., and M. M. Teeter., 1992, Phospholipid binding to plant toxins. X-ray and NMR studies of thionins Submited to **Biochemistry** March 1993.

Teeter, M. M., Ma, X.-Q., Rao ,U. & Whitlow, M., 1990, Crystal structure of a protein-toxin α1-purothionin at 2.5Å and comparison with predicted models. *Proteins: Structure, Function and Genetics*, **8**, 118-132.

Teeter, M. M., Roe, S. M. & Heo, N-H., 1993, The atomic resolution (0.83 Å) crystal structure of the hydrophobic protein crambin at 130 K. *J.Mol.Biol..*, **230**, 293-311.

Whitlow, M., Teeter, M. M., 1985, Energy mininization for tertiary structure prediction of homologous proteins: α_1-purothionin and viscotoxin-A3 models from crambin. *J. Biochem. Struct. Dynam.*, **2**, 831-48.

Yashimura, T., Goto, Y. and Aimoto, S., 1992, Fusion of vesicles induced by an amphiphilic model peptide: close correlation between fusogenicity and the hydrophobicity of the peptide in an alpha helix. *Biochemistry*, **31**, 6119-26.

BIOSYNTHESIS OF GLYCOSYL-PHOSPHATIDYLINOSITOL PROTEIN ANCHORS IN AFRICAN TRYPANOSOMES

Kenneth G. Milne, Robert A. Field and Michael A. J. Ferguson

Department of Biochemistry
University of Dundee
Dundee DD1 4HN
Scotland

INTRODUCTION

Covalent linkage to a glycosyl-phosphatidylinositol (GPI) has been widely recognised as a mode of anchoring proteins to eukaryotic cell surfaces (Ferguson and Williams, 1988; Low, 1989; Thomas *et al.*, 1990; Cross, 1990; Ferguson, 1991). The GPI anchor can be thought of as an alternative to the hydrophobic transmembrane polypeptide domain for membrane protein anchorage. There is no clear correlation between the occurrence of a GPI anchor and protein function, examples of GPI anchored proteins include, protozoal coat proteins, hydrolases, differentiation antigens, adhesion molecules and receptors (Fig. 1). The basic structure of GPI anchors is conserved between different eukaryotes (Ferguson *et al.*, 1988; Homans *et al.*, 1988; Roberts *et al.*, 1988; Schneider *et al.*, 1990; Guther *et al.*, 1992; Deeg *et al.*, 1992), suggestive of a shared biosynthetic pathway.

The conserved GPI core region can be variously substituted with carbohydrate chains and/or ethanolamine phosphate (Fig. 1). In addition, the lipid moieties of the phosphatidylinositol (PI) group are quite variable. The trypanosome VSG anchor contains only dimyristyl-phosphatidylinositol, whereas GPIs of *Leishmania* promastigote surface protease and many mammalian GPI anchored proteins, including those on erythrocyte acetylcholinesterase, folate receptor and decay-accelerating factor contain exclusively 1-alkyl-2-acyl inositol phospholipids.

New Developments in Lipid-Protein Interactions and Receptor Function
Edited by K.W.A. Wirtz *et al.*, Plenum Press, 1993

GPI anchorage appears to be used most extensively in the protozoa. African trypanosomes have a cell surface coat consisting of 10 million copies of a 60 kDa glycoprotein molecule called variant surface glycoprotein (VSG). Each of these are anchored to the plasma membrane by a GPI anchor. The surface coat is essential for trypanosome survival in the mammalian bloodstream. It protects the parasite in two ways. First, the dense packing of the coat acts as a macromolecular diffusion barrier (Cross and Johnson, 1976; Ferguson and Homans, 1989). As such, the coat prevents access to sensitive components of the plasma

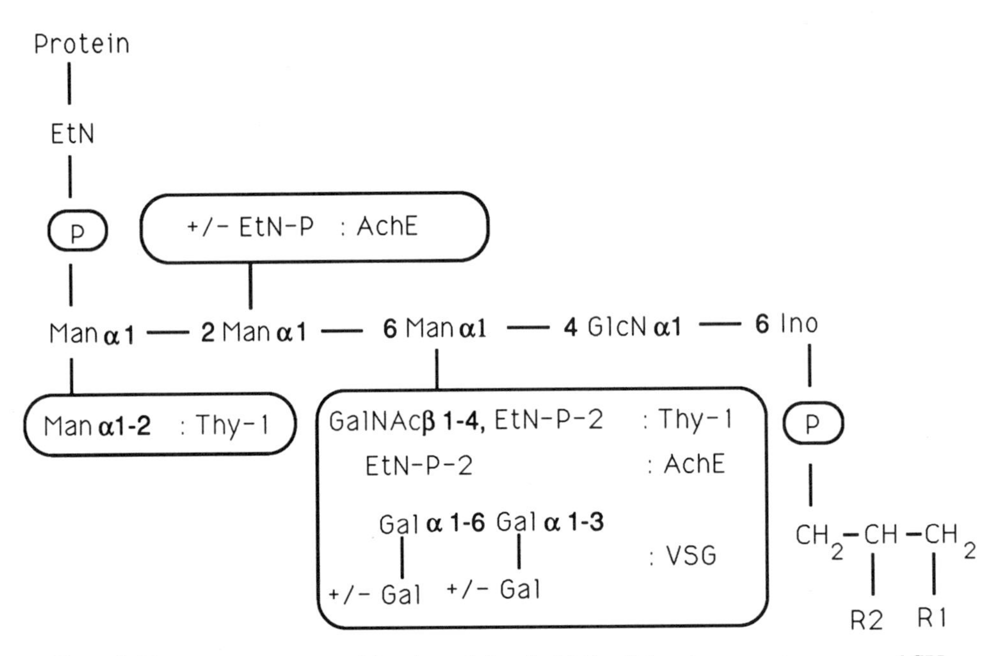

Figure 1. The consensus structure of the glycosyl phosphatidylinositol anchor suggests a conserved GPI biosynthetic pathway. VSG, Trypanosome variant surface glycoprotein; AchE, Human erythrocyte acetylcholinesterase; Thy-1, Rat brain Thy-1; EtN-P, Ethanolamine-phosphate; R1 and R2, diacyl (VSG) and alkyl-acyl lipid moeity (Thy-1, AchE).

membrane which would otherwise trigger lysis of the parasite by the alternative complement pathway (Ferrante and Allison, 1983). Secondly, the trypanosome surface coat undergoes antigenic variation (Boothroyd, 1985; Cross, 1990). Each trypanosome has up to 1000 different genes encoding antigenically distinct VSGs. Only one of these VSG genes is expressed by a trypanosome at any time. Hence, by switching from expressing one VSG gene to another, a competely different surface coat is produced. In this way the parasite eludes the humoral immune response of the mammalian host.

The relative abundance of trypanosome VSG has made it a model system for the study of GPI biosynthesis. The GPI anchor is synthesised as a precursor and attached to newly synthesised VSG by an amide linkage to the carboxyl-terminal amino acid of the protein. The kinetics of the GPI addition to protein are extremely rapid, and most likely occur in the lumenal surface of the endoplasmic reticulum (Bangs *et al.*, 1985; Ferguson *et al.*, 1986; Mayor *et al.*, 1991). Studies using a cell-free system, based on washed trypanosome membranes (Masterson *et al.*, 1989, 1990; Doering *et al.*, 1989; Menon *et al.*, 1990a,b), have led to the delineation of the GPI precursor biosynthetic pathway (Fig. 2).

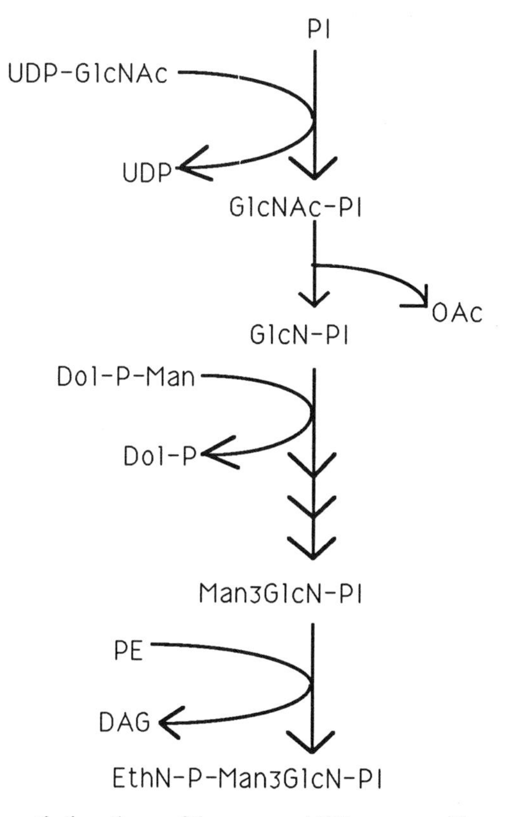

Figure 2. Biosynthetic pathway of the conserved GPI precursor. The topology of the pathway is not known, but it may occur mostly on the lumenal face of the endoplasmic reticulum. PI, phosphatidylinositol; PE, phosphatidylethanolaine; Dol-P-Man, dolichol-phosphate-mannose. In mammalian cells the extra EtN-P moieties are added before the addition of the EtN-P bridge.

The first step in the pathway is the transfer of N-acetyl glucosamine (GlcNAc) from UDP-GlcNAc to PI, to form GlcNAc-PI which is then rapidly de-*N*-acetylated to form GlcN-PI (Doering *et al.*, 1989). Subsequently, three α-mannose residues are transferred from dolicholphosphate-mannose (Menon *et al.*, 1990a) to form the intermediate Man_3GlcN-PI.

Ethanolamine phosphate is transferred from phosphatidyl-ethanolamine (Menon and Stevens, 1992) to the terminal mannose residue to form EtN-P-Man_3GlcN-PI (known as glycolipid A'). This species then undergoes a complex series of fatty-acid-remodelling reactions (Masterson, *et al.*, 1990) which are unique to trypanosomes. The 2-*sn* fatty acid of glycolipid A' is removed to form a *lyso* species called glycolipid θ, which is then myristoylated to form glycolipid A". The 1-sn fatty acid is then removed from glycolipid A" and replaced by myristic acid to form the GPI precursor, glycolipid A.

CHARACTERISATION OF THE GPI BIOSYNTHESIS ENZYMES

GlcNAc transferase

GlcNAc transferase is the first enzyme in the formation of the GPI precursor. This enzyme is responsible for the transfer of GlcNAc, from UDP-GlcNAc to endogenous PI to form GlcNAc-PI (Doering *et al.*, 1989). In a trypanosome cell-free system the addition of low concentrations of sulphydryl alkylating reagents resulted in the inhibition of this GlcNAc transferase (Milne *et al.*, 1992). This is consistent with the selective alkylation of sulphydryl groups in the transferase active site. N-ethylmaleimide (NEM) irreversibly inhibits this transferase in a time dependant manner which reaches 90% after 5 min (Fig. 3A). Membranes incubated with UDP-GlcNAc before the addition of NEM formed significant amounts of radiolabelled glycolipids (Fig. 3B), indicating that the GlcNAc transferase was partially protected from inactivation. The protection of the transferase activity was dependent on the concentration of the substrate UDP-GlcNAc, tending to a maximum of approximately 30% protection at 0.1 mM UDP-GlcNAc (Fig. 3C)

The transferase can also be protected from NEM inactivation by incubation with UDP-sugars and nucleotides. The authentic donor substrate, UDP-GlcNAc, gave the best protection. However, UMP, UDP and UTP were all more effective than other UDP-sugars, suggesting that the nucleotide portion (rather than the sugar portion) of the donor plays the greater role in enzyme binding and sulphydryl protection. Interestingly, GDP gave similar levels of protection to the uridine nucleotides, while CDP and ADP gave poor protection. The pyrimidine uracil and the purine guanine possess a C=0 instead of a C-NH_2 group found in the equivalent position in adenine and cytosine. One could speculate that the sulphur atom of the putative free sulphydryl group in the transferase active site may attack the carbonyl (C=0 group) in UDP and GDP to form a reversible covalent bond, thus preventing the irreversible binding of NEM. In contrast to the result with GDP, GDP-Man gives no protection. This result suggests that the nucleotide-sugar binding site can not tolerate an axial hydroxyl group at C2 of the hexose ring.

Man (α 1-2) transferase

Mannosamine (2-amino-2-deoxy D-mannose) has been shown to block the incorporation of GPI into mammalian and trypanosomal GPI anchored proteins. It was found to exert its effect in procyclic (insect stage) trypanosomes by inhibiting the incorporation of [^{3}H]ethanolamine into GPI anchors and thus preventing anchor attachment to the protein

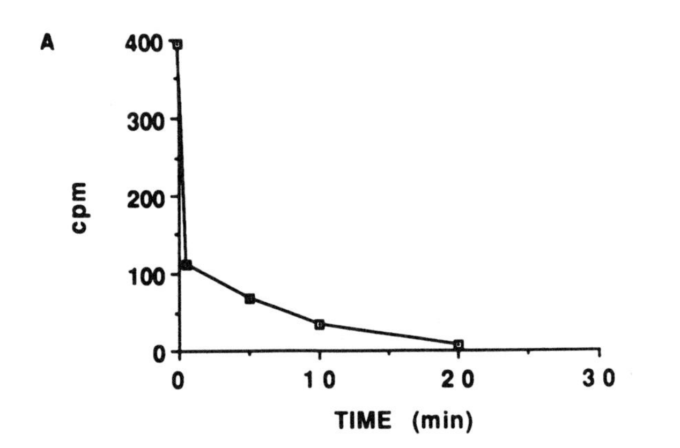

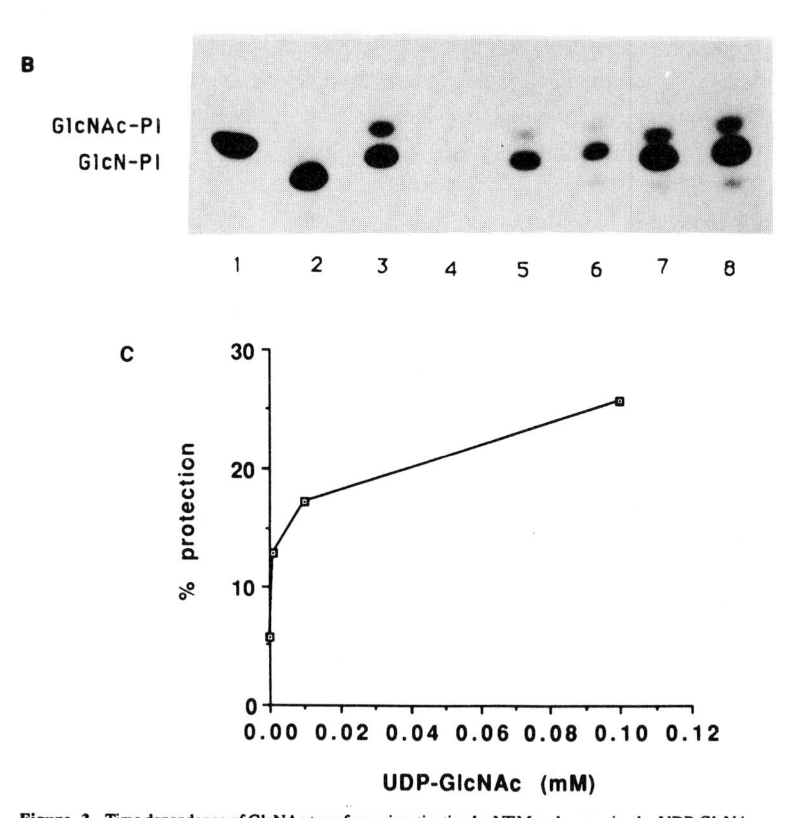

Figure 3. Time dependence of GlcNAc transferase inactivation by NEM and protection by UDP-GlcNAc. A) Time dependent inactivation of GlcNAc-transferase by 0.2mM NEM at 0°C. The y axis shows radioactivity incorporated into both GlcNAc-PI and GlcN-PI. B) Autofluorograph of a TLC showing the protection of GlcNAc transferase by UDP-GlcNAc from NEM inactivation. Membranes were assayed for GlcNAc transferase activity after incubation with 0.2mM NEM in the absence (lane 4) or presence of 1 and 10mM UDP-GlcNAc (lanes 5 and 6 respectively). Control assays were performed without NEM in the absence (lane 3) or presence of 1 and 10mM UDP-GlcNAc (lanes 7 and 8). Lanes 1 and 2, GlcNAc-PI and GlcN-PI standards, respectively. C) Protection of GlcNAc transferase by UDP-GlcNAc from NEM inactivation is concentration dependent. Membranes were assayed for GlcNAc transferase activity after incubation with 0.2mM NEM in the presence of 0, 1, 10 and 100mM UDP-GlcNAc. Transferase protection was quantitated by autofluorography of the TLC plate followed by densitometry of the autofluorograph.

(Lisanti *et al.*, 1991). By studying bloodstream form trypanosomes incubated in the presence of mannosamine, we observed a 80% reduction in the formation of glycolipid A. The inhibition of the GPI pathway occurs in the formation of the Manα1-2Man linkage resulting in an accumulation of $ManNH_2$-Man-GlcN-PI (Milne, unpublished observations) (Fig. 4). This is consistant with the inhibition of the Man (α 1-2) transferase which is responsible for the transfer of mannose from dolicholphosphate-mannose, to Man_2GlcN-PI to form the GPI intermediate Man_3GlcN-PI.

Little is currently known about the mechanism of glycosyl transferases. Hindsgaul's group, using a series of deoxy-sugars as substrate analogue-inhibitors have shown that for a number of transferases there is a critical interaction between the hydroxyl group of the acceptor sugar and a basic group on the enzyme. This interaction may increase the nucleophilic character of the acceptor hydroxy oxygen atom, thus enhancing its rate of reaction with the donor directly, with an oxocarbonium ion derived from the donor or with a donor-derived glycosyl enzyme intermediate. An adaptation of Hindgaul's model applied to the mannosyl transferase suggests a possible interaction between the mannosaminide amino group and the enzyme active site base, implying that the mannosaminide $ManNH_2$-Man-GlcN-PI may act as a competitive inhibitor of the GPI pathway Man (α1-2) Man transferase.

Phosphoethanolamine transferase

Phosphoethanolamine transferase is responsible for the transfer of phosphoethanolamine from phosphatidylethanolamine to Man_3GlcN-PI to form glycolipid A'. The serine esterase inhibitor, phenylmethanesulphonyl fluoride (PMSF) inhibit phosphoethanolamine incorporation into the GPI precursor resulting in an accumulation of a Man_3GlcN-PI intermediate (Masterson and Ferguson, 1991). PMSF exerts this effect both in living trypanosomes and in trypanosome derived cell-free system. Inhibition of the phosphoethanolamine transferase by PMSF suggests that this enzyme might belong to the general class of enzymes with an active-site serine. PMSF is known to sulphonylate the hydroxyl group of the active-site serine of chymotrypsin, the best characterised enzyme of this class. Further support for this suggestion comes from the observed accumulation of Man_3GlcN-PI in the cell-free system in the presence of di-isopropyl fluorophosphate, a reagent which specifically phosphorylates activated serines. In contrast, several other serine protease inhibitors had no effect at high concentration, ruling out the possibility that PMSF inhibits proteolytic activation of the phosphoethanolamine transferase. Given these observations, we propose a two-step reaction mechanism for phosphoethanolamine transfer in GPI biosynthesis (Masterson and Ferguson, 1991). In the first step, the active serine of the transferase would react with phosphatidylethanolamine forming an enzyme-phosphoethanolamine intermediate and releasing diglyceride. In the second step, the phosphoethanolamine would be transferred onto the C-6 hydroxyl group of the terminal mannose of Man_3GlcN-PI, with the enzyme being released for a further round of transfer.

Fatty acid remodelling enzymes

Trypanosomes undergoe an unique series of reactions where the fatty acids of glycolipid A' are removed and replaced by myristic acid to form the GPI precursor, glycolipid A. This strict myristate specificity is particularly striking because of the great quantities of VSG

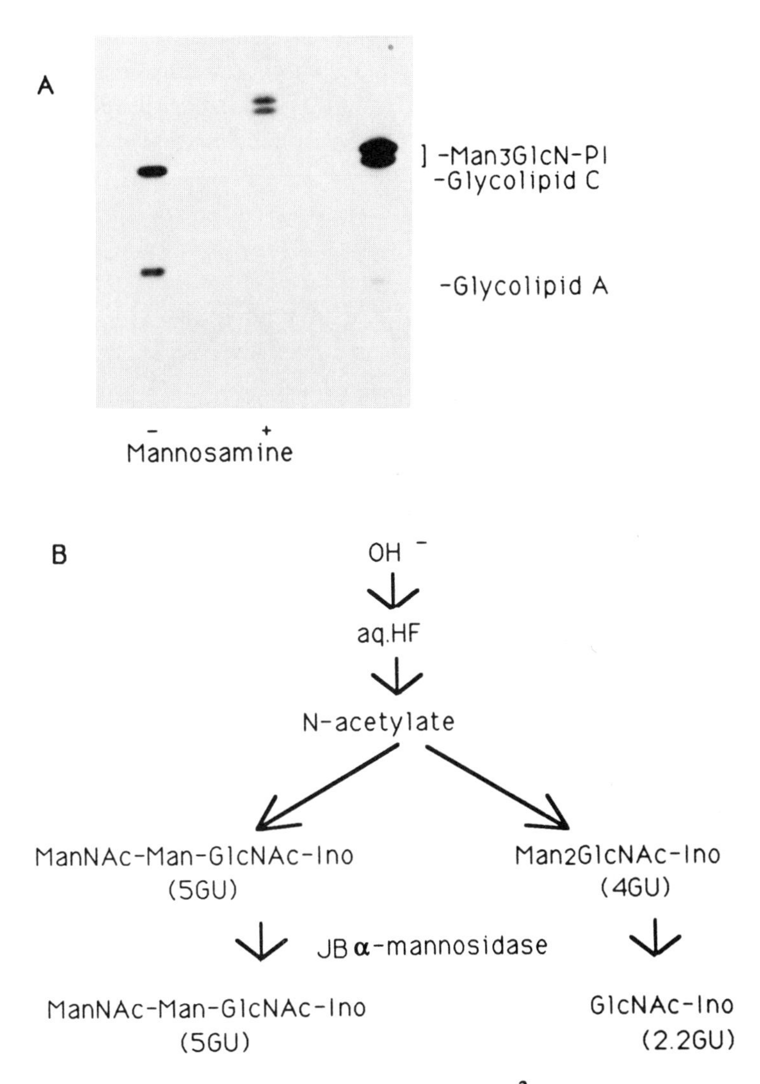

Figure 4. Bloodstream form trypanosomes were labelled with [^{3}H]glucosamine in the presence or absence of mannosamine, and analysed by HPTLC (panel A). Radiolabelled lipids accumulating in the presence of mannosamine were extracted from the silica and neutral, N-acetylated glycans were analysed using Bio Gel P4 before and after alpha-mannosidase digestion. Their elution positions relative to glucose oligomers (GU) were consistant with the structures shown above (panel B). The Man_3GlcN-PI standard (panel A) runs as 2 bands on HPTLC due to fatty acid heterogeneity.

made by this parasite, which lacks the ability to synthesise fatty acids de novo. Trypanosomes must import these compounds from the host bloodstream, where myristate comprises only 1.5% of the fatty acids. This suggests that trypanosomes employ efficient mechanisms for the uptake of myristate and concommitant GPI biosynthesis. The utilisation of heteroatom-containing analogues of myristate in the cell-free system and *in vivo* have been studied (Doering *et al.*, 1991).These anologues, with oxygen substituted for one methylene group, are similar to myristate with respect to predicted geometry but are more hydrophilic. For example, a myristate anologue with oxygen substituted for the eleventh carbon (10-(propoxy)decanoic acid, termed O-11) (Fig. 5) has hydrophathic properties similar to those of decanoic acid (10:0).

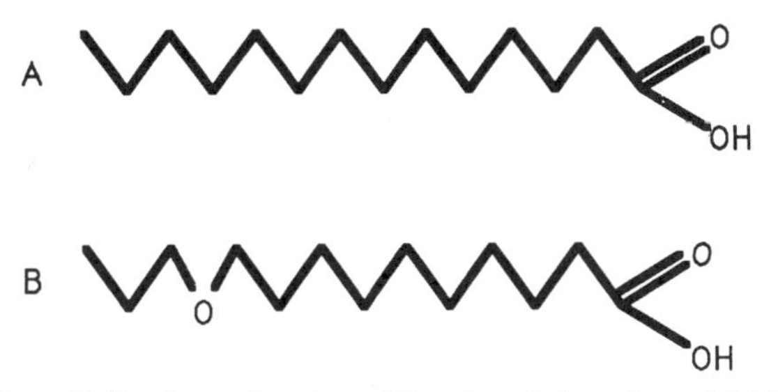

Figure 5. Structures of myristate (A) and myristic analogue 0-11 (B).

O-11 was incorporated into glycolipid A by the acyl exchange reaction even more efficiently than myristate. In contrast, an oxygen substituted anologue of palmitate, 12(propoxy)dodecanoic acid, with hydrophobicity comparable to that of myristate, are not utilised by the remodelling process. The results indicated that the specificity of fatty acid remodelling depends on chain length rather than on hydrophobicity. The anologue, O-11, was found to be highly toxic to trypanosomes in culture at 0.01mM concentration although it is nontoxic to mammalian cells (Doering *et al.*, 1991).

Acylation of glycolipid A

This enzyme is responsible for the addition of palmitic acid to the inositol of glycolipid A, to form glycolipid C. Glycolipd C may have a role as a GPI precursor store or as a step in the pathway for the degradation of excess GPI precursors. This enzyme has also been observed to have been inhibited by serine proteinase inhibitor, PMSF (Masterson, Guther and Ferguson, unpublished observations). Unlike the inhibition of phosphoethanolamine addition which was inhibited 90% by PMSF, the acylation of glycolipid A is 100% inhibited.

The identification of inhibitory agents against these transferases has provided some information on the mechanism of action of the GPI biosynthetic enzymes. It is hoped that this information will lead to the design of inhibitors of the GPI pathway for use as chemotherapeutic agents against parasitic protozoa.

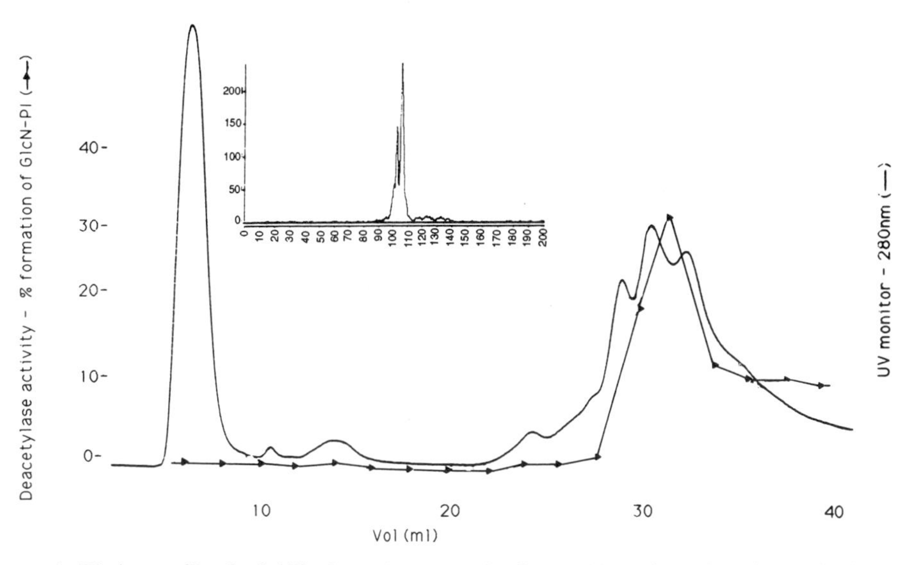

Figure 6. Elution profile of solubilised membrane proteins from DEAE column. Proteins retained at pH 8.0 in 25mM Na-HEPES, 10% glycerol and 2mM Zwittergent 3-14 were eluted with a linear gradient from 0 to 0.7M NaCl and collected in 1ml fractions which were assayed for deacetylase activity. The *inset* shows a RITA TLC analyses of the deacetylase assay for fraction 32.

INITIAL PURIFICATION OF GPI ENZYMES

In order to better understand the enzymes of GPI biosynthesis it is desirable to purify them and to establish *in vitro* assays.

An early step in the GPI biosynthetic pathway involves GlcNAc-PI deacetylase. This is crucial for the formation of the non N-acetylated glucosamine, a typical component of all GPI structures.

Deacetylase activity was assayed using an exogenous substrate of [^{3}H]myristate labelled GlcNAc-PI. This substrate was prepared from Man_3GlcN-PI (which accumulates in PMSF treated cells) which was treated with jack bean α-mannosidase and then N-acetylated to form GlcNAc-PI. Trypanosome membranes (a washed cell lysate) were incubated with this substrate and the labelled product and substrate resolved from each other by TLC. These membranes were then solubilised above the critical micelle concentration of detergent (2mM Zwittergent 3-14) in the presence of 10% glycerol to stabilise the solubilised enzyme and 10mM NEM to inactivate cysteine proteinases. The solubilised deacetylase was recovered in a supernatant fraction after ultracentrifugation (100,000g) and converted approximately 20% of the substrate to the GlcN-PI product when assayed at 37°C for 3hr.

The solubilised deacetylase has a broad pH range (6-8.5) with a pH optimum of approximately pH7.4. The first purification step was preformed using a DEAE anion exchange column at pH8, eluted with a linear gradient of NaCl (Fig. 6). The deacetylase activity was eluted at approximately 0.6M NaCl and found to have a greater activity than the solubilised membranes suggestive of the removal of an inhibiting factor. Fractions 29-33, which contained the deacetylase activity, were pooled and desalted using a Centricon-30 microconcentrator and finally suspended in a pH6.5 buffer.

The next step was gel permeation chromatography on a TSK-HW50(S) column. As shown in Fig. 7, the activity migrates in the void volume suggestive of a molecular weight in excess of 100kD.

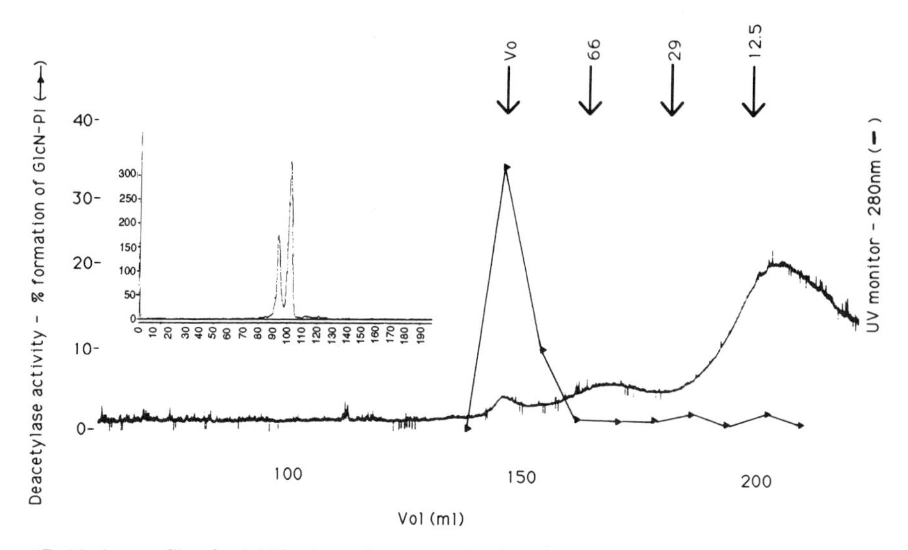

Figure 7. Elution profile of solubilised membrane proteins from gel permeation chromatography (TSK-HW50). Proteins were eluted with 25mM MES pH6.5, 0.3M NaCl, 10% glycerol and 2mM Zwittergent 3-14 at a flow rate of 0.5ml/min and 1ml fractions collected and assayed for deacetylase activity. The *inset* shows a RITA TLC analysis of the deacetylase assay for the void volume.

Fractions containing deacetylase activity were pooled and the proteins precipitated using chloroform/methanol/water and analysed by SDS-PAGE. The gel was silver stained and three faint bands migrating at 50, 75 and 80kD were detected (data not shown). We hope to scale up the purification procedure in order to purify the enzyme to homogeneity.

ACKNOWLEDGMENT

KGM thanks the SERC for a PhD studentship. This work was supported by a grant from the Wellcome Trust. We thank Malcolm McConville, Lucia Guther and Julie Ralton for helpful suggestions.

REFERENCES

Bangs, J.D., Hereld, D., Krakow, J.L., Hart, G.W. and Englund, P.T., 1985, Rapid processing of the carboxyl terminus of a trypanosome variant surface glycoprotein, *Proc. Natl. Acad. Sci. (USA)* 82: 3207-3211.

Boothroyd, J.C., 1985, Antigenic variation in african trypanosomes, *Ann. Rev. Microbiol.* 39: 475-502.

Cross, G.A.M. and Johnson, J.G., 1976, Structure and organization of the variant-specific surface antigens of *Trypanosoma brucei*, in: "Biochemistry of parasites and host parasite relationships," Van den Bosche, ed., pp. 413-420, Elsevier, Amsterdam.

Cross, G.A.M., 1990, Glycolipid anchoring of plasma membrane proteins, *Ann. Rev. Cell Biol.* 6: 1-39.

Deeg, M.A., Humphrey, D.R., Yang, S.H., Ferguson, T.R., Reinhold, V.N. & Rosenberry, T.L., 1992, Glycan components in the glycoinositol phospholipid anchor of human erythrocyte acetylcholinesterase, *J. Biol. Chem.* 267: 18573-18580.

Doering, T.L., Masterson, W.J., Englund, P.T. and Hart, G.W., 1989, Biosynthesis of the glycosyl-phosphatidylinositol membrane anchor of the trypanosome variant surface glycoprotein. *J. Biol. Chem.* 264:11168-11173.

Doering, T.L., Raper, J., Buxbaum, L.U., Adams, S.P., Gordon, J.I., Hart, G.W. and Englund, P.T., 1991, An analog of myristic acid with selective toxicity for african trypanosomes, *Science* 252: 1851-1854.

Ferguson, M.A.J., Duszenko, M., Lamont, G.S., Overath, P.O. and Cross, G.A.M., 1986, Biosynthesis of *Trypanosoma brucei* variant surface glycoproteins. *J. Biol. Chem.* 261: 356-362.

Ferguson, M.A.J., Homans, S.W., Dwek, R.A. and Rademacher, T.W., 1988, Glycosyl-phosphatidylinositol moiety that anchors *Trypanosoma brucei* variant surface glycoprotein to the membrane,*Science* 239: 753-759.

Ferguson, M.A.J. and Williams, A.F., 1988, Cell-surface anchoring of proteins via glycosyl-phosphatidylinositol structures. *Ann. Rev. Biochem.* 57: 285-320.

Ferguson, M.A.J. and Homans, S.W., 1989, The membrane attachment of the variant surface glycoprotein coat of *Trypanosoma brucei*, in: "New strategies in parasitology," McAdam, K.P.W.J., ed., pp. 121-143, Churchill Livingstone, London.

Ferguson, M.A.J., 1991, Lipid anchors on membrane proteins, *Curr. Opinion Struct. Biol.* 1: 522-529.

Ferrante, A. and Alison, A.C., 1983, Alternative pathway activation of complement by african trypanosomes lacking a glycoprotein coat, *Parasite Immunol.* 5: 491-498.

Guther, M.L.S., Cardoso deAlmeida, M.L., Yoshida, N. and Ferguson, M.A.J., 1992, Structural studies on the glycosylphosphatidylinositol membrane anchor of *Trypanosoma cruzi* 1G7 antigen, *J. Biol. Chem.* 267: 6820-6828.

Homans, S.W., Edge, C.J., Ferguson, M.A.J., Dwek, R.A. and Rademacher, T.W., 1989, Solution structure of the glycosylphosphatidylinositol membrane anchor glycan of *Trypanosoma brucei* variant surface glycoprotein. *Biochem.* 28: 2881-2887.

Lisanti, M.P., Field, M.C., Caras, I.W., Menon, A.K. and Rodriguez-Boulan, E., 1991, Mannosamine, a novel inhibitor of glycosyl-phosphatidylinositol incorporation into proteins, *EMBO J.* 10: 1969-1977.

Low, M.G., 1989, The glycosyl-phosphatidylinositol anchor of membrane proteins, *Biochim. Biophys. Acta* 988: 427-454.

Masterson, W.J., Doering, T.L., Hart, G.W. and Englund, P.T., 1989, A novel pathway for glycan assembly: biosynthesis of the glycosyl-phosphatidylinositol anchor of the trypanosome variant surface glycoprotein, *Cell* 56: 793-800.

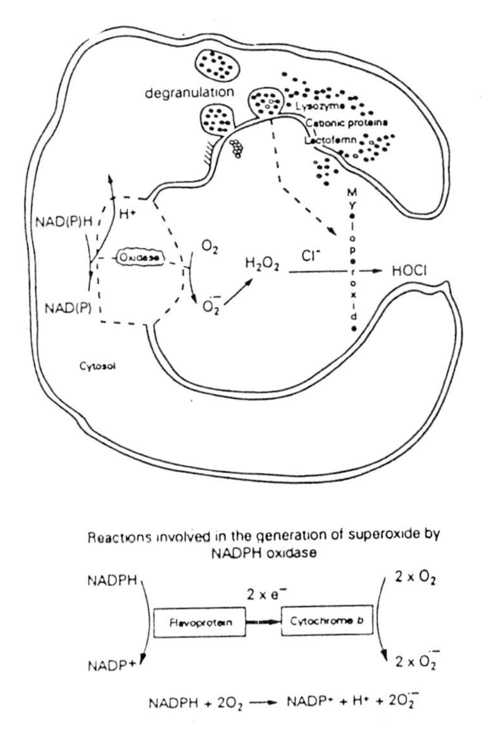

Figure 1. The respiratory burst of phagocytes

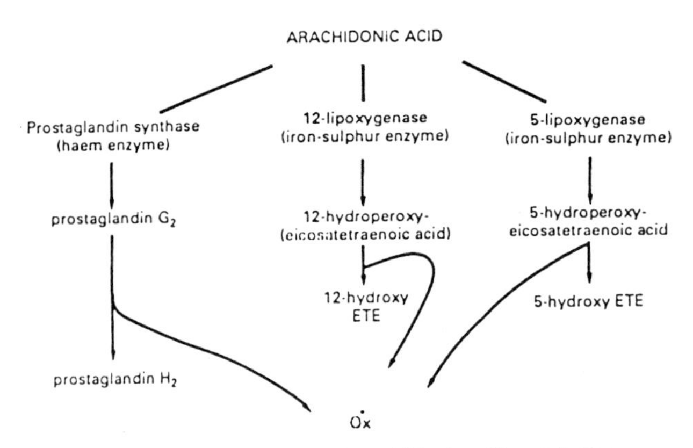

Figure 2. Eicosanoid pathways of arachidonic acid metabolism

ANTIOXIDANTS

In the normal course of events, cells and tissues have adequate antioxidant defences, both intracellularly and extracellularly. Of the range of antioxidants in the human body, those placed intracellularly are appropriate for dealing with aberrant generation of radicals; those placed extracellularly are appropriate for binding metal ions, delocalised haem proteins and for intercepting peroxidation mechanisms (reviewed in Halliwell, 1990). The range of antioxidants and radical scavengers is shown in Table 1.

A good marker of the antioxidant status of the individual is the levels in plasma of the range of antioxidants and free radical scavengers located therein.

The major antioxidants in human plasma and their effectiveness against radicals

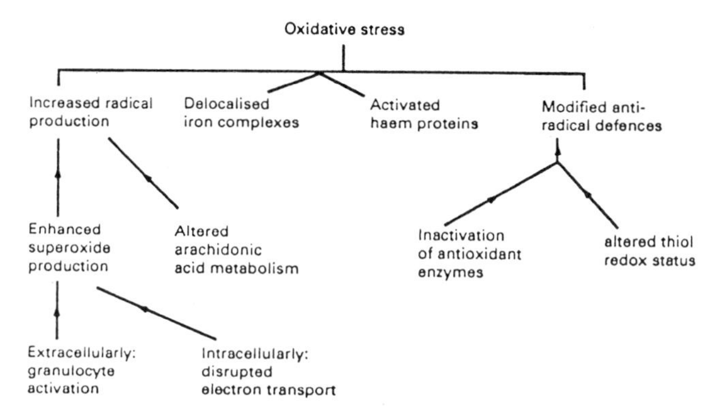

Figure 3. Factors controlling oxidative stress

Table 1. Protective antioxidant systems

NONENZYMIC
i) alpha tocopherol
- bound to membranes and in plasma bound to lipoproteins
- chain breaking antioxidant and hydroxyl radical scavenger.

ii) ascorbate
- water soluble antioxidant which acts synergistically with tocopherol.

MAJOR EXTRACELLULAR PROTECTION MECHANISMS
i) uric acid
- scavenges hydroxyl radical and singlet oxygen
- iron and copper chelator.

ii) caeruloplasmin
- acts as an antioxidant by virtue of its ferroxidase activity.

iii) transferrin
- sequesters iron (III) rendering it unavailable for catalysing the Haber-Weiss reaction, initiating lipid peroxidation or catalysing the decomposition of lipid hydroperoxides.

iv) albumin
- binds metals, especially copper but also iron weakly.

v) beta carotene

ENZYMIC ANTIOXIDANT DEFENCES
i) superoxide dismutase
- disposes of superoxide radicals.

ii) catalase
- detoxifies hydrogen peroxide.

iii) glutathione peroxidase
- detoxifies hydrogen peroxide and lipid peroxides in the presence of reduced glutathione.

generated in the aqueous phase are:
ascorbate = protein thiols>bilirubin>urate>tocopherol.
Lipid peroxidation only occurs when ascorbate is completely consumed; in ascorbate-replete plasma the lipids are completely protected (Frei *et al.*, 1989).

The major antioxidants located in the nonpolar phase, protecting the polyunsaturated fatty acyl chains of low density lipoproteins (LDL), for example, are tocopherol (7 moles/mol LDL), carotene (1/3 LDL), lycopene (1/5 LDL), other carotenoids and ubiquinol.

Recent evidence suggests that the presence of antioxidant vitamins E and C in the blood may have a protective role against cardiovascular diseases. The WHO cross-cultural epidemiological survey has demonstrated an inverse correlation between plasma tocopherol level and mortality from ischaemic heart disease (Gey *et al.*, 1991). Studies have shown that vitamin E can help normalise an atherogenic blood lipid profile and reduce hyperlipidaemia.

WHAT ARE THE FACTORS CONTROLLING THE ENDOGENOUS RELEASE OF FREE RADICALS DURING TISSUE INJURY?

1. Phagocyte recruitment and activation at the site of injury;
2. Xanthine oxidase and other superoxide-producing enzymes;
3. Disrupted mitochondrial electron transport, e.g. during ischaemia, allowing leakage of electrons onto oxygen during reperfusion;
4. A number of cell types including endothelial cells, macrophages, smooth muscle cells have been shown to be capable of producing superoxide radicals, from studies in culture.

Superoxide radical and hydrogen peroxide, the product of its dismutation, are not very cytotoxic per se, their reactivity can be amplified by interaction with specific components [Fig. 4]. Iron and haem proteins are normally protected *in vivo* from exerting pro-oxidant activities by their compartmentalisation within their functional locations. Trace transition metals, or haem proteins may be redistributed or delocalised from their normal functional locations during cell damage; haem proteins may be activated by local oxidants to the ferryl haem protein radical species which is more selective and perhaps more relevant than $^{\bullet}OH$ *in vivo*.

$O_2^{\bullet -}$ ----> H_2O_2 ----> trace transition metals ----> $OH^{\bullet}$
----> haem proteins ----> $^{\bullet}X\text{-}[Fe^{IV}{=}O]$

Figure 4. Amplification of superoxide and hydrogen peroxide reactivity

It is becoming well-recognised that reactive oxygen species [Table 2] such as superoxide radical, hydrogen peroxide, ferryl haem protein radicals and hydroxyl radicals may be important mediators of cellular and extracellular injury via the destruction of membranes, lipids, lipoproteins, or alteration of critical enzyme systems, proteins, ion channels, thus compromising cellular function and antioxidants status and amplifying the initial lesion.

Table 2. Reactive oxygen species

superoxide radical $O_2^{\bullet -}$	
hydroxyl radical	$^{\bullet}OH$
peroxyl radical	$ROO^{\bullet}$
singlet oxygen	$^1O_2$
perhydroxyl radical	$HOO^{\bullet}$
alkoxyl radical	$RO^{\bullet}$
ferryl haem protein radical	$^{\bullet}X\text{-}[Fe^{IV}{=}O]$
nitric oxide	$NO^{\bullet}$
hydrogen peroxide	H_2O_2

LIPID DAMAGE

The phospholipid component of cellular membranes is a highly vulnerable target due to the susceptibility of its poly-unsaturated fatty acid sidechains, arachidonic acid etc, to free radical attack ultimately forming lipid hydroperoxides (LOOH) (see Halliwell and Gutteridge, 1989 for review). This can induce changes in membrane permeability characteristics and in the ability to maintain transmembrane ionic gradients, altered lipid fluidity and modified lipid-protein interactions.

LOOH are quite stable under physiological conditions; their decomposition may be catalysed by transition metal- or haem protein-complexes forming alkoxyl and peroxyl radicals which can re-initiate the process and amplify the initial effects. The breakdown products of lipid peroxidation, alkenals, alkanals, hydroxy alkenals are toxic and can undergo Schiff base formation with amino groups or proteins or interaction with thiol groups, and can mediate inactivation of enzymes as well as demonstrating other cytotoxic properties, as summarised in Table 3. The physiological significance is tabulated in Table 4.

Table 3. Consequences of lipid peroxidation

Consequences
Loss of polyunsaturated fatty acids
Decrease in lipid fluidity
Altered membrane permeability
Effects on membrane-associated enzymes
Altered ion transport
Generation of cytotoxic metabolites of LOOH
Release of material from subcellular compartments, e.g. lysosomal enzymes

Table 4. Products of lipid peroxidation with physiological significance

	Product / effect
a)	LOOH ----
	stimulate prostaglandin synthesis (activate cyclo-oxygenase)
	affect cell growth
b)	EPOXY FATTY ACIDS ----
	affect hormone secretion
c)	4-HYDROXY ALKENALS ---
	affect adenylate cyclase activity
	potential anti-tumour agents
	affect DNA synthesis
	inhibit platelet aggregation
	block macrophage action
	block thiol groups

Secondary products of peroxidative events such as lipid peroxyl radicals, lipid hydroperoxides, may diffuse in the plane of the membrane before reacting further, thereby spreading the biochemical lesion. Such processes therefore not only affect the structural and functional integrity of the membrane, its fluidity and permeability, but also the breakdown products of lipid peroxidation can further damage cell function.

Apart from peroxidative dysfunction of membranes, a controlled phospholipid peroxidation could be involved in other regulatory mechanisms (Sevanian and Hochstein, 1985). Enzymic peroxidation is catalysed by different lipoxygenases active on arachidonic acid. The peroxide tone (Lands *et al.*, 1984) of membranes can control the activity of phospholipases, cyclo-oxygenase and possibly protein kinase C. The control of cell proliferation may be another major physiological effect of membrane hydroperoxides through the activity of the above-mentioned enzymes.

PROTEIN MODIFICATION

Radical generation at inappropriate sites may lead to protein destruction since they are also critical targets for free radical attack, both intracellularly and extracellularly. Proteins may be directly damaged, by specific interactions of free radicals with particularly susceptible amino acids, or may be modified by the aldehydic products of lipid peroxidation or monosaccharide oxidation. Several amino acyl constituents crucial for a protein's function are particularly vulnerable to radical damage [Sies, 1986; Rice-Evans *et al.*, 1991]. In some instances, when protein radicals are formed at a specific amino acyl site, they can be rapidly transferred to other sites within the protein infrastructure (Butler *et al.*, 1988) from:

methionine ---> tryphophan ---> tyrosine ---> cysteine.

Proteins are also particularly susceptible to attack from free radical intermediates of lipid peroxidation, alkoxyl $LO^{\bullet}$ and peroxyl $LOO^{\bullet}$ radicals. The consequences of such damage may be altered enzymic activity and altered membrane and cellular function resulting from degradation or crosslinking of receptor proteins, for example. Damage to the membrane transport proteins might affect ionic homeostasis leading to calcium accumulation. Consequently, the potential for activation of phospholipases, proteases, or the accumulation of mitochondrial calcium may lead to extensive membrane damage, cellular deterioration and gross exacerbation of the initial lesion.

Carbonyl derivatives as breakdown products of lipid peroxidation can interact with amino groups on amino acid side chains on proteins thus altering their charge and nature. This has been proposed as a contributory mechanism to the oxidative modification of low density lipoproteins, the cholesterol carrier, facilitating recognition by scavenger receptors on macrophages; the binding of malonyldialdehyde, 4-hydroxynoneral etc, secondary metabolites of the peroxidation of polyunsaturated fatty acyl chains to lysine residues on the apoB protein portion of low density lipoproteins decreases the positive charge on the surface of the LDL and limits its uptake by cells. Rather, it becomes recognisable by the scavenger receptors on macrophages, forming cholesterol-laden foam cells, contributing towards the fatty layers in atherosclerosis. Protein modification via carbonyl interaction has been described in diabetes. There are two modes of modification by glucose:

1. Glucose can oxidise when catalysed by trace amounts of transition metals, generating free radicals, hydrogen peroxide and reactive keto-aldehydes directly. The process of monosaccharide oxidation can lead to protein damage by free radicals and by covalent binding of the carbonyl products of the process.
2. Maillard reaction or nonenzymic glycation of proteins (Fig. 5): glucose is considered to be toxic by virtue of its ability to behave chemically as an aldehyde and is known to form chemically reversible early glycosylation products with protein at a rate proportional to the glucose concentration. These Schiff bases then rearrange to form the more stable Amadori-type early glycosylation products [EGP]. Protein which has been glycated *in vitro* is conformationally altered. The amount of early glycosylation products *in vivo* in diabetics, whether on Hb or basement membrane, increases when blood glucose levels are high and returns to normal after the glucose levels are normalised by treatment.

   Some of the early glycosylation products on collagen and other long-lived protein of the vessel wall do not dissociate but they undergo a slow, complex series of chemical rearrangements to form irreversible advanced glycosylation end products. A number of these irreversible end products are capable of forming covalent bonds with amino groups on other proteins, forming cross-links.

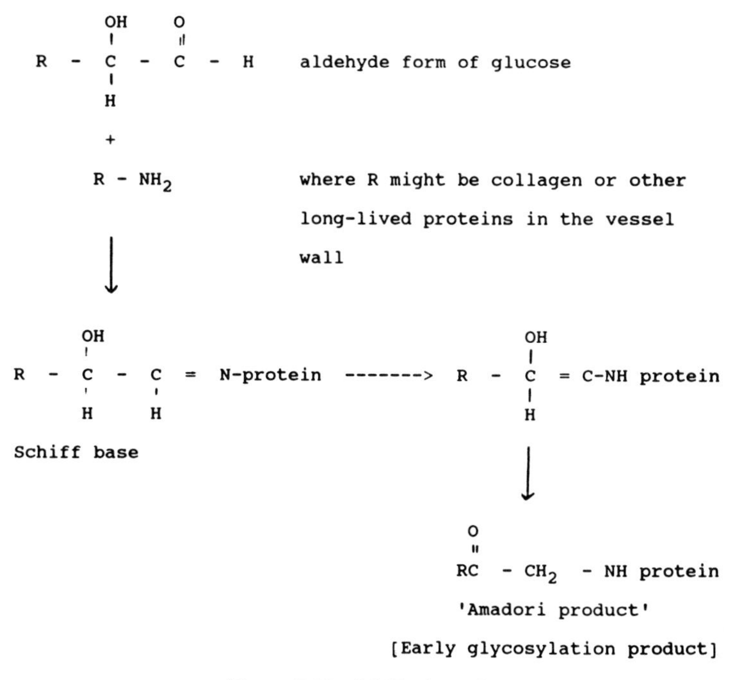

Figure 5. The Maillard reaction

INTERACTIONS BETWEEN LIPOPROTEINS AND OXYGEN RADICALS RELEASED FROM CELLS: IMPLICATIONS FOR ATHEROSCLEROSIS

The fatty streak is considered to be the earliest lesion of atherosclerosis. This lesion contains foam cells which are primarily derived from blood macrophages/monocytes. There is increasing evidence that the accumulation of cholesterol in macrophages is explained by the uptake of modified low density lipoproteins (LDL) [Fig. 6].

The mechanism of LDL oxidation in the artery wall *in vivo.* remains uncertain but *in vitro* cellular studies have been shown to involve free radical-mediated peroxidation

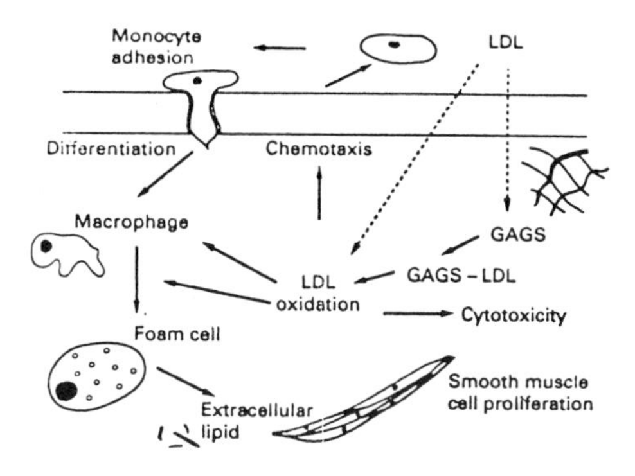

Figure 6. Early events in atherosclerosis

(Esterbauer *et al.*, 1989, 1990; Gebicki *et al.*, 1991). Evidence is accumulating that LDL can be oxidatively modified by arterial endothelial cells in culture (Henriksen *et al.*, 1981), arterial smooth muscle cells (Henriksen *et al.*, 1983) and macrophages by ruptured erythrocytes and myocytes (Parthasarathy *et al.*, 1986; Rankin and Leake, 1987; Leake and Rankin, 1990), and is subsequently recognised and rapidly taken up by the scavenger receptors on macrophages [Fig. 7]. Ruptured erythrocytes and myocytes also induce oxidation of LDL *in vitro* (Paganga *et al.*, 1992). Native LDL is not recognised by macrophage scavenger receptors.

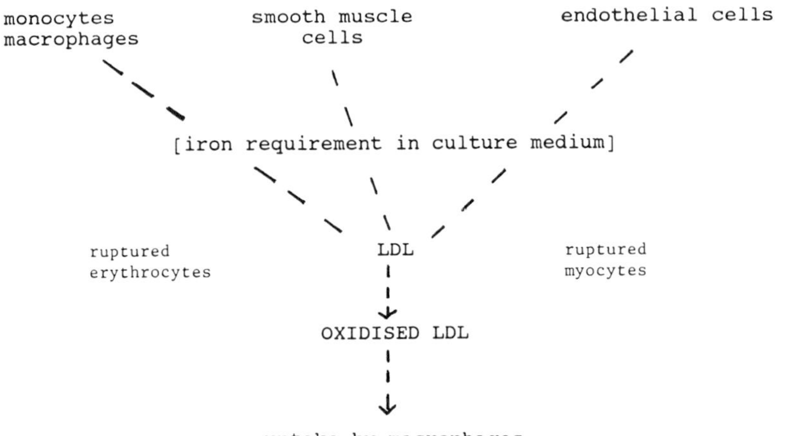

Figure 7. Cells involved in the oxidative modification of low density lipoprotein

Oxidative modification by cultured cells results in a number of compositional and structural changes to LDL including increased electrophoretic mobility, increased density, peroxidation of the fatty acyl phosphatidyl choline with increased lyso-phosphatidyl choline, derivatisation of lysine amino groups and generation of fluorescent adducts due to covalent binding of lipid oxidation products to apoB (Steinbrecher, 1987). Similar changes occur when LDL is oxidised in the absence of cells by incubation with transition metal ions such as copper and this modified form is also recognised by the macrophage scavenger receptors. Oxidative modification of LDL is inhibited in some instances by lipoxygenase inhibitors (most of which are also antioxidants) (Parthasarathy *et al.*, 1988; Rankin *et al.*, 1991) and in other instances by superoxide dismutase, implicating superoxide radical in the mechanism (Morel *et al.*, 1984; Heinecke *et al.*, 1986).

A major question is what are endothelial cells, macrophages/ monocytes, smooth muscle cells or ruptured erythrocytes, myocytes releasing such that peroxidation of LDL is mediated? The cells are all capable of generating superoxide radical. As described earlier, superoxide radical and hydrogen peroxide, the product of its dismutation, are not very reactive per se and will not initiate lipid peroxidation; their reactivity can be amplified by interaction with available transition metal ions forming hydroxyl radicals or delocalised haem proteins, producing ferryl haem protein radicals, which are more selective and perhaps more relevant *in vivo* (Rice-Evans and Bruckdorfer, 1992).

The significance of superoxide radical in the initiation of LDL oxidation in cultures of monocytes-macrophages is indicated by experiments showing inhibition of the oxidative modification by superoxide dismutase only if the antioxidant is added within a few hours of the initiation of the incubation. However, a lipid chain breaking antioxidant such as butylated hydroxy toluene was effective in inhibiting the modification when added as late as 11 h after the onset of incubation. Other studies suggest that endothelial cells can initiate the oxidation of LDL through a superoxide-independent pathway that involves lipoxygenase whereas the superoxide-dependent pathway predominates in smooth muscle cells. *In vivo* studies of Yla-Herttuala *et al.* [1990] showing co-localisation of 15-lipoxygenase mRNA and a protein with epitopes of oxidised LDL in macrophage-rich areas of atherosclerotic lesions have also suggested that cellular lipoxygenases may play a role in initiating lipid peroxidation in LDL. In support of these observations, McNally *et al.* have reported that monocyte-mediated oxidation of LDL involves monocyte lipoxygenase products which induce release of superoxide radical from the monocytes. Wilkins and Leake have obtained evidence that the NADPH oxidase of macrophages is involved in the oxidation of LDL. The mechanism by which the various cell types oxidise LDL are therefore still very unclear.

There is evidence that LDL occurs *in vivo.* LDL extracted from human or animal atherosclerotic lesions has been shown to be taken up much faster than plasma LDL by macrophages by means of their scavenger receptors (Shaikh *et al.*, 1988; Palinski *et al.*, 1990). Antibodies that recognise oxidised LDL, but not native LDL, stain human or animal atherosclerotic lesions but not the normal arterial wall (Rosenfeld *et al.*, 1990) and antioxidants that might inhibit LDL oxidation decrease the development of atherosclerosis in LDL receptor-deficient rabbits (Carew *et al.*, 1987). In addition, there is little evidence for LDL oxidation occurring to any significant extent in the circulation due to the protective effects of the plasma antioxidants; it is more likely that it occurs in the arterial wall in the locality of higher oxidative stress but lower antioxidant protection. The consensus seems to be that LDL is oxidised in areas of lesions rich in macrophages and then taken up by the cells, whereas in advanced lesions that are relatively devoid of macrophages, native and oxidised LDL, as well as oxidation products released from dying cells, are trapped in the matrix, out of the reach of those cells capable of accumulating oxidised LDL.

REFERENCES

Babior, B., 1978, The respiratory burst of phagocytes, *New England Journal of Medicine* 298:659-668.

Butler, J., Hoey, B.M. and Lea, J.S., 1988, The measurement of radicals by pulse radiolysis, *in:* Free Radicals, Methodology and Concepts (Rice-Evans, C. and Halliwell, B. eds) Richelieu Press, London pp. 457-479.

Cross, A. and Jones, O.G. 1991, Enzyme mechanisms of superoxide production, *Biochim. Biophys. Acta* 1057:281-298.

Esterbauer, H., Dieber-Rotheneder, M., Waeg, G., Striegl, G. and Jurgens, G., 1990, Biochemical, structural and functional properties of oxidised low-density lipoproteins, *Chem. Res. Toxicol.* 3:77-91.

Esterbauer, H., Rotheneder, M., Striegl, G., Waeg, G., Ashby, A., Sattler, W. and Jurgens G., 1989, Vitamin E and other lipophilic antioxidants protect LDL against oxidation. *Fat. Sci. Technol.* 91:316-324.

Frei, B., Stocker, R. and Ames, B., 1989, Antioxidant defences and lipid peroxidation in human blood plasma, *Proc. Natl. Acad. Sci.* 85:9748-9752.

Gebicki, J.M., Jurgens, G. and Esterbauer, H., 1991, *in:* Oxidative stress: oxidants and antioxidants, (H. Sies, ed) Academic Press pp 371-397.

Gey, K.F., Puska, P., Jordan, P. and Moser, U.K., 1991, Inverse correlation between plasma vitamin E and mortality from ischemic heart disease in cross-cultural epidemiology, *Am. J. Clin. Nutr.* 53:326-334S.

Halliwell, B., 1990, How to characterise a biological antioxidant, *Free Rad. Res. Comm.* 9:1-32.

Halliwell, B. and Gutteridge, J.M.C., 1989, Free Radicals in Biology and Medicine, Oxford University Press.

Henriksen, T., Mahoney E.M. and Steinberg, D., 1981, Enhanced macrophage degradation of low-density lipoproteins previously incubated with cultured endothelial cells: recognition by receptors for acetylated low-density lipoproteins, *Proc. Natl. Acad. Sci. U.S.A.* 78:6499-6503.

Henriksen, T., Mahoney, E.M. and Steinberg, D., 1983, Enhanced macrophage degradation of biologically modified low density lipoprotein, *Arteriosclerosis* 3:149-159.

Kuehl, F.A., Humes, J.L., Egan, R.W., Han, E.A., Beveridge, G.C. and Van Arman, G.G., 1977, Role of prostaglandin endoperoxide PGG_2 in inflammatory processes, *Nature* 265:270-273.

Lands, W.E.M., 1979, The biosynthesis and metabolism of prostaglandins, *Ann. Rev. Physiol.* 41:633-652.
Lands, W.E.M. *et al.*, 1984, *in:* Free radicals in Biology Vol VI (ed Pryor W) pp. 39-61, Academic Press.
Leake, D. and Rankin, S., 1990, Blood cells and ischaemia-reperfusion injury, *Blood Cells* 16:183-192.
Miyamoto, T., Ogino, N., Yanamoto, S. and Hayaishi, O., 1976, Purification of prostaglandin endoperoxide synthetase from bovine vesicular gland microsomes, *J. Biol. Chem.* 251:2629-2636.
Moncada S., 1990, From endothelium-dependent relaxation to the L-arginine:NO pathway, *Blood Vessels* 27:208-217.
Morel, D.W., DiCorleto, P.E. and Chisholm, G.M., 1984, Endothelial and smooth muscle cells alter low density lipoprotein *in vitro* by free radical oxidation, *Arteriosclerosis* 4:357-364.
Paganga G., Rice-Evans, C., Rule, R. and Leake, D., 1992, The interaction of ruptured erythrocytes with low density lipoproteins, *FEBS Lett.* 303:154-158.
Palinski, W., Yla Herrtuala, S., Rosenfeld, M.E., Butler, S.W., Socher, S.A., Parthasarathy, S., Curtiss, L.K. and Witztum, J.L., 1990, Antisera and monoclonal antibodies specific for epitopes generated during oxidative modification of low density lipoprotein, *Arteriosclerosis* 79:59-70.
Rice-Evans C. and Bruckdorfer, KR, 1992, Free radicals, lipoproteins and cardiovascular dysfunction, *Molecular Aspects of Medicine* 13:1-111.
Rice-Evans, C., Diplock A.T. and Symons, M.C.R., 1991, Techniques in Free Radical Research, Elsevier Science Publishers.
Rosenfeld, M.E., Palinski W., Yla Herttuala, S., Butler, S. and Witztum, J.L., 1990, Distribution of oxidation specific lipid-protein adducts and apolipoprotein B in atherosclerotic lesions of varying severity from WHHL rabbits, *Arteriosclerosis* 10:336-349.
Sevanian, A. and Hochstein, P., 1985, *Ann. Rev. Nutr.* 5:365-390.
Sies, H., 1986, Biochemistry of oxidative stress, *Angewandte Chemie* (Int. Ed. Engl.) 25:1058-1071.
Steinberg, D., Parthasarathy, S., Carew, T.E., Khoo, J.C. and Witztum, J.L., 1989, Beyond cholesterol - modification of low density lipoproteins that increase its atherogenicity, *New England Journal of Medicine* 320:915-924.
Steinbrecher, U.P., 1987, Oxidation of human low density lipoproteins results in derivatisation of lysine residues of apolipoprotein B by lipid peroxidation decomposition products, *J. Biol. Chem.* 262, 3603-3608.
Yla-Herttuala, S., Rosenfeld, M.E., Parthasarathy, S., Glass, C.K., Sigal, E., Witztum J.L. and Steinberg, D., 1990, Colocalization of 15-lipoxygenase messenger RNA and protein with epitopes of oxidized low density lipoprotein in macrophage-rich areas of atherosclerotic lesions, *Proc. Natl. Acad. Sci. U.S.A.* 87:6959-6963.

THE VITAMIN E ANTIOXIDANT CYCLE IN HEALTH AND DISEASE

Lester Packer

Dept. of Molecular and Cell Biology
University of California
251 Life Sciences Addition
Berkeley, CA 94720

ABSTRACT

Vitamin E is the major chain breaking antioxidant in membranes and is present in extremely low concentrations. Nevertheless it is very efficient in protection against lipid peroxidation. Vitamin E is actually a very dynamic molecule tied in intimately with the cells' metabolism and is involved in a series of antioxidant cycles. This occurs through the interactions between water- and lipid- soluble substances; by non-enzymatic and enzymatic mechanisms , which regenerate vitamin E from its radical (tocopheroxyl- or tocotrienoxyl-) form back to the initial biological antioxidant form. Under conditions where these auxiliary systems act synergistically to keep the steady state concentration of vitamin E radicals low, the loss or consumption of vitamin E is prevented. These provide efficient mechanisms for regeneration of vitamin E in membranes and low density lipoproteins (LDL). This lies at the heart of its effectiveness as a biological antioxidant. Vitamin E slows the course of degenerative diseases, and is useful for antioxidant protection in therapy in acute clinical conditions such as ischemia reperfusion injury or protection of tissues against the toxicity generated by redox cycling drugs (e.g. Doxorubicin) used to treat infections or to target tumors where the side effects on healthy tissues can be minimized by higher levels of vitamin E in the diet.

New Developments in Lipid-Protein Interactions and Receptor Function
Edited by K.W.A. Wirtz *et al.*, Plenum Press, 1993

I. OXYGEN - A DANGEROUS FRIEND

Besides our own metabolism, which during strenuous exercise may be a source of free radicals, there are other exogenous sources of free radicals that we have become exposed to during the last several hundred years . In fact, certain anti-cancer drugs are based on this principal, like adriamycin (or doxorubicin) which produce reactive oxygen species and free radicals as a means of targeting cancer. So we are constantly exposed in our environment to many different sources of free radicals. Let's consider just one of them which is ionizing irradiation. When this radiation strikes us, it reacts mostly with water, cleaving the water molecule, breaking it into a protonated electron and a hydroxyl radical. The hydroxyl radical is very reactive and will react with whatever lies next to it and will damage DNA or protein. Since it is such a highly reactive species, the protonated electron immediately dissociates into an electron and a proton, and if oxygen is present it forms superoxide. Thus we make two free radicals when ionizing radiation strikes water, the hydroxyl radical and the superoxide radical. The point is that during our own metabolism we produce the same two radicals that are produced by ionizing radiation. Oxygen should be reduced very smoothly to water, but there are side reactions of electron transport where electrons are added to oxygen to make superoxide. Superoxide is a free radical that may be potentially dangerous, so there are enzymes that remove it, like superoxide dismutase, making hydroperoxides. Hydroperoxides should not be a problem because they are very stable molecules. But if they encounter reduced species of metals (like iron or copper), then they decompose to make hydroxyl radicals (Figure I). During metabolism one is being constantly irradiated and we can expect that free radical reactions are continuously occurring during aging.

FREE RADICAL SOURCES

* SMOG (NITROGEN DIOXIDE, OZONE
* CIGARETTE SMOKE
* ENVIRONMENTAL CHEMICALS
* HALOGENATED HYDROCARBONS
* HEAVY METALS
* ALCOHOLIC BEVERAGES
* ANTI - CANCER DRUGS
* STRENUOUS EXERCISE
* RADIATION

Figure 1

To prevent free radical damage the human body developed major natural lipophilic antioxidants: vitamin E, the quinones (sometimes called coenzyme Q10), and carotenoids. These are typically located in membranes and in lipoproteins because they are hydrophobic, lipid soluble molecules. Water soluble antioxidants such as vitamin C and thiols (like glutathione which is a tripeptide), which are very important antioxidant substances in cellular and extracellular fluids, are located in the aqueous phase. Then, we have various enzymes that react with these antioxidants. There are also substances like ferritin (in cells) or transferrin (in the plasma) which bind iron. Ferritin, for example, has 4,500 binding sites for iron. It keeps iron from being free iron, which by chemical reactions can decompose hydroperoxides to make hydroxyl radicals.

It is now generally thought that probably all diseases and aging itself involve oxygen radicals at some stage in their course of occurrence. In some instances, oxygen radicals may be involved in the initiation of the disease, and certainly during the course of the disease. There are many of these you probably are aware of, e.g. cancer, cataracts, and neurological diseases. All the neurological diseases that have been looked at seem to show evidence of

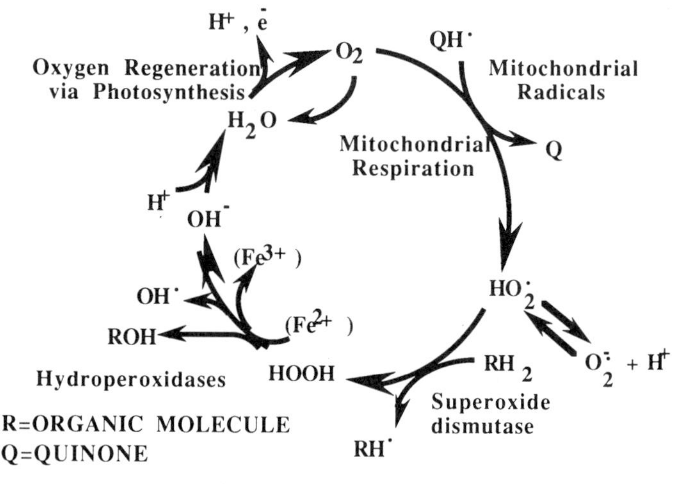

Figure 2. Oxygen cycle

the importance of antioxidants and free radicals. So, whenever free radicals are involved, antioxidants will also be involved. There is an oxidative imbalance in many nutritional diseases and diabetes, and also in cardiovascular diseases and in organ transplantation, phototoxicity with sunburn, and so on.

II. THE VITAMIN E CYCLE

Vitamin E is considered to be the major lipophilic antioxidant in membranes[1-3]. Tocopherols and tocotrienols, which are part of the vitamin E family, are composed of alpha, beta, gamma, and delta forms. Tocotrienols are present in many different plant oils, but they are not as common as the tocopherols, which we mostly concern ourselves with. The vitamin E molecule has two parts that are interesting: a chromanol nucleus, which is a cyclic aromatic ring structure, and a hydrophobic tail. The hydrophobic tail is the means by which vitamin E anchors into membranes or inserts into lipoproteins. The chromanol nucleus lies at the surface of a lipoprotein or at the surface of the membrane, and it is the phenolic hydroxyl group which lies near the surface and quenches free radicals[4]. When chromanol encounters a radical like a peroxyl radical it reacts to make a hydroperoxide and in the process a vitamin E radical is formed (Figure 3).

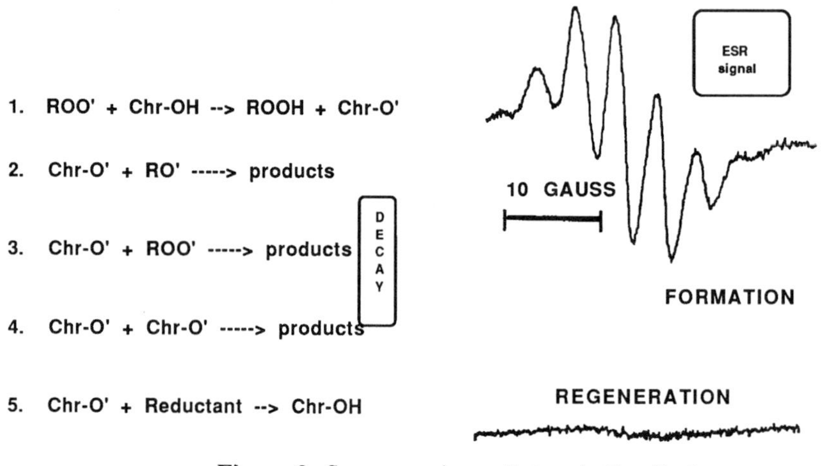

Figure 3. Some reactions of vitamin E radical

The vitamin E radical can be directly observed in biological systems, if it is present in high enough concentrations, by electron spin resonance. This radical can react with itself or with other lipid radicals to make products, which result in the loss of vitamin E. However it is very difficult to render animals deficient in vitamin E and vitamin E deficiency is seldom found in people. These remarkable properties of vitamin E may be explained by its ability to be efficiently re-reduced from its radical form to its native state[5-10] by other intracellular reductants. When the recycling take place the vitamin E free radical signal disappear and you won't see it at all by electron spin resonance. We demonstrated that tocopheroxyl radicals can be reduced by ascorbate, superoxide anion as well as by NADPH-, NADH- and succinate-dependent electron transport in microsomes and mitochondria, whereas glutathione, dihydrolipoate and ubiquinols synergistically enhance vitamin E regeneration supported by membrane electron transport enzymes (Fig. 4).

In LDL the vitamin E, phospholipid and cholesterol are located at the surface around the lipoprotein particle. In the center of the interior are the cholesterol esters, i.e. cholesterol attached to fatty acids and the trace amounts of carotenoids. It is believed that the

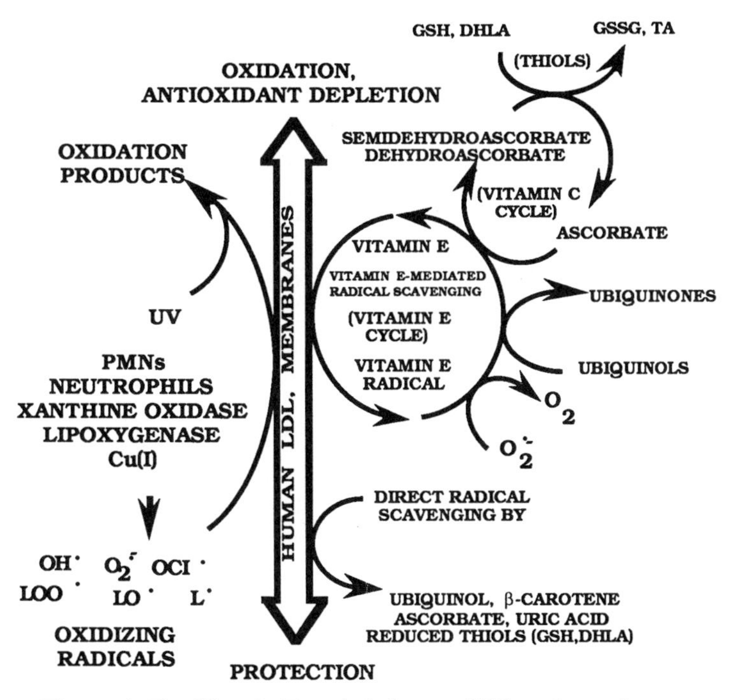

Figure 4. The Vitamin E cycle in human LDL and membranes

carotenoids are concentrated in the interior and the vitamin E is concentrated at the periphery, i.e. in the surface, of LDL particles. One can test for vitamin E radicals after oxidatively stressing LDL. UV-B irradiation is one way that will cause vitamin E to form tocopheroxyl radicals. In the presence of oxidation systems the signal from vitamin E appears and it is stable in time[11] (Fig. 5). In the presence of reductant (vitamin C) the chromanoxyl radical disappears and the ascorbyl - radical signal appears. Only when all the ascorbate is gone does the signal from vitamin E-homologue appear again. In the presence of ascorbate and thiol (dihydrolipoic acid) the time of regeneration was increased and the signal from vitamin E-homologue did not appear during the period of measurable time. Vitamin C, which is in the aqueous part of the system (comparable to plasma), approaches the surface of the lipoprotein (or a membrane) where it regenerates the vitamin E from its radical form. Thus there is an interaction between the vitamin C cycle and the vitamin E cycle. Vitamin E is usually present in nanomolar concentrations whereas vitamin C is present in micromolar intracellular concentrations. The primary preventative oxidant, the thiols (glutathione), are present in millimolar intracellular concentrations. Thiols like dihydrolipoic acid, which we are investigating, regenerate vitamin C. Vitamin C in turn regenerates vitamin E thus protecting LDL or membranes against oxidation .

Another interesting observation we have made is that the vitamin E cycle appears to be able to protect against the loss of LDL carotenoids (Table 1). We exposed LDL to irradiation to make vitamin E into a radical. Under these conditions UVB is mainly reacting

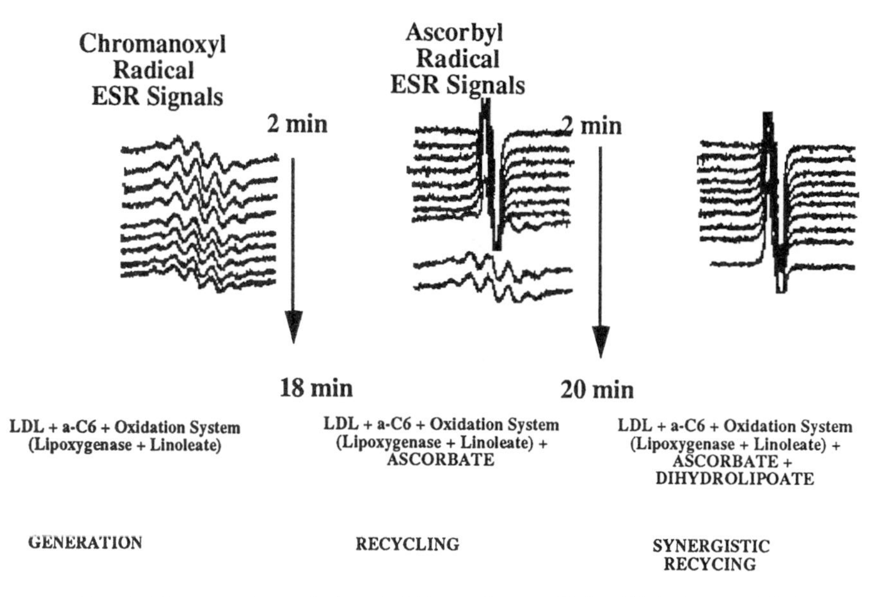

Fig. 5. Regeneration of vitamin E homologue Alpha-C6 in human LDL

with vitamin E, however we observed that the Beta-carotene was much depleted too. If the thiol dihydrolipoic acid is added a little protection is observed. Vitamin C addition also results in only a little protection. However, if both DHLA and the vitamin C are added together, there is almost complete protection against the loss of Beta-carotene. Why this is so is not clear because Beta-carotene works on a different principal than vitamin E. Vitamin E gets oxidized and reduced, and of course it can also be destroyed chemically by radical reactions. Beta-carotene however cannot be oxidized and reduced. When it reacts with radicals (not singlet oxygen), it is chemically destroyed and many products form, but you never recover Beta-carotene afterward; it is not regenerated. How then do vitamin E and Beta-carotene "talk to each other"? Many investigators believe they are located in different parts of the LDL particle, yet vitamin E cycle activity protects Beta-carotene against destruction. This is an unanswered mystery at this time. Pallozi and Krinsky[12], in studies with oxidized membrane phospholipids in the presence of carotenoids and vitamin E, have reported synergistic effects of both lipophilic antioxidants in protecting against lipid peroxidation.

In conclusion, the well known synergistic effects of physiologically important antioxidants (reductants) with vitamin E are mediated via their ability to donate electrons necessary for recycling tocopheroxyl radicals in membranes and LDL.

III. VITAMIN E EFFECTS ON ULTRAVIOLET EXPOSED SKIN

The skin is an environmentally embattled organ constantly exposed to chemicals and environmental influences that affect its health and appearance. When our skin is exposed to

Table 1 Changes in the vitamin E and beta-carotene level in LDL by different oxidation systems

	Endogenous Vitamin E (% of the control)
Control (50 min incubation)	100
+ UV	84.0 ± 6.0
+ Lipoxygenase + Linolenic acid	< 0.3
+ Lipoxygenase + Linolenic acid + Ascorbate	40.0 ± 3.0
+ Lipoxygenase + Linolenic Acid + DHLA	10.0 ± 1.0
+ Lipoxygenase + Linolenic acid + Ascorbate + DHLA	70.0 ± 5.0
	Beta - carotene (picomoles/mg protein)
Control	320
+ UV	80
+ DHLA	120
+ Ascorbate	175
+ DHLA + Ascorbate	300

UV we get wrinkling, since unexposed portions of the body wrinkle less. This is related to the degree of exposure to ultraviolet radiation. Very good studies have been made with low levels of UV irradiation, which is below what is called the minimal erythemal dose (MED). The MED is the minimum dose is that will cause your skin to redden. Over an extended period of time skin exposed to these sub-MED doses will wrinkle and this is clearly related to skin aging. Accelerated aging of the skin occurs when you are exposed to UV irradiation[13]. A recent article reports a careful study was done in Norway [14]. Norway is a country that goes from 69.5° latitude to 58.5° latitude and of course the number of people living in the north are a little less than the number of people in the very south. Most of the people live at 60° latitude where there are a million and a half people; that is where Oslo, the capitol of the country, and the city of Bergen are located. Since they keep very good health records in this country, investigators have carefully analyzed the UV dose falling on the surface at these different latitudes; it goes from 0.66 to 1.07. So the UV dose shows an increase and the UV dose correlates with a graded increase in malignant melanoma of 5.1 to 11.2, and also of non-maligmant melanoma.

Thus a very nice relationship was shown here between the dose of the UV exposure and the occurrence of skin cancer. In fact, malignant melanoma is increasing in the United States now at a rate greater than any other human cancer, except lung cancer in women (particularly in California). It is estimated that by the year 2000 that 1 in 90 people are going

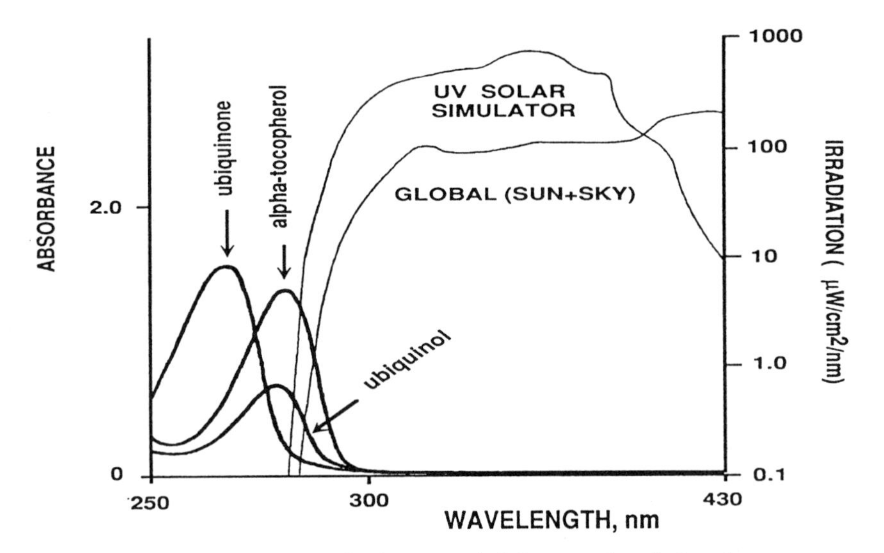

Fig. 6 Comparison of ubiquinol, ubiquinone and alpha-tocopherol absorbance spectra with the spectra of global irradiation (sun + sky) or a solar simulator.

to contract melanoma. It is a frightening statistic and although that does not mean that you are going to die from it, since few people die from skin cancers, it is of course potentially very dangerous. So, the most numerous source of cancer and the fastest growing metastatic human cancer occurs in the skin.

The components of sunlight responsible for skin damage are wavelength regions arbitrarily defined as UVB and UVA irradiation, which are in the range of 290 to 320 nm and 320 to 400 nm, respectively. UVB light normally penetrates through the epidermis and reaches the dermis but is most damaging to the epidermis, whereas UVA, although less damaging at a given dose, penetrates much deeper into the skin through the dermis and to the subcutis.

Vitamin E and quinols (quinol Q_{10}) absorb light in the UVB range (Fig.6). This means that if UVB light strikes the skin you can expect that a direct reaction is possible by the UVB irradiation with vitamin E to make it into a free radical [15]. The vitamin E concentration in the skin epidermis is among the lowest concentrations of vitamin E in any human tissue. The free radical of vitamin E (tocopheroxyl or chromanoxyl) shows a characteristic ESR spectrum.

Extensive recent studies in our laboratory have confirmed that ubiquinols are capable of quenching vitamin E radicals. Although both ubiquinols and vitamin E are lipid-soluble antioxidants in their own right, the reaction rates with lipid radicals by vitamin E are usually about 100 times greater than those of quinones[16]. However, quinones effectively act as antioxidants by being "slaves to vitamin E," i.e., helping cells to regenerate and more efficiently recycle vitamin E. At a dose of UV light equivalent to about 5 hours exposure to natural sunlight we find about a 50% reduction in the concentration of vitamin

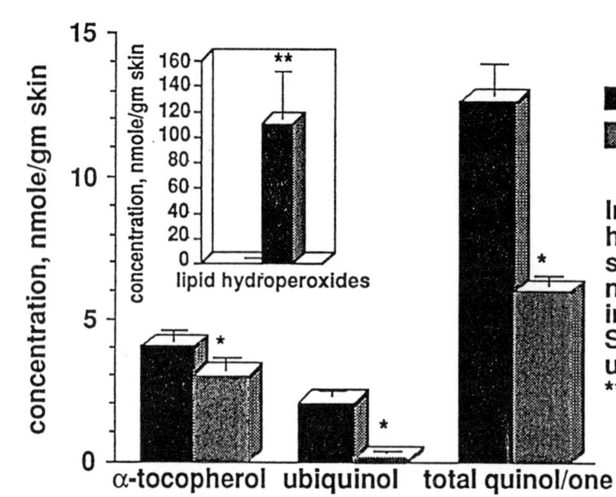

■ before irradiation
■ after irradiation

Insert: Concentration of lipid hydroperoxides in the same skin samples, irradiated and non-irradiated (none detectable in non-irradiated samples). Statistically different than unirradiated controls
**** $p<0.01$; * $p<0.05$**

Total quinol/one = ubiquinol + ubiquinone. n= 6

Fig. 7 Concentrations of cutaneous lipophilic antioxidants after a single large dose (25J/cm^2) of simulated sunlight.

E and complete depletion of ubiquinol9 (Fig. 7). At this dose of UV light we have also detected enormous increases in lipid hydroperoxides, products of lipid peroxidation, *in vivo*[17]. Levels increase from undetectable in nonirradiated skin of 106 nmole/gm skin in irradiated skin. The skin content of protein carbonyls, which are products of oxidative protein damage, increased 33.1% with irradiation of skin *in vivo* at this dose of light. It doesn't tell us which proteins are damaged, but it now gives us the possibility to find out in the future which proteins are damaged. We suspect some of these may be very important proteins for the integrity of the skin to prevent cancer in aging.

What can we do about this UVAB damage? Of course, we can feed people or animals with vitamin E. The diet of a mouse can be supplemented, for example, with a very high concentration of vitamin E (10,000 IU's per kilogram per diet). We have demonstrated that cutaneous concentrations of vitamin E can be increased 8-fold in mice fed a vitamin E-supplemented diet, but it takes four weeks of feeding very high oral supplements of vitamin E to achieve this enhanced skin vitamin E level (you can load the liver in 24h this way). In this case, one observes that the skin lipid hydroperoxide concentration after UVAB exposure is enormously decreased as a result of this loading of the tissue with vitamin E. Thus you protect the skin against UV-induced damage by orally loading the skin vitamin E ten-fold.

Protection could also be afforded by topical application. If you topically apply vitamin E to the skin as we did in two different studies (both of which came out similarly), you get an enormous increase in the amount of vitamin E in the skin, but now you can load skin in 24 hours, so in just one day you can load up vitamin E in the skin .

It is quite clear that if the skin is loaded with vitamin E, there is protection against the loss of vitamin E by exposure to a given dose of UV irradiation. Of course, there is still a large loss of vitamin E by UV exposure in the vitamin E loaded skin, but when the experiment is over, there is much more vitamin E present in the skin than remained in the control. By applying vitamin E to the skin you also protect the quinones. After treatment with alpha-tocopherol, the total amount of quinones was slightly elevated. After irradiation there is no decrease in the quinones. Quinols and vitamin E do seem to talk to each other.

V. SOME WAYS IN WHICH VITAMIN E MAY BE INVOLVED IN CHRONIC DISEASES AND ACUTE CLINICAL CONDITIONS

Appreciating how all the different redox based antioxidants, vitamin C, the thiols and the ubiquinones, can interact with the vitamin E cycle you could perhaps use this knowledge to predict how we might derive clinical benefits. With this objective we have been testing several model systems. One has been to see which antioxidants protect against ischemia/reperfusion injury in the isolated perfused animal heart. We subject a rat heart to fourty minutes of global ischemia, and then afterward reoxygenate for twenty minutes, and then test for recovery of mechanical activity (contractility) and many other parameters indicative of function and molecular markers of damage.

We have performed this kind of experiment now in animals supplemented with either ubiquinone (Q_{10}), vitamin E (as alpha-tocopherol or as palm oil vitamin E [POE]--45% tocopherol and 55% alpha-tocotrienol), or thiols (alpha lipoic acid) and in most cases we get protection[18]. One typical experiment is a case where we fed animals with a high vitamin E diet for 6 weeks. The isolated hearts from control animals, after forty minutes of global ischemia, exhibited 20 to 25% recovery of mechanical activity and a similar small recovery at their ATP. There is also leakage of enzymes (LDH, CPK) and the oxidation of proteins

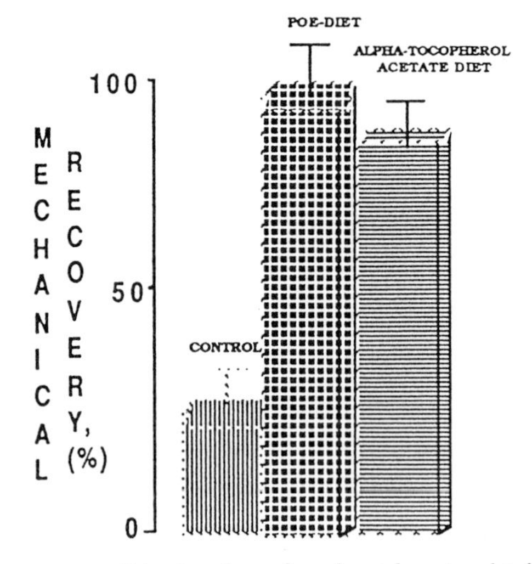

Fig. 8 Mechanical recovery of isolated perfused rat hearts obtained from animals fed different types of diet and subjected to 10 min perfusion + 40 min ischemia + 20 min reperfusion. Alpha-tocopherol acetate content was 60 mg/kg in the control and 20g/kg in the supplemented diet. POE-supplemented diet contained 7 g/kg palm oil vitamin E. N=6.

and lipids that occur after reperfusion of these control hearts. In the hearts from the vitamin E fed animals, almost complete recovery of all these injurious events is observed (Fig. 8).

So although we realize the potential benefits in chronic diseases of antioxidant protection, it is also important to appreciate that in acute clinical conditions like acute exposure to UV radiation to the skin, or suffering a myocardial infarct, you have to have the protection in place beforehand because it takes a long time to load tissues with lipophilic antioxidants like vitamin E and carotenoids which do not move into tissues immediately. Often you need many weeks to load up and to saturate the tissue levels. You want that kind of protection in place ahead of time because you do not always know beforehand when you are going to experience an acute clinical situation.

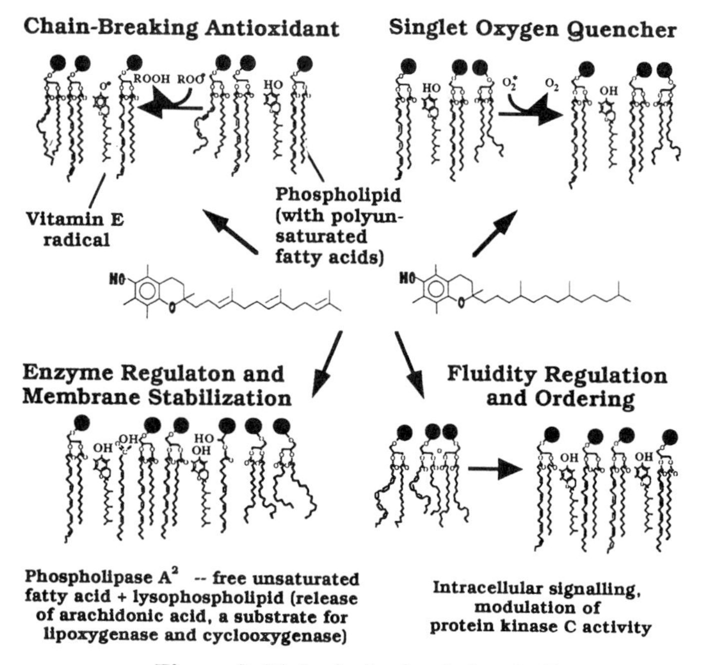

Figure 9. Biological role of vitamin E

In nutrition there is a whole range of antioxidant deficiency/sufficiency, but we do not know what is sufficient or normal for humans. Clearly there is considerable epidemiological evidence that protection, repair, and prevention may be benefited by supplementation with vitamin E alone and other antioxidants. Fig. 9 summarized our knowledge's for biological activity of vitamin E[19-24].

Moreover, there is now much evidence emerging that gene expression is regulated by oxygen radicals. Therefore antioxidants which regulate oxygen radicals have an effect on regulating gene expression. This is a whole new field of knowledge that is now beginning to emerge and we should witness very exciting findings in the future.

We need further molecular biological, biochemical, and physiological and clinical experiments going on so that we understand on a conceptual basis how vitamin E works. With such knowledge we can predict how we can design better experiments to vitamin C and other antioxidant nutrition interventions for optimizing human health and for protection against aging and disease. With such knowledge we should be able to desing more perfect future experiments important to adding "life to your years" and even perhaps "years to your life".

REFERENCES

1. Machlin, L.J., Ed. Vitamin E. New York: Marcel Dekker; 1989.
2. Packer L., Landvick S. In: Diplock A.T., Machlin L.J., Packer L., Pryor W.A., Eds. Vitamin E Biochemistry and Health Implications. New York: New York Academy of Science 570, 1-6, 1989.
3. Serbinova, E.A., Kagan, V.E., Han, D., Packer, L. Free Radical Biol. Med. 10, 263-275, 1991.
4. Packer, L. In: Fernandez-Gomez, J., Ed. Ed. Advances in Biomembranes and Biomaterials. Basel: Birkhauser Verlag, 1991 (in press).
5. Kagan, V., Serbinova, E., Packer, L. *Arch.Biochem.Biophys.,* 280, 33; 1990.
6. Packer, L., Maguire, J.J., Mehlhorn, R.J., Serbinova, E.A., Kagan, V.E. *Biochem.Biophys.Res.Comm.*, 159, 229; 1989.
7. Hiramatsu, M., Packer, L. In: Ozawa, T., Ed. New Trends in Biological Chemistry. Tokyo: Japan Science Society Press, Berlin: Springer Verlag, 323-331, 1991.
8. Hiramatsu, M., Velasco, R.D., Packer, L. Free Radical Bio. Med. 9, 459-464, 1990.
9. Maguirre, J.J., Wilson, D.S., Packer, L. J. Biol. Chem. 264, 21462-21465, 1989.
10. Mehlhorn, R.J., Sumida, S., Packer, L.J. Biol. Chem. 264, 13448-13452, 1989.
11. Kagan, V., Serbinova, E., Forte, T., Scita, G., Packer, L., *J. of Lipid Research* , 33, 385-397; 1992.
12. Pallozi, P., Krinsky, N., *Arch. Biochem. Biophys.*, 1992 (in press)
13. Council on Scientific Affairs; Harmful effects of ultraviolet radiation; J. Amer. Medical. Assoc. 262 , 380-384, 1989.
14. Henriksen, T., Dahlback, A., Larsen, S., Moan, J., *Photochemistry and Photobiology,* 51, 579-582, 1990.
15. Kagan, V.E.; Witt, E.H.; Goldman, R.; Scita, G.; and Packer, L.; Free. Rad. Res. Comm. 1991 (in press).
16. Kagan, V.E.; Serbvinova, E.A.; Packer, L., Biochem. Biophys. Res. Comm. 169 , 851- 857, 1990.
17. Witt, E.H.; Motchnik, P.A.; Han, D.; Ames, B.; and Packer, L.;.; J. Invest. Dermatol., 96 585 -587, 1991.
18. Serbinova, E.A., Khwaja, S., and Packer, L., *Nutrition Research,* 12, suppl. 1, S203-S215, 1992.
19. Mahoney, C.V., Azzai, A. Biochem. Biophys. Res. Commun. 154, 694-697, 1988.
20. Boscoiboinik, B., Szweczyk, A., Hensey, C.E., Azzai, A., J. Biol. Chem. 266, 6188-6194, 1991.
21. Douglas, C.E., Chan, A.C., Choy, P.C., Biochim. Biophys. Acta. 876, 639-645, 1986.
22. van Kuijk, FJGM, Sevanian, A.,Handelman, G.J., Dratz, E.A. TIBS 12, 31-34, 1987.
23. Reddanna, P., Rao, M.K., Reddy, C.C. FEBS Lett. 194, 39-43, 1985.
24. Bakalova, R.A., Nekrasov, A.S., Lankin, V.Z., Kagan, V.E., Stoychev, T.S., Evstigneeva, R.P. Proc. Natl. Acad. Sci. USSR 299, 1008-1011, 1988.
25. Halevy, O., Sklan, D. Biochim. Biophys. Acta. 918, 304-307, 1987.

CONTRIBUTORS

A

Avila, M.A.

B

Batty, I.
Becker, K.
Benusiglio, E.
Beyaert, R.
Boelens, R.
Breton, M.

C

Carter, A.N.
Casey, P.J.
Chap, H.
Ciardo, A.

D

De Valck, D.
De Groot, R.P.
Delmas, V.
Devaux, P.F.
Downes, C.P.

E

Eib, D.
Eichholtz, T.
Eppenberger, H.M.
Estevez, F.

F

Ferguson, M.A.J.
Field, R.A.
Fiers, W.
Foulkes, N.S.

G

Gerhardt, C.C.
Gil, B.
Grohovaz, F.
Grondin, P.
Guinebault, C.

H

Heffron, G.J.

J

Jalink, K.
Jastorff, B.

K

Kaldis, P.
Kaptein, R.
Katahira, M.
Kits, K.S.
Knegtel, R.M.A.
Kuvichkin, V.

L

Laoide, B.M.
León, Y.
Lewis, K.A.
Lodder, H.J.

M

Markman, O.

INDEX